Primate Social Conflict

Primate Social Conflict

Edited by
WILLIAM A. MASON
and
SALLY P. MENDOZA

State University of New York Press

Published by
State University of New York Press, Albany

© 1993   State University of New York

For information, address State University of New York
Press, State University Plaza, Albany, N.Y. 12246

Production by Bernadine Dawes
Marketing by Bernadette LaManna

Library of Congress Cataloging-in-Publication Data

Primate social conflict / edited by William A. Mason and Sally P.
   Mendoza.
        p.   cm.
      Includes index.
      ISBN 0–7914–1241–5 (alk. paper) : $59.50. — ISBN 0–7914–1242–3
   (alk. paper : pbk.) : $19.95
      1. Aggressive behavior on animals. 2. Social hierarchy in
   animals. 3. Primates—Behavior. I. Mason, William A. , 1926–
   II. Mendoza, Sally P.
   QL758.5.P77   1993
   599 . 8 ' 0451—dc20
                                                      91–39372
                                                         CIP

10 9 8 7 6 5 4 3 2 1

Contents

CHAPTER ONE

Primate Social Conflict: An Overview of Sources, Forms, and Consequences

WILLIAM A. MASON AND SALLY P. MENDOZA

Social order and social conflict always are continuing, simultaneous aspects of interaction among individuals and groupings in every society. There is no society without some conflict. Without some order, there is no society.

If one can adequately explain conflict, he will also have advanced our knowledge of order; if he can adequately explain social order, he will have helped us to understand conflict.

(Williams 1970)

Virtually all primates live in organized social groups, the forms of which vary widely across species and much less within species. Some species are characteristically found in groups containing a single adult male together with several females and their offspring. Other species live in groups that include multiple adult males and females and immature animals at all stages of development. A few species favor social

William A. Mason—Department of Psychology and California Regional Primate Research Center, University of California, Davis

Sally P. Mendoza—Department of Psychology and California Regional Primate Research Center, University of California, Davis

Preparation of this manuscript was supported by Grant RR00169 from the Division of Research Resources of the National Institutes of Health.

systems in which a single pair does all the mating, while all other adults are excluded from the group or participate in caring for the young.

A central and continuing concern in the study of primate social behavior is to understand the sources of order in these groups. What processes produce the differences among species in social organization? How are groups formed? How do they maintain themselves as social entities? How capable are they of accommodating to demographic changes brought about by births, deaths, defections, or the perturbations caused by untoward environmental events? What manner of adaptive benefits are conferred by a given form of social organization, and how are they achieved?

Although such questions call for different approaches and make use of different explanatory models, they share the assumption that social organizations are the emergent products of dynamic social processes. The collective order is in some manner created and maintained by interactions among the individual constituents. Together, they form the society. Together, they determine who will be included or excluded as members of the group, and they establish the place which each individual will occupy in the dynamic web of social relationships. In this ongoing pattern of mutual influence, adjustment, and control, the individual is, at once, agent and object, acting while being acted upon.

SOURCES OF SOCIAL CONFLICT

Given this basic feature of primate social systems and the high degree of behavioral autonomy of the individual primate, the likelihood is great that conflict will occur. The potential sources of conflict are many. Conflict may occur whenever two individuals are seeking the same thing at the same time and both cannot be satisfied, or because one individual wants or anticipates something from another who is unwilling or unable to comply (Coombs 1987; Hand 1986). Social conflict may be the result of actions that are ambiguous, moods or intentions that are disregarded or misinterpreted, or motives that are ambivalent or fundamentally incompatible (see Mason in this volume). The potential for conflict is multiplied as group size increases, not only because of the additional number of participants, but also because the possibilities for the formation of factions and coalitions increase (see Itoigawa in this volume; Noë, 1986; Silk in this volume).

Awareness of the prevalence of social conflict in the daily life of nonhuman primates has been sharpened by the growing body of information on the life-span development of known individuals within established social groups. For examples, see Altmann and Altmann 1970; Altmann et al. 1985; Goodall 1975; Itoigawa in this volume; Rawlins and Kessler 1986; Rhine 1986. These data have drawn attention to the number and types of social problems that an individual is likely to encounter at the level of daily intercourse throughout its life, and in more intense form as it negotiates passage through major social transitions from weaning and the birth of younger siblings, adolescence, mating and the care of young, and extending into late maturity and senescence. For examples, see Abbott in this volume; Andrews et al. in this volume; Anzenberger in this volume; Hinde 1983; de Waal in this volume; Mason et al. in this volume; Mendoza and Mason 1986; Menzel in this volume; Norton 1986; Rawlins and Kessler 1986).

Another development that has helped to make social conflict more salient is an improved understanding of the subtlety and complexity of social relationships and their powerful effects on the participants (Anderson and Mason 1974, 1978; Hinde 1983; Mendoza in this volume; Mendoza and Mason 1989; Mendoza et al. 1991; Sapolsky in this volume). The dynamic processes involved in the formation and maintenance of relationships can readily put individuals in a place where their interests do not coincide. Indeed, given sufficient time, the elements of rejection, resistance, or indifference are bound to emerge in some form in any active relationship—even those in which the level of mutual satisfaction is high (Anzenberger in this volume; Menzel in this volume). Moreover, primates rely extensively on their knowledge of other individuals in the regulation of their social lives (Anderson and Mason 1974, 1978; Cheney et al. 1986; de Waal in this volume). Such information, however it is acquired, is an imperfect and fallible guide to social conduct and can easily lead to conflict (Mason in this volume).

A critical element in our current understanding of the sources of social conflict has been provided by sociobiological theories. Sociobiology is based on the assumption that natural selection has favored individuals who act so as to maximize their own fitness. From this perspective, social conflicts are anticipated whenever the genetic self-interests of interacting organisms diverge. As sociobiological arguments have emphasized, this condition is quite common. The potential for conflict exists, not only with respect to competition for access to mates and essential resources—which has always been recognized as conducive to conflict—but also in relationships in which conflict may seem intuitively unlikely, such as between breeding pairs, between siblings, and between parents and offspring (Maynard Smith 1974; Tiger in this volume; Trivers 1972, 1974). From a sociobiological perspective natural selection should also favor the evolution of strategies to avoid, mitigate, or resolve social conflicts. Such strategies are particularly likely to be manifest in relationships in which the participants have a stake in tempering or eliminating the conflict because, for example, they are close genetically, are mutually dependent, or risk being injured.

In spite of the seminal contribution of sociobiological theory, the utility of any evolutionary approach to empirical studies of primate social conflict is necessarily limited. A major source of this limitation is that evolutionary strategies are abstractions—generalized scenarios referring to the actuarial properties of populations—rather than the actual moment-to-moment behavior of individuals. Instances of social conflict are not generated by incompatible evolutionary strategies, but by interacting animals who operate in the here and now, who lack prescience, and who do not always act in their own best interests. An essential complement to sociobiological theory is information on the proximate sources of conflict and the factors that determine its frequency, intensity and form.

FORMS OF SOCIAL CONFLICT

In view of the ubiquity of social conflict, the richness and variety of its sources, and its powerful influence on social life, it is remarkable that it has so rarely been treated as a distinct and important topic in animal research. This is not to say that all forms of social conflict have been neglected. Aggression, in particular, has received a huge amount of attention in both the scientific and popular literature. Indeed, in many quarters, aggression is treated as though it were synonymous with social conflict. It is not. Although aggression is a salient aspect of social conflict, it is only one part of a much broader range of phenomena, and not necessarily the most important part at that. One suspects that the preoccupation with aggression says more about our understandable concern with the extravagant human capacity for violence and mayhem than about the actual frequency of aggression or its significance in the social world of nonhuman primates.

Conflict, rather than aggression, is the generic concept. It is an inherent and protean accompaniment of life in organized social groups. Our common vocabulary for describing social interactions is replete with terms referring to different forms or degrees of social conflict. Individuals resist, coerce, and impose upon one another; they clash, compete, work at cross purposes, disagree, and have falling-outs; they may also negotiate, resolve, compromise, or reconcile their differences. Although such terms usually refer to human behavior, many are applicable to the behavior of nonhuman primates, as the chapters in this volume amply illustrate.

Conflict may also occur within an individual. This form of conflict is not specifically social. As the early ethologists pointed out, however, intraindividual conflict is frequently present in a social context (Tinbergen 1952). This is the case, for example, when an individual is simultaneously motivated by two incompatible drives such as flight and attack. Intraindividual conflict has also been investigated extensively in experimental animal behavior laboratories (Brown 1971; Miller 1944, 1982). Generally speaking, cognitive complexity is conducive to intraindividual conflict. The greater the diversity of stimuli that an organism is able to register, the larger its repertoire of responses for acting on such stimuli, and the more elaborate its cognitive processes, the more frequently it will encounter decision-making conflict (Berlyne 1960). Small wonder that conflicts are so common among group-living primates!

CONSEQUENCES OF SOCIAL CONFLICT

Responses to conflict are as diverse as one might expect, given the variety of its sources and forms, and they can have important biosocial consequences. We know from our own experience that being involved in social conflict often produces a state of acute discomfort or distress. Apparently, this is also the case with monkeys and apes. An animal may respond to the unpleasantness of conflicts by quitting the field, intensifying its original pursuit, accommodating to the resistance which it encounters, or aggressing against another. Its options are many. Whatever the immediate response to social conflict may be, however, it can have significant long-range consequences for all participants, both individually and in respect to their future relations with one another.

Social conflicts may be brief, one-time episodes, or they may be prolonged or repeated frequently. It is reasonable to assume that the effects of chronic or recurrent social conflict can be cumulative and potentially harmful to the individual, owing to persistent activation of the catabolic processes associated with the stress response (Weiss 1972). Experimental evidence demonstrating pervasive and damaging physiological effects of chronic social conflict is available for rodents and tree shrews (Henry and Stephens 1977; von Holst 1977, 1985). With respect to primates, the indications are clear that periods of social instability—during which conflict is probably frequent—are associated with persistent and potentially pathogenic changes in some individuals (Kaplan et al. 1983; Manuck et al. 1988; Sapolsky 1987, and in this volume).

The destructive potential of social conflict has presumably contributed to the evolution of ways to forestall or mitigate its deleterious consequences. Conspicuous among these are various species-typical ritualized behaviors shown by all primates. Such behaviors are particularly likely to be associated with conflict. In social situations they apparently convey intolerance, submission, friendly intent, reassurance, and so on. Ritualized behaviors tend to elicit appropriate complemental responses. This has given rise to the idea that certain social conventions or rules govern interactions during conflict and serve to establish or preserve potentially valuable relationships (de Waal in this volume; de Waal and Yoshihara 1983; Kummer 1975; Maxim 1976, 1982; Mendoza in this volume; Mendoza and Barchas 1983).

An intriguing possibility is that some social situations in which the potential for conflict is great may reduce this potential by the adaptive strategy of causing changes in physiological regulatory systems. For example, the physiological suppression of reproduction in young adult or subordinate members of social groups of talapoins or marmosets could effectively eliminate sexual competition between potential rivals (Abbott in this volume; Hansen et al. 1980). Likewise, physiological differentiation among males during initial formation of dominance relations may be an evolved strategy that reduces competition and conflict by altering metabolic processes and behavioral tendencies among otherwise similar animals (Mendoza in this volume and 1984; Mendoza and Mason 1989).

Conflict may also contribute to social order as a creative and positive force (Lyons in this volume). The possibility that conflicts can promote psychological development and strengthen social bonds has been recognized by a number of influential theorists. A common ingredient in this notion is the idea that conflict generates tension or energy that is an impetus to growth and innovation. John Dewey described conflict as "the gadfly of thought" (Dewey 1930, 200). Jean Piaget hypothesized that peer conflict caused cognitive conflict, which was a major impetus to cognitive development (Shantz 1987). Erik Erikson characterized human psychosocial development as a sequence of stages, each involving a nuclear conflict that "...adds a new ego quality, a new criterion of accruing human strength" (Erikson 1963, 270). The social theorist, Georg Simmel, viewed social conflict as a form of socialization and a necessary factor in maintaining group integrity (Coser 1956). In contrast to these theorists, students of animal behavior have tended to emphasize the negative consequences of social conflict. There are some exceptions, however. M. R. A. Chance, certainly among the first to call attention to the unique importance of conflict in primate societies, suggested that social conflict played a significant role in the evolution of primate intelligence (Chance 1966; Chance and Mead 1953). Konrad Lorenz also recognized the positive role of conflict as a force in evolution and believed that "...the personal bond, love, arose in many cases from intraspecific aggression, by way of ritualization of a redirected attack or threatening" (Lorenz 1966, 217). At a more proximate level, primatologists have hypothesized that conflict makes a positive contribution to social order by increasing interpersonal attraction, creating social bonds, and strengthening group cohesion (Chance 1955, 1963; de Waal 1986; de Waal and Yoshihara 1983; Kummer 1971; Mason 1964).

CONCLUSIONS

Our view of conflict as an important factor in the social life of animals has been enlarged by primate research. Beyond the occasional dramatic clash between antagonists, the concept has been extended to include the commonplace. Conflict is seen as a normal and recurrent feature of social life, varied in its causes and manifestations, but nonetheless consequential for the participants.

As the chapters in this volume illustrate, the study of social conflict has much to offer toward understanding the sources of order in primate societies. Conflict differentiates roles within organized groups. It leads to changes in reproductive potential and physical well-being. It is a daily accompaniment of such mundane processes as selecting travel routes, choosing to lead or to follow, and deciding where or when to settle down for the night. Most likely, conflict contributes to the evolution of rituals and rules of social conduct, and gives rise to the phenomena of interpersonal attraction, charisma, and social bonds. It is difficult to imagine a stage in the life of the individual primate into which social conflict does not intrude.

REFERENCES

Altmann, J.; Hausfater, G; Altmann, S. A. (1985) Demography of Amboseli baboons. *American Journal of Primatology* 8:113–125.

Altmann, S. A.; Altmann, J. (1970) "Baboon Ecology: African Field Research." Chicago: University of Chicago Press.

Anderson, C. O.; Mason, W. A. (1974) Early experience and complexity of social organization in groups of young rhesus monkeys *(Macaca mulatta)*. *Journal of Comparative and Physiological Psychology* 87:681–690.

Anderson, C. O.; Mason, W. A. (1978) Competitive social strategies in groups of deprived and experienced rhesus monkeys. *Developmental Psychobiology* 11:289–299.

Berlyne, D. E. (1960) "Conflict, Arousal, and Curiosity." New York: McGraw-Hill.

Brown, J. S. (1971) Principles of intrapersonal conflict. In Smith CG (ed.), "Conflict Resolution: Contributions of the Behavioral Sciences." Notre Dame, Ind.: University of Notre Dame. 88–97.

Chance, M. R. A. (1955) The sociability of monkeys. *Man* 176:1–4.

Chance, M. R. A. (1963) The social bond of the primates. *Primates* 4:1–22.

Chance, M. R. A. (1966) Resolution of social conflict in animals and man. In de Reuck AVS, Knight J (eds.), *Ciba Foundation Symposium on Conflict in Society* 16–35. London, J & A Churchill, Ltd.

Chance, M. R. A.; Mead, A. P. (1953) Social behaviour and primate evolution. *Society for Experimental Biology Symposia* 7:395–439.

Cheney, D.; Seyfarth, R.; Smuts, B. (1986) Social relationships and social cognition in nonhuman primates. *Science* 234:1361–1366.

Coombs, C. H. (1987) The structure of conflict. *The American Psychologist* 42:355–363.

Coser, L. (1956) "The Functions of Social Conflict." London: The Free Press.

de Waal, F. B. M. (1986) The integration of dominance and social bonding in primates. *Quarterly Review of Biology* 61:459–479.

de Waal, F. B. M.; Yoshihara, D. (1983) Reconciliation and redirected affection in rhesus monkeys. *Behaviour* 85:224–241.

Dewey, J. (1930) "Human Nature and Conduct: An Introduction to Social Psychology." New York: The Modern Library.

Erikson, E. H. (1963) "Childhood and Society." New York: W. W. Norton.

Goodall, J. (1975) Chimpanzees of the Gombe National Park: Thirteen years of research. In Kurth G, Eibl-Eibesfeldt I (eds.), "Hominisation und Verhalten." Stuttgart: Fischer.

Hand, J. L. (1986) Resolution of social conflicts: Dominance, egalitarianism, spheres of dominance, and game theory. *Quarterly Review of Biology* 61:201–220.

Hansen, S.; Keverne, E. B.; Martensz, N. D.; Herbert, J. (1980) Behavioural and neuroendocrine factors regulating prolactin and LH discharges in monkeys. In "Non-Human Primate Models for Study of Human Reproduction." *Satellite Symposium of the 7th Congress of the International Primatological Society*, Bangalore 1979. Basel: Karger. 148–158.

Henry, J. P.; Stephens, P. M. (1977) "Stress, Health and the Social Environment: A Sociobiologic Approach to Medicine." New York: Springer-Verlag.

Hinde, R. A. (ed.) (1983) "Primate Social Relationships: An Integrated Approach." Sunderland, Mass.: Sinauer Associates, Inc.

Kaplan, J. R.; Manuck, S. B.; Clarkson, T. B.; Lusso, F. M.; Taub, D. M.; Miller, E. W. (1983) Social stress and atherosclerosis in normocholesterolemic monkeys. *Science* 220:733–734.

Kummer, H. (1971) "Primate Societies: Group Techniques of Ecological Adaptation." Chicago: Aldine Publishing Company.

Kummer, H. (1975) Rules of dyad and group formation among captive gelada baboons *(Theropithecus gelada)*. In Kondo, S., Kawai, M., Ehara, A., Kawamura, S. (eds.), "Proceedings from the Symposia of the Fifth Congress of the International Primatological Society, Nagoya, Japan, August 1974." Tokyo: Japan Science Press.

Lorenz K (1966) "On Aggression." New York: Harcourt, Brace & World, Inc.

Manuck, S. B.; Kaplan, J. R.; Adams, M. R.; Clarkson, T. B. (1988) Studies of psychosocial influences on coronary artery atherogenesis in cynomolgus monkeys. *Health Psychology* 7:113–124.

Mason, W. A. (1964) Sociability and social organization in monkeys and apes. In Berkowitz L (ed.), "Advances in Experimental Social Psychology," vol. 1. New York: Academic Press, Inc.

Maxim, P. E. (1976) An interval scale for studying and quantifying social relations in pairs of rhesus monkeys. *Journal of Experimental Psychology: General* 105:123–147.

Maxim, P. E. (1982) Contexts and messages in macaque social communication. *American Journal of Primatology* 2:63–85.

Maynard Smith, J. (1974) The theory of games and the evolution of animal conflicts. *Journal of Theoretical Biology* 47:209–221.

Mendoza, S. P. (1984) The psychobiology of social relationships. In Barchas PR, Mendoza, S. P. (eds.), "Social Cohesion: Essays Toward a Sociophysiological Perspective." Westport, Conn.: Greenwood Press. 3–29.

Mendoza, S. P.; Barchas, P. R. (1983) Behavioral processes leading to linear status hierarchies following group formation in rhesus monkeys. *Journal of Human Evolution* 12:185–192.

Mendoza, S. P.; Mason, W. A. (1986) Parental division of labour and differentiation of attachments in a monogamous primate *(Callicebus moloch)*. *Animal Behaviour* 34:1336–1347.

Mendoza, S. P.; Mason, W. A. (1989) Primate relationships: Social dispositions and physiological responses. In Seth, P. K., Seth, S. (eds.), "Perspectives in Primate Biology," vol. 2. New Delhi: Today & Tomorrow's Printers and Publishers. 129–143.

Mendoza, S. P.; Lyons, D. M.; Saltzman, W. (1991) Sociophysiology of squirrel monkeys. *American Journal of Primatology* 23:37–54.

Miller, N. E. (1944) Experimental studies of conflict. In Hunt, J. McV. (ed.), "Personality and the Behavior Disorders," vol. 1. New York: The Ronald Press Company. 431–465.

Miller, N. E. (1982) Motivation and psychological stress. In Pfaff, D. W. (ed.), "The Physiological Mechanisms of Motivation." New York: Springer-Verlag. 409–432.

Noë, R. (1986) Lasting alliances among adult male savannah baboons. In Else,

J. G., Lee, P. C. (eds.), "Primate Ontogeny, Cognition and Social Behaviour." Cambridge: Cambridge University Press. 381–392.

Norton, G. W. (1986) Leadership: Decision processes of group movement in yellow baboons. In Else, J. G., Lee, P. C. (eds.), "Primate Ecology and Conservation." Cambridge: Cambridge University Press. 145–156.

Rawlins, R. G.; Kessler, M. J. (ed.) (1986) "The Cayo Santiago Macaques: History, Behavior and Biology." Albany, N.Y.: State University of New York Press.

Rhine, R. J. (1986) Ten years of cooperative baboon research at Mikumi National Park. In Else, J. G., Lee, P. C. (eds.), "Primate Ontogeny, Cognition and Social Behaviour." Cambridge: Cambridge University Press. 13–22.

Sapolsky, R. M. (1987) Stress, social status, and reproductive physiology in free-living baboons. In Crews, D. (ed.), "Psychobiology of Reproductive Behavior: An Evolutionary Perspective." Englewood Cliffs, N.J.: Prentice-Hall. 291–322.

Shantz, C. U. (1987) Conflicts between children. *Child Development* 58:283–305.

Tinbergen, N. (1952) "Derived" activities: Their causation, biological significance, origin, and emancipation during evolution. *Quarterly Review of Biology* 27:1–32.

Trivers, R. L. (1972) Parental investment and sexual selection. In Campbell, B. (ed.), "Sexual Selection and the Descent of Man." Chicago: Aldine Publishing Company. 136–179.

Trivers, R. L. (1974) Parent-offspring conflict. *American Zoologist* 14:249–264.

von Holst, D. (1977) Social stress in tree-shrews: Problems, results, and goals. *Journal of Comparative Physiology* 120:71–86.

von Holst, D. (1985) Coping behaviour and stress physiology in male tree shrews *(Tupaia belangeri)*. In Holldobler, B., Lindauer, M. (eds.), "Experimental Behavioral Ecology and Sociobiology." New York: Gustav Fischer Verlag.

Weiss, J. M. (1972) Influence of psychological variables on stress-induced pathology. In "Physiology, Emotion, and Psychosomatic Illness." *Ciba Foundation Symposium 8*. Amsterdam: Elsevier. 253–265.

Williams, R. M., Jr. (1970) Social order and social conflict. *Proceedings of the American Philosophical Society* 114:217–225.

CHAPTER TWO

The Nature of Social Conflict:
A Psycho-Ethological Perspective

WILLIAM A. MASON

In primate societies the potential for social conflict is high. Primate groups are complex, loosely coordinated, and characterized by a high degree of interdependence among members. The preservation of social order is largely based on the exchange of information among members of the group, each of which has primary responsibility for its own self-maintenance and for obtaining the social satisfactions that contribute to its comfort and well-being. It is not surprising in these circumstances that the individual primate often finds itself in a conflicted state or acting at cross-purposes with other members of its group. Conflict from this perspective is an inevitable accompaniment of social life. Its sources are many, as are the responses which it engenders.

A useful distinction can be made between conflicts that occur within the individual (intraindividual) and conflicts that occur between individuals (interpersonal). Both forms of conflict participate closely in the processes contributing to social stability and social change. They are conceptually distinct, however.

Intraindividual conflict is present whenever two or more incompatible tendencies are aroused simultaneously. The hallmark of intraindividual conflict in a social context is that the actor is in a state of ambivalence or uncertainty with respect to the perceived consequences of its actions. Extreme or prolonged intraindividual conflict is an aversive state and can have harmful consequences.

William A. Mason—Department of Psychology and California Regional Primate Research Center, University of California, Davis

Preparation of this chapter was aided by grants HD06367 and RR00169 from the United States National Institutes of Health. I thank D. M. Lyons for his comments.

In contrast to intraindividual conflict, interpersonal conflict is inherently social; it is an emergent property within a social interchange. The defining characteristic of interpersonal conflict is incompatibility between interacting individuals. The efforts of one individual to achieve some desired outcome are thwarted because another individual resists or interferes with these efforts, withdraws from the situation, or in other ways fails to comply. Interpersonal conflict may occur because the participants have different priorities at the moment, or as the result of misunderstanding, fear, ignorance, or mere indifference of one individual to the desires or intentions of another. The interplay between intraindividual conflict and interpersonal conflict is intimate. It is also complex and indirect.

Current treatments characteristically emphasize competition as the cause of social conflict and aggression as its principal consequence. This simple formulation is seriously incomplete. It fails to convey the richness, the subtlety, the pervasiveness of conflict, and its powerful influence on the social dynamics of organized primate groups.

Everyone is familiar with the idea of conflict, for it is deeply rooted in our ways of looking at and thinking about the world. We see it everywhere. Conflict is a common theme in myth, religion, philosophy, and literature (Rapoport 1974). Freud's (1943) psychoanalytic theory is based on three conflicting aspects of human personality. The progressive resolution of conflicts between a person's needs and the demands of society characterizes Erikson's (1963) theory of human development. In biological thought, conflict is widely viewed as a driving force in organic evolution, as implied, for example, in the time-worn metaphor, the struggle for existence (Cloudsley-Thompson 1965; Darwin 1872). Clearly, if we are to make headway in the scientific understanding of so broad a topic—and one so congenial to our intuitive perceptions of the world—some distinctions are required.

LEVELS OF CONFLICT

The term conflict is used in several senses in scientific writing, as well as in ordinary discourse. The core meaning implicit in all uses, however, is the idea of some sort of clash, interference, or incompatibility between elements or forces. In other words, conflict assumes at least two identifiable parties or agents. Differences in the scientific approach to conflict are found chiefly in the way these agents are characterized, in the levels of description and analysis that are deemed appropriate, and in the aims of the investigation. For example, the conflicting agents may be assumed to be goal-directed or not; the language for describing and analyzing conflict may be specific and concrete, or highly abstract; and the investigation may be concerned with such matters as the sources of conflict, ways of forestalling or resolving it, or its consequences in the short- or long-term.

In the social sciences, the approaches to conflict most remote from present concerns are at a level where the conflicting agents are highly abstract, impersonal, or multipersonal entities—such as social classes, commercial enterprises, political factions, nation states, or ideologies. The individual constituents are unknown to each other and are treated as anonymous single-minded participants in a larger organization that may be characterized in terms of its goals, tactics, and the resources it commands. The processes that lead to conflict between such entities, and the modes in which conflict is expressed or resolved are usually quite beyond the control of single individuals.

An analogous level in the biological sciences is exemplified by the Competitive Exclusion Principle, in which the parties in conflict are conceptualized as two or more noninterbreeding populations living in the same habitat and competing for the same limiting resource (Gause 1934). Here, as in the parallel examples from the social sciences, individual organisms are treated as anonymous, unidimensional, and individually incapable of influencing the sources of the conflict or its outcome. The successful population survives and becomes more numerous in relation to the losing population, which either becomes extinct or evolves a new niche (Hardin 1960).

The present essay is about social conflict at a level below the anonymous population, namely, individual primates in face-to-face interactions, who either live together as familiars or recognize each other as strangers. At this level, social conflict is most often manifest as an interruption or perturbation in the ordered flow of social events. Agonism (threats, attacks, flight, gestures of fear or submission) is one of the most conspicuous signs of conflict. More generally, however, social conflict exists whenever the expectations or behavior of two or more interacting individuals are incompatible. One individual's efforts to achieve some outcome are thwarted because another individual resists or interferes with these efforts, withdraws from the situation, or, for whatever reason, is unwilling or unable to comply.

CONTRIBUTING FACTORS

Anyone who observes group-living primates regularly will see evidence of social conflict. It occurs between adult and young, between animals on familiar terms, and between strangers; it occurs between males and females, and between members of the same sex. There is virtually no social context in which conflict does not occasionally emerge.

The frequent occurrence of social conflict among group-living primates appears to be attributable to characteristics shared by most primate species. None of these general characteristics relates directly to social conflict. Together, however, they constitute a configuration of contributing factors that seem to make social conflict virtually inevitable. They include the following:

1. Gregariousness or sociality. Most primates spend their entire lives in close association with other animals. All major life events take place within a social context. To appreciate the potential for social conflict it is critical to recognize that powerful forces keep the individual within a social system. Were it not for this compelling attraction to life in groups, the occasions for social conflict would obviously be reduced. A solitary life free of social conflict, however, is seldom a realistic possibility for the individual monkey or ape. We do not fully understand why the majority of primate species are so strongly committed to being social. The proximate sources of gregariousness are probably many. Even when individuals leave a group—which is not uncommon in some species—they usually do so with a small party of other group members, or, if alone, they attempt to join another group. The truly solitary individual is an exception in most species and,

if solitude occurs at all, it is restricted to brief periods within the total life span.

2. Individualism. Beyond infancy, each member of a primate group has primary responsibility for its own self-maintenance and well-being, and is capable, in principle, of acting as an independent agent. For this reason, competition and conflicts of interest are an inherent feature of life in primate groups. Although in the course of evolving toward their current levels of sociality, some species have come up with various forms of "mutual aid" or "cooperation," these share an uneasy coexistence with a resolute individualism.

3. Looseness and complexity of social organization. Each primate species tends toward a characteristic or modal form of social organization that is created and maintained by patterns of interaction among the members of the group. Members collectively determine who will be in the group, who will be excluded, and the place that each member occupies in the network of social relationships. The actual social processes through which these outcomes are achieved are extraordinarily dynamic and complex. They include elements of coercion and restraint, solicitation and support, gratification and denial. At any given moment, the individual may figure simultaneously as the generative agent of social events and the object upon which they impinge.

4. Dependence on exchange of information. Exchange of information is an essential source of order in primate social life. Members of a primate social group are regularly and frequently confronted with the need to adjust to those with whom they interact—to take into account or assess in some fashion, their idiosyncracies, knowledge, desires, moods, expectations, intentions, and so on. All these qualities are subject to change and to errors of interpretation. Change may be progressive (as in aging), cyclic (as in reproduction), or a matter of the moment (as in irritability). If change is the rule in primate social life, information is the currency through which it is recognized and dealt with. This poses problems in interpersonal relations which humans find complex

and demanding, and it is safe to assume that the individual monkey or ape is no better endowed than we are with the general cognitive skills it needs to solve them

COMPETITION, AGGRESSION, AND SOCIAL CONFLICT

The common calculus of animal social conflict treats competition as cause and aggression as effect. Indeed, aggression is often used as a synonym for social conflict, and, if aggression occurs, competition is assumed, even when the object of competition is not evident. This simple formulation has diverted attention from the analysis of three potentially distinct phenomena: competition, aggression, and social conflict.

Competition, of course, can be a source of social conflict. If two individuals want the same commodity at the same time, and there is enough for only one of them, their actions toward each other are likely to be incompatible. The specific occasion may be food, a prospective mate, or even a comfortable place in the sun. Social conflicts also emerge, however, when there is no hint of competition over some limited commodity. For example, two individuals may be seeking different things in circumstances in which both cannot be satisfied (Coombs 1987; Hand 1986). They may have different priorities at the moment, or be engaged in different social agendas. Social conflict may also be the result of misunderstanding, fear, ignorance, or mere indifference of one individual to the desires or intentions of another.

Similarly, the relation of aggression to competition and to social conflict is neither simple nor direct. Although there are many other ways of responding to competition, competition may cause aggression. Conversely, aggression can result from diverse causes. It may be provoked by fear, frustration, pain, and unfamiliarity. It may even occur in the absence of any apparent external cause. The target of an attack may have had nothing to do with precipitating it, as in the familiar case of redirection or scapegoating. In such circumstances, aggression is the cause of social conflict, rather than a response to it.

In summary, competition and aggression are not tightly linked to each other nor to social conflict. Competition can provoke aggression, but it need not do so. Aggression may occur in the absence of competition. Incompatibility between interacting individuals—the hallmark of social conflict—can occur for many reasons that do not involve competition or aggression.

Clearly, the received view of social conflict in which competition and aggression stand as cause and effect is incomplete. As a first concern, descriptive studies of conflict should focus on the problems which animals encounter in their dealings with one another. The relevant questions concern the nature of these problems. Under what circumstances is conflict likely? How do these situations present themselves to the individuals involved and when do they occur? How does the broader social setting influence the nature of conflict between individuals? Can different forms of conflict be distinguished? How are conflicts expressed and resolved?

SOURCES OF SOCIAL CONFLICT

Situational Sources

Consider first the social circumstances that are conducive to conflict. For descriptive purposes, five types of intergrading situations can be identified in which social conflict is likely. A common theme in these situations is a discrepancy between what an individual is seeking or expects to encounter, and what the actual situation provides or portends.

Probably the most common of source of conflict occurs when one individual is anticipating some form of direct social satisfaction from another and fails to achieve it. For example, one animal may attempt to have sex with another, or engage it in some seemingly benign and mutually agreeable activity such as grooming or playing, and is rebuffed because, for whatever reason, the other is presently disinclined to participate. Instances of this type of conflict are often seen between mother and infant, but it must occur frequently in any established relationship in which interdependence between the participants is high. Obviously, the frequency with which this sort of conflict appears in a particular relationship will play a significant part in the development of general expectations in each participant concerning the probable reactions of the other to its social overtures. This situation grades imperceptibly into the next, which is also quite common.

In the second situation, conflict occurs because established expectations regarding another are no longer reliable. Often this is the result of progressive endogenous changes in one or both participants. Examples of such changes include the seasonal onset of breeding readiness, the birth of an infant, a fulminating illness, or puberty. Although most of these phenomena have been investigated extensively, their implications for creating conflicts in established relationships have been largely ignored.

A third situation associated with social conflict is the establishment of new relationships when expectations are in the process of being formed. Although this occurs in natural populations as the result of individuals changing from one group to another or forming new groups, it has been investigated most thoroughly in captive groups or in arranged paired encounters between strangers.

A fourth situation that is conducive to social conflict poses a threat to an established and valued relationship. Most often this involves a third party that is perceived as intruding upon and disrupting the relationship or as offering some harm to one of the participants. Specific examples include the intervention of mothers of many primate species in response to perceived danger to their infants, and the reactions of male Hamadryas baboons and monogamous titi monkeys to intruders.

The fifth situation that engenders social conflict is face-to-face competition over access to some tangible incentive, most often food or mates. In baboons, macaques, and chimpanzees, competition is often accompanied by threats, chases, or attacks. In many other species, agonism is less prominent.

Status Conflicts

In addition to these five situations, a potential source of conflict that must be considered critically is the agonistic contest over social dominance or status. On the surface, this appears to be another example of competition, and is readily accommodated within the traditional calculus of social conflict. In contrast to the foregoing situations, however, competition is, in this case, presumed to be over social status. The implicit assumption is that superior status is a goal. In its strong form, a specific dominance drive is hypothesized (Maslow 1935). Although contemporary scientists seldom refer to a dominance drive, their interpretations of social conflict in the more aggressive primates (macaques, baboons, and chimpanzees) make frequent reference to "status striving," "challenges" to individuals of higher rank, "dominance contests," and "fights over status." All such terms imply a goal-directed process centering around the achievement and maintenance of dominance.

In the context of the present chapter, this implication needs to be examined critically, for it has an important bearing on descriptive studies of social conflict. The assumption that social dominance is sought for its own sake transforms the concept of status into something akin to a resource or commodity. Individuals can be said to compete for it, to engage in contests to achieve it, and to defend it against challenges. Animals of lower status are expected to show upward mobility, and to be perennially on the lookout for ways of gaining another rung on the social ladder. The descriptive literature on primate social dynamics is replete with such phrases. Instead of serving as a descriptive term for phenomena that require more refined analysis, social dominance assumes the status of an explanatory motivational construct that readily accounts for all sorts of agonistic social conflicts.

A simple alternative formulation that does not assume that dominance-striving is an independent process or distinct social motive may be described as the *minimax model*. According to this model individual primates strive to have their own way wherever they can—that is, to move freely within the group; to interact with whomever they choose, whenever they choose; to have unimpeded access to the good things in life; and to accomplish this with as little pain and travail as possible. In this respect, they are not fundamentally different from other mammals, although the details of what is satisfying or aversive will, of course, vary with species. According to this view, the individual monkey or ape is not looking for ways of advancing its social status, but for ways of maximizing its satisfactions while minimizing its discomforts and frustrations. This agenda is mostly carried out within a social context. When the activities of one individual impinge on the interests of another, conflict is likely to result and aggression may be one of its accompaniments.

Aggression is not incompatible with a minimax model. Aggression is viewed as one response to events that affect the individual directly and interfere with or threaten its hedonic pursuits (Craig 1921). This is consistent with the finding that the principal elicitors of aggression in many mammalian species are pain, frustration, direct competition, unfamiliarity, uncertainty, and encroachment on personal space (Archer 1976; Marler 1976; Scott 1977). There is good reason to suppose that such conditions are abundantly present within the setting of primate social groups.

From the standpoint of a minimax model, a clear distinction is made between the causes of aggression and its consequences. For example, among the macaques, the frequency and intensity of aggression varies with species, and the more aggressive species also tend to show more strongly developed (or clearly expressed) relations of dominance and subordination (Hawkes 1970; Lahiri and Southwick 1966; Rosenblum, Kaufman and Stynes 1964; Simonds 1965; Sugiyama 1971; Thierry 1985, 1986). Differences between species in aggressive behavior and in the prominence of the asymmetrical relations of dominance and subordination appear to reflect fundamental attributes of temperament. Macaque species with the stronger disposition to respond aggressively show this tendency in many different contexts, including those in which the establishment or maintenance of dominance and subordination are minimally involved, if at all, as in interactions between consort pairs, between mothers and infants, and responses to predators and human caretakers. Furthermore, differences are not specifically social. They are seen in distinctive behavioral and physiological profiles over a broad range of novel or unexpected nonsocial and social situations and events (Clarke and Mason 1988; Clarke, Mason, and Moberg 1988a, 1988b; Kling and Orbach 1963). It would thus appear that the critical antecedent condition differentiating these macaque species is not the degree to which they strive to achieve dominance, but their generalized modes of responding to provocative events (Mason 1964, 1976).

The number of potentially provocative events within a social group are many. Provocation can lead to an aggressive interchange, the outcome of which can be construed in terms of *winner* and *loser* or dominant and subordinate. Although this interpretation may convey something important about one aspect of the relationship between the participants, it does not imply that a struggle or contest has taken place over the abstract prize of social dominance. Status differentiation is a consequence of the agonistic encounter, not its cause.

The distinction between cause and consequence is particularly difficult to keep clear because strong tendencies toward aggression and well-developed relations of dominance and subordination are usually associated. Rhesus monkeys and several other species of macaques are prime examples. When two adult males from these species encounter each other for the first time, an exchange of threats or a fight is likely. In this event, it is also likely that one of the participants will show submissive signals. Dominance/subordination is now included as an aspect of their developing relationship. It does not follow, of course, that the agonistic interaction was caused (or motivated) by a desire to achieve higher status. In fact, sudden encounters between strangers are typically arranged by experimenters, and many factors are present in these situations that are conducive to aggression. The setting is usually unfamiliar, space for maneuvering may be limited, and each participant is likely to experience some degree of fear, ambiguity, ambivalence, and uncertainty with respect to the other. Given the ample provocations to aggression of such arranged encounters, it is noteworthy that dominance and subordination are often settled quickly, frequently in the absence of serious fighting, even when multiple animals are involved (Barchas and Mendoza 1984; Bernstein and Mason 1963; Hawkes, 1970; Maxim 1976, Mendoza and Barchas 1983; Thompson 1967). Although this outcome clearly suggests that some species are strongly disposed to incorporate dominance and subordination as elements in their social relationships, there is no need to assume a specific motive to become dominant, any more than to assume a specific motive to become subordinate. Clearly, the causes of the outcome need to be examined separately from the outcome itself.

Also contrary to a dominance-striving hypothesis is the finding that rhesus males joining a new group (whether freely or as the result of external manipulations) are generally ranked near the bottom of the status hierarchy and do not contest this position. A study by Bernstein and Gordon (1980) is particularly instructive because the two highest ranking males from each of two groups were alternately placed in the other group. In both instances, the newly introduced males were subordinate to the resident males over whom they were dominant in their own groups, and, in fact, they became the lowest ranking animals in the host group. When returned to their own groups, they immediately resumed their original ranks. Vessey and Meikle (1987) reviewed the findings on male migration among free-living rhesus monkeys. They found that males joining a group were at or very near the bottom of the hierarchy and moved up only as the higher ranking males left or died. There is no suggestion in their report of ongoing struggles to achieve high status.

Another type of observation that seemingly supports a dominance-striving hypothesis is described as a challenge, overthrow, or takeover, implying that the critical determinant is a quest for higher status. The phenomenon suggesting this interpretation is recurrent aggression directed by a lower ranking animal toward another of higher rank, which may lead to a reversal in status relationships between the participants. Here, too, the minimax model offers a plausible alternative interpretation that does not invoke the concept of status-striving.

The essential feature of this interpretation is that it looks to the quality of the total relationship between the two individuals for an explanation of the event. The social primates are characterized by a high degree of interdependence among many members of the group. Interdependence is based to a large extent on the historical details of a relationship, and reflects each individual's expectations regarding the probable behaviors of those with whom it interacts. Interdependence thus engenders constraints on each participant in a relationship. In species showing strong aggressive tendencies and sharp status differentiation, one type of constraint concerns the expectations of the subordinate that some of its actions are likely to have aversive consequences—for example, elicit an aggressive response. Such constraints can be oppressive. If so, individuals are expected to take whatever steps they can to eliminate them or mitigate their effects. If the circumstances are right this may lead to aggressive social conflict.

What constitutes the right set of circumstances is a matter of conjecture at this point. It does not appear to be the perceived loss of fighting ability by the individual of higher rank, for there are many reports of older or debilitated individuals of higher status maintaining their positions in spite of the presence of younger or more vigorous animals of lower rank (de Waal 1975). I suggest that differences in power, rank, or fighting ability between participants in a relationship are constraints that need not be strongly oppressive. The critical issue is not social status per se, but the extent to which the relationship is oppressive and the type of satisfactions it affords. When the oppressive constraints in a relationship consistently exceed the satisfactions for one of the participants, the potential for conflict is high. Aggression against the oppressor may be one response, but other options are available. Some form of conciliation or active accommodation to the oppressor are possibilities, as are avoidance, interacting selectively with individuals that offer a better combination of minimal oppressive constraints and maximal social satisfactions, and, in field settings, emigration to another group. On occasion, all these options are used.

In conclusion, there is no compelling evidence that the achievement of high status exists as an end in itself. Primates generally act so as to maximize their personal freedom and mobility under demanding circumstances. They compete for mutually desired incentives. They thwart each other in myriad ways, and they must deal with uncertainty, ambiguity, and change on a daily basis. Provocations to conflict are many, as are the modes of responding to them. In some species agonistic behavior is a common reaction, and it is expressed in a broad range of conflictual situations. It also appears that it is these species that show the most clear-cut differentiation of status. Although status differences may be established or reinforced in conflictual situations, it is argued here that the conflict is not about a contention for status.

Social Context

Another complexity that must be dealt with in any discussion of the proximate sources of social conflict is the immediate social setting (Maxim 1982). It is frequently the case among group-living primates that conflicts do not occur as spatially isolated dyadic events, but in proximity to other animals. The number of individuals present, their arrangements in space, and their interrelationships can influence the occurrence, the form, and the outcome of social conflict.

The descriptive literature contains many illustrations of the diverse influences of context on social conflict. Whether an individual chooses to approach another and whether it initiates affiliative behavior or displays threatening or appeasing gestures will often depend on the other animals that are in the vicinity and their relationships with the actor and the recipient. Bystanders may intrude on and disrupt ongoing affiliative interactions, such as sex or grooming, even when they have no special relationship with the participants (harassment). They may intercede spontaneously or as the result of recruitment on behalf of one of the participants in an agonistic conflict (agonistic aiding, alliances, coalitions, or incitement). They may also cause social conflict to extend to uninvolved bystanders (redirected aggression, scapegoating, or social contagion).

In dealing with the diverse and complex effects of context on social conflict, it may help to think of them in terms of conditional probabilities in a tripartite situation. The three elements in the situation are the actor (A), the target (B) and the social context (C). Under this formulation, C must be construed broadly. It may include the mere proximity of other animals, such as a dominant male, kin of A or B, or virtually any other set of bystanders with whom neither A nor B are presently interacting. The essential question concerns the differential effects of the added element (C) on the actor's response to the target animal—and vice versa.

Social Disorganization

The last factor to be considered as a potential source of social conflict is more pervasive and indirect in its effects than those described thus far. An organized primate group is characterized by interconnectedness or interdependence among its members. Animals move together, sleep together, and carry out their daily routines together. Organization is also seen in the highly structured patterns of interindividual relationships, expressed in spatial arrangements, and in who does what to whom, how often, and under what circumstances. Social organization is never static, of course, but it is usually sufficiently stable to provide experienced members of the group with a shared core of knowledge about the environment and each other, and to create in each of them some sense that their social world is orderly and predictable.

When this sense is violated at the same time in many members of a social group, the existing organization is disrupted, and the likelihood of social conflict is increased. Major perturbations in the social order can be brought about either through changes within the group or by events originating outside the group that impinge on all or most of its members. Disturbances arising within the group may be the result of such factors as the loss of a particularly influential individual or subset of individuals, an increase in the size of the group to the point at which coordination of group movements and other activities breaks down, or age-related changes in behavior associated with the attainment of reproductive maturity or the onset of senescence. Disturbances originating outside the group include environmental factors, such as a reduction in food supplies, space, and other important resources owing to drought or habitat destruction. They may also result from increases in predation pressure or from harassment by conspecifics living outside the group.

FORMS OF CONFLICT

Beyond recognizing the diversity of sources of social conflict, an integrated perspective needs to consider the structure, content, and significance of specific instances of conflict. This task requires a distinction between two types of conflict, one occurring within the individual participants, and the other generated by the differences between them. To illustrate, consider the following episode: One monkey is eating in a choice patch of food when it notices a larger and stronger member of its group approaching. The smaller individual continues to eat. When the larger one threatens it, the smaller one stops eating but does not move. The larger animal slaps it, and the smaller monkey runs away. Obviously, at some point during this episode, the victim was in a state of *intraindividual* conflict. It had to decide whether or when to abandon the food. It is also clear from the actions of both animals that an *interpersonal* conflict had taken place, and that the outcome of the episode was potentially significant for both participants.

The intimate interplay between intraindividual and interpersonal conflict in this hypothetical episode is characteristic. Nevertheless, the two forms of conflict are conceptually distinct. They require different operational criteria and different formal models. Each form of conflict will be considered separately in this chapter.

Intraindividual Conflict

Intraindividual conflict is a hypothesized state within the individual. It is not limited to social settings, and occurs whenever two or more incompatible courses of action of roughly equivalent strength are aroused simultaneously. A simple and useful paradigm for thinking about intraindividual conflict is presented in figure 2.1 (Brown 1971; Miller 1944). Intraindividual conflict occurs when a stimulus complex or situation (S) has a high probability of arousing two different tendencies (T1 and T2) to perform incompatible actions (R1 and R2).

Conflicts are classified into three major groups, based on the conditions in which the conflict occurs.

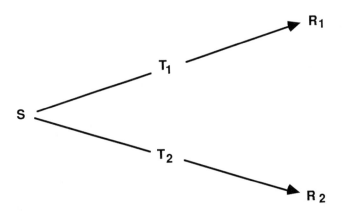

Figure 2.1 Schematic representation of intraindividual conflict. Conflict occurs when a situation or stimulus complex (S) simultaneously arouses two different tendencies (T_1 and T_2) to perform incompatible actions (R_1 and R_2).

1. Spatial conflict. Three types of spatial conflict are distinguished. In a spatial *approach-avoidance* conflict, the organism is simultaneously attracted to and repelled from the same locus or object in its environment. The tendencies to approach and avoid both increase as the individual draws closer to the goal, but the tendency to avoid increases more sharply than that of approaching, causing the individual to stop short of the goal. In a spatial *avoidance-avoidance* conflict, the individual, while attempting to avoid one object, must move toward another that is equally threatening. In a spatial *approach-approach* conflict, the individual is simultaneously drawn toward two objects that occupy separate places.

2. Uncertainty-induced conflicts. Conflicts that occur when an individual is required to make a discrimination between cues which are similar but lead to different outcomes—such as reward or punishment—are described by Brown as discrimination-induced conflicts (Brown 1971; Miller 1944). This is only one step removed from the condition in which the actor is simply uncertain as to whether the outcome of its response to a specific stimulus on any given occasion will be positive or negative, and it is therefore conflicted (Berlyne 1960; Garner 1962).

3. Temporal conflict. Temporal conflict occurs when the strength and degree of equality between competing tendencies increases as a function of nearness in time to the event that elicits these tendencies.

    In a social context the presence of intraindividual conflict can usually be inferred from the following criteria:

1. Situational. At least two distinct and mutually exclusive alternatives are afforded by the situation, or there are grounds for assuming that the situation is perceived as containing an element of ambiguity or uncertainty.

2. Behavioral. The conflicted individual exhibits signs of ambivalence. For example, it freezes, vacillates between two different

and mutually incompatible courses of action, and shows overt indications that it is agitated, aroused, or distressed.

3. Physiological. When physiological measures are possible, they are expected to indicate high levels of arousal as reflected, for example, in autonomic activity—such as heart rate—and measures of pituitary-adrenal responsiveness—as in cortisol and catecholamines.

Thus, intraindividual conflict is a state of psychological disequilibrium, a source of tension, or emotional arousal. This state need not be wholly aversive. On the contrary, inasmuch as moderate increments in arousal have positive motivational effects, individuals will seek out or place themselves in situations that are mildly conflictual (Berlyne 1960; Mason 1964, 1971; Menzel 1964, 1966; Sackett 1965). This seems to contribute to the attraction to novel objects, to playful social interactions, and to a variety of events that fall somewhat outside the usual range of experience. In contrast to the positive qualities of mild conflict, experimental research shows that intense or sustained intraindividual conflict is a purely aversive state, both behaviorally and physiologically disruptive, and carrying great potential for harmful effects on the conflicted individual (Brown 1971; Frankenhaeuser 1983; Henry 1975; Liddell 1944; Masserman 1972; Masserman and Pechtel 1953; Miller 1982; Weinberg and Levine 1980; Weiss 1984).

For obvious reasons, few, if any, intraindividual conflicts within established natural groups reach the degree of severity demonstrated in the laboratory. There is space to maneuver, the animals are known to each other, and individuals usually have the option of choosing those with whom they will interact or avoid.

This process of assessment and selection of companions is no doubt influenced by many factors. Among the more important of these are the perceived attributes of the partner, which may be termed its *social valence* (Carpenter 1945; Hinde 1976; Kummer 1978; Mason 1971, 1978). Social valence is a form of social attribution. It refers to the positive or negative qualities of one individual from the standpoint of another. These qualities depend on such general characteristics as gender, age, size, and social comportment—also called *bearing* or *style*. Most particularly, however, they depend on the history of the relationship between the individuals.

Valences provide a means of characterizing one aspect of a dyadic relationship with reference to each participant. Valences may be similar between participants, or they may differ widely in strength or direction. Strongly positive valence is characterized by the general tendency to move toward, and is often described in such terms as *affiliation, attraction, attachment, friendship*, and the like. B's positive valence for A is inferred when A acts as though B is able to provide social commodities or satisfactions which A seeks, such as sex, play, grooming, warmth, reassurance, and similar activities. Strongly negative valence is characterized by one of two contrasting general behavioral tendencies. One is *movement against*. In this case, negative valence of B for A is inferred when A shows intolerance for B, as expressed in physical attacks, threats, and chases—responses that are indicative of A's hostile, aggressive, or antagonistic stance toward B. The other general behavioral tendency associated with negative valence is *movement away*, or fear. In this case, negative valence of B for A is reflected in B's ability to displace A, to disrupt A's ongoing behavior, or to deter or deflect it from some intended course of action. Fear is also manifest in some species in a variety of distinctive postures, facial gestures, and vocalizations.

Most often, valences are mixed. Established relationships usually involve some mingling of positive and negative features for both participants, and their influence can vary with circumstances. Thus, B's perceived value for A reflects not only the historical qualities of their relationship, but also A's current motivational state, and what it is seeking to gain from interacting with B. Naturally, the valence of either participant may be changed significantly by the quality of any interaction between them. It is appropriate, however, to consider most moment-to-moment variations in social valence as transient departures from some norm or baseline value that remains fairly stable over long periods. This appears to characterize relationships among matrilineal kin in macaques; between some dominants and subordinates in chimpanzees, baboons and macaques; between some pairs of adult males and females in baboons and macaques; and between most pairs in monogamous primates, such as *Callicebus*.

To the extent that individuals have generalized expectancies as to how specific others are likely to respond to them, they may be said to be aware of the valence which these others attribute to them. It also seems likely that in some species the valence of a single individual may be similar for many animals within the social group (Chance 1967). Members of the group may even be cognizant of this consensus (Anderson and Mason 1974, 1978; Bernstein 1964).

The concept of valence is critical to the application of Brown's classification of intraindividual conflict to social settings. The most common forms of conflict in such settings are probably spatial approach-avoidance conflict and uncertainty-induced conflict. Ambivalence is engendered because an individual is simultaneously drawn and repelled from the same region in space or seeks to perform some action, but is unsure as to the positive or aversive consequences of its actions. In naturalistic social settings, it is often difficult to distinguish between approach-avoidance and uncertainty-induced conflict. For example, intraindividual conflict might occur because A is attracted to B, who is sitting in proximity to C who is feared, or because A is simultaneously attracted to and fearful of the same individual. These might be construed as instances of spatial approach-avoidance. It may also be argued, however, that, in the foregoing situation and many like it, there is inevitably an element of uncertainty. The actor has no completely reliable basis for predicting how another will respond to it on any given occasion and frequently must rely on subtle cues indicating the other's current mood or probable reactions. Common examples in which uncertainty appears to be prominent include an infant approaching its mother during the weaning process (Will she reject it on this occasion or permit it to nurse?); an estrus female soliciting sex from a seemingly indifferent and potentially aggressive male (Will he attack or copulate with her?); and a subordinate animal approaching a food source that is near a more dominant individual (Is he satiated?).

Most of the other types of conflict described by Brown may be less frequent in a naturalistic social context. Spatial avoidance-avoidance conflict requires that the actor has no choice but to approach one of two equally-feared individuals. This may be relatively uncommon in natural social groups where the actor usually has the option of quitting the field. It may be imagined, however, that there are occasions when this option is not available. This might occur, for example, when two feared individuals move into proximity to the actor, and it can only withdraw from the situation by approaching one or the other. Spatial approach-approach conflict is exemplified in any situation in which two equally attractive and spatially separate alternatives are available—to seek grooming or to seek food; to join a recently formed play group; or to follow one's mother. Presumably, such conflicts are seldom severe or prolonged. Temporal conflicts may be fairly common in primate societies, inasmuch as social groups tend toward cyclicity throughout the day and year. Temporal conflicts may arise in a male with respect to a particular female (whom he fears) when she approaches her time of maximal receptivity and sexual attractiveness. (In the same circumstances, temporal conflicts may also arise in a female with respect to a male whom she fears and usually avoids.) These conflicts may also occur in less dramatic form, as in the individual who continues to feed at a particularly rich source while its group moves on. At what time do the negative effects of separation override the attraction to food?

Interpersonal Conflict

In contrast to intraindividual conflict, interpersonal conflict is uniquely social. It represents an emergent property within a social interchange, indicated by some element of incompatibility, mismatching or resistance in the interbehaviors of the participants.

All but the simplest dyadic social interchanges can be viewed as distinct and temporally bounded episodes with a beginning, a middle, and an end. The interchange is initiated when one individual attempts to alter its current association with another. The middle comprises a series of mutually contingent responses organized around the attempted change. The end of the interchange occurs with the initiator either achieving the sought-after change or failing to do so. This formulation implies that the participants are engaged in a transactional process concerning an outcome of interest to at least one of them, more often both (Mason 1979). Interpersonal conflict can emerge at any point in this process.

Virtually any behavior in an individual's repertoire that impinges significantly on another can precipitate interpersonal conflict. Responses to interpersonal conflict are also highly variable. Because of this diversity, the form of behaviors shown during a conflictual interchange is, per se, an unreliable guide to the source or structure of the social conflict. For example, a threat face may function in macaques to deter a competitor, recruit an ally, redirect an incipient attack by another, or sexually incite the other (Altmann 1962; Anderson and Mason 1974, 1978; de Waal, van Hooff, and Netto 1976; Zumpe and Michael 1970). Whether this behavior serves as an inducer of interpersonal conflict, a manifestation of conflict in progress, or a way of resolving conflict depends on the immediate social context, the history of the relationship between the participants, and their current state. This is by no means an isolated example. It is generally the case in primate social interactions that diverse behaviors may be functionally equivalent, and a given species-specific behavior pattern may serve various functions. Thus, it can be seen why interpersonal conflict is an emergent property within a social transaction. The specific behaviors of the participants are relatively uninformative when viewed outside the context of the transaction in which they occur. Their mere presence does not indicate that a transaction is conflictual nor does it show where the conflict emerges. The critical characteristic is the element of interpersonal incompatibility in the transaction itself.

The emphasis on interpersonal conflict as an emergent phenomenon within a transaction is different from the prevailing approach, which focuses on aggression as the principal form of social conflict. The transactional orientation is broader and is explicitly concerned—as the prevailing approach is not—with the structure of interpersonal conflict. It is meant to distinguish among the presumptive sources of the conflict, the behavioral manifestations of conflict in each of the individuals involved, and the ways that they deal with their interpersonal conflict.

Transactions are most often initiated by an approach. The expectations of the participants and their motivational/emotional state will have an important influence on the ensuing interchange, but do not determine it directly. For example, one or both of the participants may be ambivalent toward the other. A state of ambivalence is inherently unstable with respect to its effects on behavior, and it increases the likelihood that interpersonal conflict will occur.

Smuts describes an incident involving a young female baboon in her first sexual swelling who was strongly ambivalent toward the males she approached. She backed slowly toward a couple of males, glancing nervously over her shoulder while presenting her perineum. Both males ignored her. The next male she approached indicated some interest. He leaned forward and sniffed at her rear, whereupon she "leaps away with a cry, and then turns to look...from a few feet away, her tail up in alarm and her mouth curved into a fear-grin" (Smuts 1985, 205).

Interpersonal conflict is particularly likely if both participants are ambivalent. No doubt this contributes to the high incidence of conflict when animals encounter each other for the first time (Kummer 1975; Maxim 1976, 1978; Wade 1977).

Ambivalence existing at the outset of a transaction may also be resolved as the interchange proceeds. Maxim describes the initial thirteen minutes of interaction between an adult male and female rhesus (Maxim 1982). During the first few minutes the female stared at the male while he repeatedly lipsmacked and presented to her. After about four minutes of this, she began to approach the male. In the ninth minute, she made her first sexual present to him, whereupon he mounted. The last few minutes consisted of additional presents by the female, mountings by the male, and grooming. Maxim describes this sequence as one of "progressive negotiation," resulting in the establishment of a consort relationship (1982, 81).

Interpersonal conflict may also occur, of course, when there is no prior state of ambivalence in either participant. A monkey soliciting sex may be ignored by the object of its attentions, an instance of interpersonal conflict that presumably is also frustrating for the soliciting monkey. It may respond to this indifference with some hostile act, thus terminating the transaction and inducing intraindividual conflict in the recipient. It may also respond to the passivity of the recipient with renewed efforts that meet with success, thus effectively resolving the conflict, presumably to the satisfaction of both participants. A final possibility is that the recipient actively rejects the solicitor by an aggressive act, to which he responds in kind, whereupon the transaction is terminated leaving both participants in a state of intraindividual conflict. All transactions in these simple scenarios contain an element of interpersonal conflict, but the structure and outcome vary with the nature of the interchange between the participants.

Figure 2.2 presents a classificatory scheme for describing the patterns that a single conflictual transaction can assume. Although the relationship between intraindividual and interpersonal conflict in this scheme is intimate, it is also complex and indirect. In formal terms, the transaction is viewed as part of a sequence that includes the following components:

1. The state of participants immediately before the transaction begins. (+ = presence of intraindividual conflict, 0 = absence of conflict). One [+,0], both [+,+], or neither [0,0] of the participants may be in a state of conflict at the beginning of a transaction. Presumably, an aggressive mood at the outset of a transaction often reflects a state of intraindividual conflict induced, for example, by ambivalence, uncertainty, frustration, or fear (Archer 1976).

2. The transaction. By definition, a transaction requires (at a minimum) one response from each participant and is therefore represented as two moves, M_1 and M_2 (Maxim 1976; Mendoza and Barchas 1983). A transaction need not be conflictual ($M_1O \rightarrow M_2O$), or interpersonal conflict (C) may emerge at any point during the interchange. Conflict may also be resolved within the interchange itself ($M_1C \rightarrow M_2O$).

3. The outcome of the transaction. The transaction may end without conflict for either party [0,0], or it may leave one or both of participants in a state of intraindividual conflict with respect to the other ([+,0], [+,+]).

CONFLICTUAL TRANSACTIONS (DYADIC)

1. Immediately Prior State of Participants	2. Structure of Transaction	3. Possible Outcomes for Participants
	M1 → M2	
[+,+] [+,0] [0,0]	C → C	[+,0] [+,+]
[+,+] [+,0] [0,0]	C → O	[0,0]
[+,+] [+,0] [0,0]	0 → C	[+,0] [+,+]

+ = intraindividual conflict

0 = no conflict

C = interpersonal conflict

M1→ M2 = "Moves"

Figure 2.2.   Structural model of social conflict. At the beginning of a conflict-ual transaction, each participant may or may not be in a state of intraindivid-ual conflict, indicated by + or 0. The three possibilities are represented within brackets. A conflictual transaction may be conflictual initially and terminate in conflict [C→C], be conflictual initially and terminate with the conflict resolved [C→0], or be without conflict initially, but terminate in conflict (0→C). When-ever the outcome of a transaction is conflictual, at least one participant is left in a state of intraindividual conflict (ambivalence, uncertainty) with respect to the other participant ([+,0], [+,+]).

It is clear from figure 2.2 why the relationship between intraindi-vidual and interpersonal conflict is neither simple nor direct. Interper-sonal conflict may precede and provide the occasion for intraindivid-ual conflict. In this case, events occurring during the transaction are the source of conflict (as in the sequence of Prior [0,0], Transaction [M_1C→M_2C], Outcome [+,+] or [+,0]). This might happen if each par-ticipant incorrectly anticipates the mood or intent of the other. Because of differences in the participants' expectations, the interchange engen-ders interpersonal conflict, and the transaction terminates with one or both participants left in a state of intraindividual conflict with respect to the other. If both participants are in a state of intraindividual con-flict before the transaction begins, interpersonal conflict is likely. How-ever, because it is not inevitable, the initial behavior of each participant could immediately resolve the conflict in the other (as in the sequence of Prior [+,+], Transaction [$M_1O \rightarrow M_2O$], Outcome [0,0]). A more probable scenario is that intraindividual conflict existing prior to the transaction will result in interpersonal conflict, although this could be resolved as the transaction proceeds. Thus, a mother in the gradual process of weaning her offspring may vacillate on a particular occasion between acceding to the infant's attempts to gain the nipple or preventing it from suckling. If she resolves her intraindividual conflict by denying the infant, she creates interpersonal conflict in the transaction and intraindividual conflict in the infant. This may be resolved within the transaction if the mother capitulates [$M_1C \rightarrow M_2O$]. It may also be left unresolved [$M_1C \rightarrow M_2C$]. These outcomes—[0,0] or [0,+]—presumably have different implications for subsequent transactions and for the relationship between the two participants.

CONCLUSION

The primary concern of this essay is with the identification, description and analysis of the proximate sources of social conflict within organized primate groups. I describe the approach as *psychoethological*. It is psychological because it views the behaving organism as mainly concerned with extracting information from its surroundings, comparing this with its own needs, values, and expectancies, and with acting, or at least preparing itself to act on the basis of this epistemic and evaluative process (Mason 1985). Although these information-processing activities are necessarily covert, a great deal is known about their behavioral and physiological correlates and their functional characteristics. Such information provides strong sanctions for the use of psychological constructs to describe the behavior of nonhuman primates. At the same time, it places definite constraints on how they are used (Roitblat 1987). The approach is ethological because it is focuses on the everyday behavior of animals in naturalistic social contexts, with the problems they create for each other, how they deal with them, and the consequences of their actions.

Conflict implies opposition, resistance, or incompatibility. A basic distinction is required between conflicts that occur within an individual and those that occur between individuals. Intraindividual conflict is not specifically social. It is most clearly exemplified by an individual confronted with a situation in which two opposing or incompatible tendencies of approximately equal strength are aroused simultaneously. More generally, however, it exists whenever an individual is indecisive, ambivalent or uncertain with respect to environmental events, or about the probable consequences of its actions.

Among primates, a common indication of intraindividual conflict in a social context is the occurrence of ritualized displays. These are among the most conspicuous and dramatic behaviors in a species' repertoire. They include changes in posture and facial expressions, distinctive movements, and characteristic vocalizations. Such displays are most frequent and intense during initial encounters between strangers, in response to ambiguous stimuli, or during environmental perturbations when events deviate from their normal and expected course. They are often accompanied by spatial vacillations, and other actions that bespeak a conflicted state. Although the names given to such displays (such as threat or fear-grin) imply that the displayer is providing unequivocal information about its internal state (namely, anger or fear), and about its behavioral intentions (such as attack or withdraw), the matter is not so simple. No one doubts that displays are specialized to serve a communicative function, but they do not always provide reliable information about an animal's affective state (such as fear or anger) or its probable responses (Mason 1985; Simpson 1973). Furthermore, they are only a few among the many sources of information available to individuals in the multivalent social process of communication (Hinde 1985; Owings and Hennessy 1984; Smith 1985).

In contrast to intraindividual conflict, interpersonal conflict, can occur only when animals are interacting socially. Most social interactions are episodic. These episodes, called *transactions*, are viewed as the primary structural and functional unit in the conduct of interpersonal relations. Both forms of conflict may be manifest in a transaction. However, they are conceptually and operationally distinct. The elements of resistance, opposition, or incompatibility *between* participants distinguishes interpersonal from intraindividual conflict.

Transactions are carried out mainly through the interchange of information about some matter of interest to at least one of the participants. An individual initiating a transaction is assumed to be seeking certain outcomes. Information about its current state and intentions may be conveyed to the recipient by its posture, the speed and regularity of its movements, its orientation, and by ritualized displays (Andrew 1972; Moynihan 1982; Smith 1985). Further information potentially available to the recipient is provided by the social context, and by the recipient's historical relationship with the initiator, which includes any immediately preceding transactions. The initiator has similar information available to it regarding the probable response of the recipient to the social overture.

In spite of the many sources of information available, the potential for novelty and surprise is always present. The transaction is an ongoing social process in which each participant is making and acting on predictions about the intentions and probable behavior of the other. It is a process of mutual assessment and control—inferential, probabilistic, and subject to error. In spite of the occasional error, many transactions probably proceed without overt interpersonal conflict. Some errors do, however, lead to interpersonal conflict. The likelihood that this will occur is greatest at the outset of the transaction. The initiator may have a poor basis for anticipating the recipient's response, or the recipient may be similarly uncertain regarding the intended actions of the initiator. If either participant is ambivalent toward the other, the most innocuous actions are likely to be misconstrued and give rise to interpersonal conflict.

As a transaction proceeds, it tends to reduce ambiguity and reliance on communication. Each participant's uncertainty with respect to the other decreases—but is never entirely eliminated—as perceived intentions are transformed into social acts of grooming, nursing, embracing, sex, simple propinquity, and spatial displacement. Any of these acts may precipitate interpersonal conflict. Resistance can be shown by either party; it may take the form of passive noncompliance or forthright rejection. In spite of the emergence of interpersonal conflict, some degree of mutual accommodation may be achieved as the transaction proceeds. However, it may end on a discordant note.

Presumably, what is carried forward by the participants are memories of the overall quality and the outcome of their transactions rather than a recollection of specific behavioral details. Quality and outcome are critical factors in the historical process of forming interpersonal relationships. Every transaction provides each participant with a specific occasion to develop, confirm, or revise its expectancies regarding the other. Transactions are largely responsible for the development and maintenance of social valence, and are, thus, major determinants of the who, the how, and the why of future interactions.

Transactions are a primary source of information about other members of the group, but not necessarily the only one. Primates tend to watch each other closely. By monitoring social interactions in which they are not directly involved, it is likely that individuals in some species gain information about social relations among members of their group and about their probable actions in different circumstances (Anderson and Mason 1974, 1978; Cheney and Seyfarth 1980, 1986; de Waal 1986; Kummer 1982). Although such information may serve as a basis for action, it is presumably meager and unreliable, as compared to that obtained directly through the transactional mode.

Given the commitment of the monkey or ape to sociality, its individualism, and its heavy reliance on information in the conduct of its social life, it is in no way surprising that the individual often finds itself in a conflicted state or acting at cross-purposes with other members of its group. Conflict from this perspective is an inevitable accompaniment of life in groups. Its sources are many, as are the responses it engenders.

This is not the traditional view in which competition and aggression are treated virtually as defining characteristics of social conflict. The following quotation is concerned with human behavior, but it comes close to the perspective many primatologists seem to accept. "Social conflict may be defined as a struggle over values or claims to status, power, and scarce resources, in which the aims of the conflicting parties are not only to gain desired values, but also to neutralize, injure, or eliminate their rivals" (Coser 1968, 232). Insofar as the nonhuman primates are concerned, this is a biased and impoverished metaphor that does not come close to conveying the richness, the subtlety, the pervasiveness of conflict, and its powerful influence on the social dynamics of organized groups.

REFERENCES

Altmann, S. A. (1962) A field study of the sociobiology of rhesus monkeys, *Macaca mulatta*. *Annals of the New York Academy of Sciences* 102:338–435.

Anderson, C. O.; Mason, W. A. (1974) Early experience and complexity of social organization in groups of young rhesus monkeys (*Macaca mulatta*). *Journal of Comparative and Physiological Psychology* 87:681–690.

Anderson, C. O.; Mason, W. A. (1978) Competitive social strategies in groups of deprived and experienced rhesus monkeys. *Developmental Psychobiology* 11:289–299.

Andrew, R. J. (1972) The information potentially available in mammal displays. In Hinde, R. A. (ed.), "Non-verbal Communication." Cambridge: Cambridge University Press. 179–206.

Archer, J. (1976) The organization of aggression and fear in vertebrates. In Bateson, P. P. G., Klopfer, P. (eds.), "Perspectives in Ethology," vol. 2. New York: Plenum Press, 231–298.

Barchas, P. R.; Mendoza, S. P. (1984) Emergent hierarchical relationships in rhesus macaques: An application of Chase's model. In Barchas, P. R. (ed.), "Social Hierarchies: Essays toward a Sociophysiological Perspective." Westport, Conn.: Greenwood Press. 81–95.

Berlyne, D. E. (1960) "Conflict, Arousal, and Curiosity." New York: McGraw-Hill.

Bernstein, I. S. (1964) Role of the dominant male rhesus monkey in response to external challenges to the group. *Journal of Comparative and Physiological Psychology* 57:404–406.

Bernstein, I. S.; Gordon, T. P. (1980) The social component of dominance relationships in rhesus monkeys (*Macaca mulatta*). *Animal Behaviour* 28:1033–1039.

Bernstein, I. S.; Mason, W. A. (1963) Group formation by rhesus monkeys. *Animal Behaviour* 11:28–31.

Brown, J. S. (1971) Principles of intrapersonal conflict. In Smith, C. G. (ed.), "Conflict Resolution: Contributions of the Behavioral Sciences." Notre Dame, Ind.: University of Notre Dame. 88–97.

Carpenter, C. R. (1945) Concepts and problems of primate sociometry. *Sociometry* 8:56–61.

Chance, M. R. A. (1967) Attention structure as the basis of primate rank orders. *Man* 2:503–518.

Cheney, D. L.; Seyfarth, R. M. (1980) Vocal recognition in free-ranging vervet monkeys. *Animal Behaviour* 28:362–367.

Cheney, D. L.; Seyfarth, R. M. (1986) The recognition of social alliances by vervet monkeys. *Animal Behaviour* 34:1722–1731.

Clarke, A. S.; Mason, W. A. (1988) Differences among three macaque species in responsiveness to an observer. *International Journal of Primatology* 9:347–364.

Clarke, A. S.; Mason, W. A.; Moberg, G. P. (1988a) Differential behavioral and adrenocortical responses to stress among three macaque species. *American Journal of Primatology* 9:347–364.

Clarke, A. S.; Mason, W. A.; Moberg, G. P. (1988b) Interspecific contrasts in responses of macaques to transport cage training. *Laboratory Animal Science* 38:305–309.

Cloudsley-Thompson, J. L. (1965) "Animal Conflict and Adaptation." Chester Springs, Pa.: Dufour Editions.

Coombs, C. H. (1987) The structure of conflict. *The American Psychologist* 42:355–363.

Coser, L. (1968) Conflict: Social Aspects. In International Encyclopedia of the Social Sciences, vol. 3. 232–236.

Craig, W. (1921) Why do animals fight? *International Journal of Ethics* 31:264–278.

Darwin, C. (1872) "The Expression of the Emotions in Man and Animals." Reprinted in "The Origin of Species and The Descent of Man." New York: The Modern Library.

deWaal, F. B. M. (1975) The wounded leader: A spontaneous temporary change in the agonistic relations among captive Java-monkeys (*Macaca fascicularis*). *Netherlands Journal of Zoology* 25:529–549.

deWaal, F. B. M. (1986) Class structure in a rhesus monkey group: The interplay between dominance and tolerance. *Animal Behaviour* 34:1033–1040.

deWaal, F. B. M.; van Hooff, J.; Netto, W. J. (1976) An ethological analysis of types of agonistic interaction in a captive group of Java-monkeys (*Macaca fascicularis*). *Primates* 17:257–290.

Erikson, E. H. (1963) "Childhood and Society," 2nd ed. New York: W. W. Norton.

Frankenhaeuser, M. (1983) Human psychophysiology. The sympathetic-adrenal and pituitary-adrenal response to challenge: Comparison between the sexes. In Dembroski, T. M., Schmidt, T. H., Blumchen, G. (eds.), "Biobehavioral Bases of Coronary Heart Disease." Basel: Karger. 91–105.

Freud, S. (1943) "A General Introduction to Psychoanalysis." Garden City, N.Y.: Garden City Publishing Co.

Garner, W. R. (1962) "Uncertainty and Structure as Psychological Concepts." New York: John Wiley.

Gause, G. F. (1934) "The Struggle for Existence." Baltimore: Williams and Wilkins.

Hand, J. L. (1986) Resolution of social conflicts: Dominance, egalitarianism, spheres of dominance, and game theory. *Quarterly Review of Biology* 61:201–220.

Hardin, G. (1960) The competitive exclusion principle. *Science* 131:1292–1297.

Hawkes, P. N. (1970) Group formation in four species of macaques in captivity. Doctoral dissertation, University of California at Davis.

Henry, J. P. (1975) The induction of acute and chronic cardiovascular disease in animals by psychosocial stimulation. *International Journal of Psychiatry in Medicine* 6:147–158.

Hinde, R. A. (1976) On describing relationships. *Journal of Child Psychology and Psychiatry* 17:1–19.

Hinde, R. A. (1985) Expression and negotiation. In Zivin, G. (ed.), "The Development of Expressive Behavior: Biology-Environment Interactions." New York: Academic Press. 103–116.

Kling, A.; Orbach, J. (1963) Plasma 17-hydroxycorticosteroid levels in the stumptailed monkey and two other macaques. *Psychological Reports* 13:863–865.

Kummer, H. (1975) Rules of dyad and group formation among captive gelada baboons (*Theropithecus gelada*). In Kondo, S., Kawai, M., Ehara, A., Kawamura, S. (eds.), "Proceedings from the Symposia of the Fifth Congress of the International Primatological Society, Nagoya, Japan, August 1974." Tokyo: Japan Science Press. 129–159.

Kummer, H. (1978) On the value of social relationships to nonhuman primates: A heuristic scheme. *Social Science Information* 17:687–705.

Kummer, H. (1982) Social knowledge in free-ranging primates. In Griffin, D. R. (ed.), "Animal Mind–Human Mind." Berlin: Springer-Verlag. 113–130.

Lahiri, R. K.; Southwick, C. H. (1966) Parental care in Macaca sylvana. *Folia Primatologica* 4:257–264.

Liddell, H. S. (1944) Conditioned reflex method and experimental neurosis. In Hunt, J. M. V. (ed.), "Personality and the Behavior Disorders." New York: Ronald Press. 389–412.

Marler, P. (1976) On animal aggression: The roles of strangeness and familiarity. *The American Psychologist* 31:239–246.

Maslow, A. H. (1935) Individual psychology and the social behavior of monkeys and apes. *International Journal of Individual Psychology* 1:47–59.

Mason, W. A. (1964) Sociability and social organization in monkeys and apes. In Berkowitz, L. (ed.), "Recent Advances in Experimental Social Psychology." New York: Academic Press. 277–305.

Mason, W. A. (1971) Field and laboratory studies of social organization in *Saimiri* and *Callicebus*. In Rosenblum, L. A. (ed.), "Primate Behavior: Developments in Field and Laboratory Research," vol. 2. New York: Academic Press. 107–137.

Mason, W. A. (1976) Primate social behavior: Pattern and process. In Masterton, R. B., Bitterman, M. E., Campbell, C. B. G., Hotton, N. (eds.),

"Evolution of Brain and Behavior in Vertebrates." Hillsdale: Lawrence Erlbaum Assoc. 425–455.

Mason, W. A. (1978) Ontogeny of social systems. In Chivers, D. J., Herbert, J. (eds.), "Recent Advances in Primatology," vol. 1. London: Academic Press. 5–14.

Mason, W. A. (1979) Ontogeny of social behavior. In Marler, P., Vandenbergh, J. G. (eds.), "Handbook of Behavioral Neurobiology" vol. 3, "Social Behavior and Communication." New York: Plenum Press. 1–28.

Mason, W. A. (1985) Experiential influence on the development of expressive behaviors in rhesus monkeys. In Zivin, G. (ed.), "The Development of Expressive Behavior: Biology-Environment Interactions." New York: Academic Press. 117–152.

Masserman, J. H. (1972) Psychotherapy as the mitigation of uncertainties. *Archives of General Psychiatry* 26:186–188.

Masserman, J. H.; Pechtel, C. (1953) Conflict-engendered neurotic and psychotic behavior in monkeys. *Journal of Nervous and Mental Diseases* 118:408–411.

Maxim, P. E. (1976) An interval scale for studying and quantifying social relations in pairs of rhesus monkeys. *Journal of Experimental Psychology: General* 105:123–147.

Maxim, P. E. (1978) Quantification of social behavior in pigtail monkeys. *Journal of Experimental Psychology: Animal Behavior Processes* 4:50–67.

Maxim, P. E. (1982) Contexts and messages in macaque social communication. *American Journal of Primatology* 2:63–85.

Mendoza, S. P.; Barchas, P. R. (1983) Behavioral processes leading to linear status hierarchies following group formation in rhesus monkeys. *Journal of Human Evolution* 12:185–192.

Menzel, E. W., Jr. (1964) Patterns of responsiveness in chimpanzees reared through infancy under conditions of environmental restriction. *Psychologische Forschung* 27:337–365.

Menzel, E. W., Jr. (1966) Responsiveness to objects in free-ranging Japanese monkeys. *Behaviour* 26:130–149.

Miller, N. E. (1944) Experimental studies of conflict. In Hunt, J. M. V. (ed.), "Personality and the Behavior Disorders." New York: Ronald Press. 431–465.

Miller, N. E. (1982) Motivation and psychological stress. In Pfaff, D. W. (ed.), "The Psychological Mechanisms of Motivation." New York: Springer-Verlag. 409–432.

Moynihan, M. (1982) Why is lying about intentions rare during some kinds of contests? *Journal of Theoretical Biology* 97:7–12.

Owings, D. H.; Hennessy, D. F. (1984) The importance of variation in sciurid visual and vocal communication. In Murie, J. O., Michener, G. R.

(eds.), "The Biology of Ground-Dwelling Squirrels: Annual Cycles, Behavioral Ecology, and Sociality." Lincoln: University of Nebraska Press. 169–200.

Rapoport, A. (1974) "Conflict in Man-Made Environment." Harmondsworth, England: Penguin Books, Ltd.

Roitblat, H. L. (1987) "Introduction to Comparative Cognition." New York: W. H. Freeman.

Rosenblum, L.A.; Kaufman, I.C.; Stynes, A. J. (1964) Individual distance in two species of macaque. *Animal Behaviour* 12:338–342.

Sackett, G. P. (1965) Effects of rearing conditions upon the behavior of rhesus monkeys (*Macaca mulatta*). *Child Development* 36:855–868.

Scott, J. P. (1977) Agonistic behavior: Adaptive and maladaptive organization. In McGuire, M. T., Fairbanks, L. A. (eds.), "Ethological Psychiatry: Psychopathology in the Context of Evolution Biology." Orlando, Fla.: Grune & Stratton. 193–209.

Simonds, P. E. (1965) The bonnet macaque in South India. In DeVore, I. (ed.), "Primate Behavior: Field Studies of Monkeys and Apes." New York: Holt, Rinehart & Winston. 175–196.

Simpson, M. J. A. (1973) Social displays and the recognition of individuals. In Bateson, P. P. G., Klopfer, P. H. (eds.), "Perspectives in Ethology," vol. 1. New York: Plenum Press. 225–279.

Smith, W. J. (1985) Consistency and change in communication. In Zivin, G. (ed.), "The Development of Expressive Behavior, Biology-Environment Interactions." New York: Academic Press. 51–76.

Smuts, B. B. (1985) "Sex and Friendship in Baboons." New York: Aldine.

Sugiyama, Y. (1971) Characteristics of the social life of bonnet macaques (*Macaca radiata*). *Primates* 12:247–266.

Thierry, B. (1985) Patterns of agonistic interactions in three species of macaque (*Macaca mulatta, M. fascicularis, M. tonkeana*). *Aggressive Behavior* 11:223–233.

Thierry, B. (1986) A comparative study of aggression and response to aggression in three species of macaque. In Else, J. G., Lee, P. C. (eds.), "Primate Ontogeny, Cognition, and Social Behaviour." Cambridge: Cambridge University Press. 307–313.

Thompson, N. S. (1967) Some variables affecting the behavior of irus macaques in dyadic encounters. *Animal Behaviour* 15:307–311.

Vessey, S. H.; Meikle, D. B. (1987) Factors affecting social behavior and reproductive success of male rhesus monkeys. *International Journal of Primatology* 8:281–292.

Wade, T. D. (1977) Complementarity and symmetry in social relationships of nonhuman primates. *Primates* 18:835–847.

Weinberg, J.; Levine, S. (1980) Psychobiology of coping in animals: The effects of predictability. In Levine, S., Ursin, H. (eds.) "Coping and Health." New York: Plenum Press. 39–59.

Weiss, J. M. (1984) Behavioral and psychological influences on gastrointestinal pathology: Experimental techniques and findings. In Gentry, W. D. (ed.), "Handbook of Behavioral Medicine." New York: The Guilford Press. 174–221.

Zumpe, D.; Michael, R. P. (1970) Redirected aggression and gonadal hormones in captive rhesus monkeys (*Macaca mulatta*). *Animal Behaviour* 18:11–19.

CHAPTER THREE

The Evolution of Social Conflict among Female Primates

JOAN B. SILK

A female's ability to reproduce successfully can be limited by the availability of essential nutritional resources. In group-living primates, this may mean that females are often in competition for the same food. This chapter develops the thesis that natural selection can be expected to favor the evolution of traits that enable females to gain access to the resources they require for successful reproduction.

These traits may include aggressive qualities that enhance a female's ability to compete effectively with other females in contests over food. They may include cognitive abilities that allow a female to weigh the relative value of a contested resource against the cost of obtaining an acceptable alternative. They may also include social skills that permit a female to establish and maintain alliances with other females who will cooperate with her in defending resources.

Thus, from an evolutionary perspective, the same ecological factors—limiting resources essential to a female's reproductive success—create selection pressures that can lead in one direction toward traits that increase the likelihood of overt social conflict, aggression, and success in direct competition, and, in another direction, toward the elaboration of cognitive and cooperative skills that mitigate aggressive conflict or avoid it.

Darwin recognized that competition and conflict are fundamental elements of the evolutionary process. In the *Origin of Species* he wrote:

Joan B. Silk—Department of Anthropology, University of California, Los Angeles

A struggle for existence inevitably follows from the high rate at which all organic beings tend to increase. Every being, which during its natural lifetime produces several eggs or seeds, must suffer destruction during some period of its life, and during some season or occasional year, otherwise, on the principle of geometrical increase its numbers would quickly become so inordinately great that no country could support the product. Hence, as more individuals are produced than can possibly survive, there must in every case be a struggle for existence, either one individual with another of the same species, or with the individuals of distinct species, or with the physical conditions of life. (Darwin 1859, 53)

If population growth is ultimately limited by the availability of resources that are needed for survival and reproduction, natural selection is likely to favor morphological or behavioral traits that confer success in the "struggle for existence." In contemporary terms, this means that individuals who are able to find, process, or exploit resources most efficiently, and those who are able to defend access to resources most effectively are likely to leave more progeny than will their peers. If access to critical resources is determined by the outcome of competitive encounters, conflict is likely to become an integral element of the behavioral strategies of many organisms.

The primary goal of this chapter is to apply an evolutionary approach to conflict among female primates, and to explore the functional significance of aggression, competition, and conflict among them. I focus upon females for several reasons.

First, my own research is principally concerned with factors that influence the life histories of *Cercopithecine* primate females, and conflict plays an important role in their lives. Second, a general consensus is emerging among primatologists that, for most primate species, the behavioral strategies of females largely determine the distribution of primate males. This, in turn, influences the extent of competition and conflict that males encounter, and shapes male reproductive strategies (Wrangham 1980). This gives logical precedence to consideration of the evolution of conflict among females. Finally, discussions of competition and conflict among primates have typically emphasized aggression among males, neglecting the empirical reality and evolutionary significance of conflict among females (Hrdy and Williams 1983). This chapter represents one of many recent efforts to offset this imbalance, among which are volumes edited by Small (1984a) and Wasser (1983a).

WHAT DO PRIMATES COMPETE OVER?

In most mammalian species, females compete primarily over food, while males compete principally over females. This disparity in competitive strategies arises from the pronounced asymmetry in the reproductive roles of mammalian males and females (Trivers 1972), and has profound implications for the evolution of competitive strategies among male and female primates (Wrangham 1980).

Primate infants cannot survive unless they receive considerable care from their mothers. The fetus is carried internally for several months after conception. Then, the infant suckles for a substantial period after birth. Mammalian reproductive physiology demands that these tasks be performed exclusively by females. While infants have other needs that could conceivably be met by males, the extent of male participation in raising young is quite variable (Whitten 1987). In some monogamous and polyandrous species, fathers carry, groom, provision, and play with their offspring more often than do mothers (Chivers 1974; Goldizen 1987; Leighton 1987; Robinson et al. 1987; Wright 1984). However, monogamy is relatively uncommon among primates (Jolly 1985). Thus, in most primate species, females bear the burdens of feeding, carrying, defending, protecting, and nurturing their offspring with little help from conspecifics. Since maternity involves a number of energetically demanding activities (Portman 1970), a female's ability to invest in her offspring is strongly influenced by her ability to obtain access to nutritional resources.

In contrast, males' reproductive success is limited principally by their ability to obtain access to receptive females. Males are likely to encounter competition over mating opportunities for two reasons. First, males are capable of inseminating many females in fairly rapid succession, and females become receptive at relatively long intervals. Second, female primates tend to aggregate in cohesive groups, perhaps in response to the distribution of resources (Wrangham 1980, 1987) or the threat of predation (van Schaik 1983; Terborgh 1983). Since a single male can potentially inseminate many females, females may represent economically defendable resources for males, as suggested by Brown (1964). The intensity of competition among males over access to females is likely to be a function of the number of females per male in social groups. Comparative studies indicate that the number of males in primate groups is correlated with the number of females (Andelman 1986) and the extent of seasonality in female receptivity (Ridley 1986). Moreover, the extent of sexual dimorphism in body weight is positively correlated with the number of females per male in social groups (Clutton-Brock et al. 1977). In monogamous species, males and females are monomorphic. In species that form groups composed of one male and many females, males are much larger than females. This is generally believed to be a result of sexual selection (Clutton-Brock et al. 1977; Gaulin and Sailer 1984; but see Leutenegger and Cheverud 1982).

CONFLICT AMONG PRIMATE FEMALES

Conflict is expected to be an important element of female behavioral strategies if access to resources limits females' reproductive success. Observations in both natural populations and provisioned groups suggest that (1) the growth of primate populations is often limited by the availability of food, and (2) access to food affects female reproductive success. These data are reviewed briefly in this section.

At several sites, changes in the availability of food resources were associated with changes in the size and composition of free-ranging, unprovisioned primate populations. As a consequence of a long period of environmental deterioration in Amboseli National Park, Kenya, the size and number of groups of baboons declined dramatically. In 1963–1964 Altmann and Altmann (1970) counted 2,500 baboons in 51 groups. In 1979, there were only 123 baboons in five groups (Altmann et al. 1985). During this period, female fecundity declined and mortality among immature animals increased, producing substantial changes in the age structure of the population (Altmann et al. 1985). Sympatric vervet monkeys in Amboseli were also adversely affected by the deterioration of the environment. The size of vervet groups declined substantially, although the number of groups remained stable. During this period, the mortality of infant and juvenile vervet monkeys increased and fecundity declined (Struhsaker 1973, 1976).

A drought had similar effects upon toque macaques in Sri Lanka, where group size declined and rates of mortality among infants and juveniles rose (Dittus 1977, 1979, 1980). Periodic shortages of important dietary items may also be responsible for cyclic declines in population size among howler monkeys on Barro Colorado Island (Milton 1982). Strum and Western (1982) report that decreases in the ungulate biomass at Gilgil, Kenya, increased the amount of food available to baboons, and corresponded to increases in female fecundity. Finally, Cheney and colleagues (1988) have found that differences in the quality of adjacent vervet monkey territories are reflected in variations in the reproductive success of females who live in them. Residents of a poor territory matured more than a year later and gave birth at shorter intervals than did residents of two better territories.

At several other sites, substantial increases in population size have been observed after artificial provisioning has been initiated. Many groups of Japanese macaques that visit feeding areas have grown rapidly (Fujii 1975; Koyama et al. 1975; Kurland 1977; Masui et al. 1975; Sugiyama and Ohsawa 1975). At Takasakiyama, for example, there were two hundred monkeys when provisioning began. Twenty-four years later, the group numbered twelve hundred individuals (Kurland 1977). In these populations, growth rates and sexual maturation were accelerated, interbirth intervals after surviving infants were shortened, natality was increased, and mortality among infants and juveniles was reduced (Mori 1979; Takahata 1980).

Provisioned, semi-free-ranging populations of rhesus macaques established on Cayo Santiago Island and La Parguera Island in the Carribbean also sustained rapid expansion. Between 1959 and 1964, the annual rate of growth on Cayo Santiago Island was 16% (Koford 1965, 1966), and, between 1976 and 1983, the annual rate of growth was 13% (Rawlins and Kessler 1986). A group of Japanese macaques, transported intact to Texas and partially provisioned, grew at an average annual rate of 8.6% (Gouzoules et al. 1982). Rapid growth rates of provisioned populations are partly due to the rapid development of infants and early onset of sexual maturity in females. On Cayo Santiago and La Parguera, female rhesus monkeys give birth to their first infant at 3.5 to 4 years, and produce infants annually (Drickamer 1974; Koford 1965). In contrast, unprovisioned female rhesus monkeys in the Himalayas give birth to their first infant when they are 4.5 to 5.5 years old, and give birth at two or three year intervals (Melnick and Pearl 1987). This difference is likely to be due at least in part to differences in food availability.

The availability of food may also influence litter size in some species. Among *Callitrichids*, litter size ranges from one to four infants (Evans and Hodges 1984). In one captive colony of cotton-top marmosets, the frequency of triplets rose from zero to one-third after the protein content of their diet was increased (Kirkwood 1983).

These data suggest that population growth and female reproductive performance are often influenced by the availability of food. This does not mean that food is the only factor that limits population growth or that the availability of food resources is of equal importance in all populations or species. Predation (Cheney and Wrangham 1987; van Schaik 1983; Terborgh 1983), disease (Coelho et al. 1977; Teas et al. 1981), parasites (Mori 1979), climate (Ohsawa and Dunbar 1984), intraspecific competition for mates (Dunbar and Sharman 1983), intergroup competition for food (Cheney et al. 1988; Dittus 1986), and interspecific competition for biotic resources (Strum and Western 1982; Terborgh 1983) are thought to regulate some primate populations. We know little about the factors that influence female reproductive performance regulation in most arboreal forest-dwelling Old World monkeys, apes, prosimians, and many New World monkeys. Nonetheless, it seems clear that access to food resources is one of the factors that affects the growth of many primate groups, and is likely to influence female reproductive performance in all primate species.

If access to food limits female reproductive performance, what types of traits do we expect will evolve among primate females? First, we expect selection to favor morphological traits that enable females to increase their ability to find, process, and digest resources efficiently. Second, we expect selection to favor behavioral traits that enable females to increase their share of available resources. There are several ways in which females might accomplish the latter.

Competition and Grouping

One way for females to increase their access to resources may be to form cooperative alliances with other females. Wrangham (1980, 1987) has argued that competition over access to food provides the primary impetus for sociality among female primates. He points out that many primates feed preferentially upon fruit, a resource that is often clumped and patchily distributed. He suggests that females who band together to defend clumps of fruit will obtain more food than will females who feed alone. Further, since females increase their inclusive fitness by helping kin, females are likely to recruit kin as allies in contests over food. If access to food is challenged frequently, alliances are likely to be formed often. Females may therefore benefit by travelling with their allies in stable parties. Cohesion may be reinforced by exchanging affiliative gestures, such as grooming. As a result, female-bonded kin groups may be established.

In all primate species, members of one or both sexes routinely leave their natal groups at puberty (Pusey and Packer 1987). If females benefit from establishing permanent alliances with kin, they are unlikely to leave their natal groups. In fact, female philopatry and male dispersal characterize the majority of nonmonogamous primate species in which dispersal patterns have been documented (Pusey and Packer 1987). Female philopatry and male dispersal occur in all but three of the Old World monkey species and in the social prosimians, capuchin species, and some squirrel monkeys. In some species of howler monkeys, gorillas, and hamadryas baboons, members of both sexes disperse from their natal groups. Female dispersal and male philopatry are known only among chimpanzees and are suspected to occur in red colobus monkeys and some squirrel monkeys (Mitchell et al. 1991).

Intergroup Conflict

If females form groups to ensure access to resources, conflict between groups is expected to be observed. In nature, competition often occurs between members of different groups over resources located inside territorial boundaries and sometimes occurs over resources located within areas of home range overlap. In territorial species, females often take active roles in intergroup encounters (Cheney 1987). In some nonterritorial species, females participate aggressively in intergroup encounters while, in other nonterritorial species, males are the principal aggressors. Cheney (1987) speculates that females may be most active when access to food resources is threatened, and males may be most active when defense of mates is jeopardized.

Intergroup encounters in territorial species normally favor residents over intruders (Cheney 1987). In species that do not defend rigid boundaries, the size of competing groups seems to play an important role in determining the outcome of intergroup encounters. All other things being equal, members of large groups are frequently able to supplant members of smaller groups from resources (Cheney 1987).

Moreover, when group size declines, autonomy is threatened. In Sri Lanka, the home range of one small group of toque macaques was absorbed into the home range of a larger neighboring group (Dittus 1986). The female members of the smaller group occupied the lowest positions in the dominance hierarchy, and did not reproduce successfully after the takeover. In Amboseli, where the size of the primate groups has declined substantially since the early 1960s, small groups have also fused. In 1972, two groups of baboons joined to form one group. For the next ten years, members of the larger original group dominated most members of the smaller group (Hausfater et al. 1982; Samuels et al. 1987). The remnants of pairs of larger groups of vervet monkeys have also fused to form new groups (Hauser et al. 1986).

Intragroup Conflict in Female-Bonded Groups

The advantages of grouping for females do not necessarily eliminate intragroup competition over access to resources. Van Schaik (1983) has pointed out that, as group size increases, female fertility often declines, presumably because females in large groups encounter higher levels of competition over access to food resources. In the Virunga Mountains, there are fewer infants per female in gorilla groups that are larger than the median size as compared to groups that are smaller than the median (Stewart and Harcourt 1987). While van Schaik (1983) argues that such results contradict predictions derived from Wrangham's model—and support his own hypothesis that predation provides the primary impetus for sociality among primates—this result is also consistent with a broader interpretation of Wrangham's model.

It is easy to imagine that there is some intermediate size at which individual reproductive success is maximized. When groups are smaller than the optimal size, residents are expected to encourage recruitment into the group. They may do so by behaving altruistically toward—and thereby enhancing the fitness of—other residents, or they may actively recruit new group members. This may account for the relaxation of xenophobic responses to female members of other groups, which permits small groups to fuse. When groups are larger than the optimal size, members of these groups are expected to discourage recruitment by behaving spitefully toward—and thereby reducing the fitness of—other residents, while actively discouraging the recruitment of new group members (Boyd 1982). This may be one of the reasons that large groups undergo fission (Chepko-Sade and Sade 1976; Furuya 1969; Koyama 1970; Malik et al. 1985; Missakian 1973; Nash 1976). It also might explain why resident females are often hostile to unfamiliar females (Cheney 1987; Shopland 1982; Southwick 1967).

If this argument is correct, then conflict should be more common among females when groups exceed the optimal size, and increases in reproductive success should follow when groups fuse and divide. There is little evidence that changes in the frequency or intensity of aggression and conflict within groups correspond directly to changes in group size or changes in the availability of food resources. However, Teas and colleagues (1981) report that the frequency of aggression among free-ranging rhesus macaques in Kathmandu was elevated during winter months when food was scarce. The reproductive success of females also increased after several successive divisions of a group of rhesus monkeys in India (Malik et al. 1985).

Tactics of Intragroup Conflict in Female-Bonded Groups

Intragroup conflict takes several different forms among female primates. First, females compete directly with other members of their groups over access to food and other resources. Second, females participate in aggressive interactions that seem to be mainly concerned with establishing and maintaining their dominance status. Finally, conflict over group membership is observed.

Conflict over resources. Intragroup competition over access to communally held resources is a common cause of conflict among females (Walters and Seyfarth 1987). The forms of conflict range from ritualized displays to physical contests in which females shove, grapple, bite, and sometimes wound one another. However, escalated aggression is not common, and physical fights over food are rare (Walters and Seyfarth 1987). More often, females simply retreat from food sites when they are approached by others, or distance themselves from other members of their groups while they are feeding. Low-ranking female long-tailed macaques leave the main body of their social groups and forage in small parties to avoid competition for resources (van Noordwijk and van Schaik 1987a). Female gorillas (Stewart and Harcourt 1987) and Japanese macaques (Furiuchi 1983) maintain greater distances from other group members while they are feeding than while they are engaged in other activities.

The passive nature of these responses to competition seems somewhat paradoxical. If access to limited food resources limits female reproductive success, then why do females cede access to resources with so little resistance? The answer may be related to the relative costs and benefits associated with defending the resource (Clutton-Brock and Harvey 1976). A female who defends a food item expends energy and risks injury. A female who abandons a resource when challenged must find a new food item. If the costs of defense are greater than the costs of finding new resources, females are expected to abandon resources without further conflict (Maynard Smith 1982). Since injuries sustained in escalated fights among females can debilitate them, it may not often be worthwhile for females to defend access to resources (Clutton-Brock and Harvey 1976; Walters and Seyfarth 1987).

Regardless of their intensity, conflicts over resources among primate females tend to have highly predictable outcomes. Females usually retreat when they are approached by higher ranking females, that is, females who have previously defeated them in aggressive interactions that did not involve resources. Conversely, they often ignore the approaches of lower ranking females. Escalated conflicts produce similarly predictable results. It is important to emphasize that dominance rank is defined here in terms of the outcome of aggressive and submissive interactions that do not directly involve access to resources.

Conflict over dominance rank. Much of the aggression observed among primates seems to be concerned with establishing or maintaining dominance relationships (Walters and Seyfarth 1987). Although this seems sensible, given the correlation between dominance rank and the outcome of competitive encounters, the structure and regularity of dominance relationships among primate females are difficult to explain from an evolutionary perspective (Silk 1987).

As with many other animals, primate females often form linear dominance hierarchies. A linear dominance hierarchy is formed when all triadic relationships are transitive—that is, when the following condition is met: If A defeats B, and B defeats C, then A defeats C. When this condition is met for all possible subgroups of three animals, a perfectly linear hierarchy is produced (Chase 1980).

In some groups, the extent of linearity is remarkable. For example, Hausfater (1975) observed 1,638 dyadic agonistic interactions among female baboons during a 14-month period. Only six of these 1,638 acts of aggression were directed toward higher ranking females. In three groups of vervet monkeys observed over a two-year period, Cheney and colleagues (1981) observed 2,399 approach-retreat interactions among females. The outcomes of approximately 98% of these interactions were consistent with the established dominance rank order. The majority of remaining interactions involved two females who switched ranks with one another during the study period. In one group of rhesus monkeys, Missakian (1972) observed only 61 reversals in 5,159 interactions among females observed over a 24-month period. Smuts (1985) recorded 32 reversals in 3,799 interactions among female olive baboons in a 15-month period. Females in many groups of macaques, baboons, vervets, mangabeys, and capuchin monkeys establish dominance hierarchies with comparable degrees of linearity (Loy 1985; Melnick and Pearl 1987; Robinson and Janson 1987; Silk et al. 1981a).

It should be noted that there are some species, such as squirrel monkeys (Baldwin and Baldwin 1981), red colobus and black-and-white colobus monkeys (Struhsaker and Leland 1987), and guenons (Cords 1987) in which dominance hierarchies among females have not been detected. There are conflicting reports about the existence and stability of linear dominance hierarchies among the langurs (Hrdy 1977; Struhsaker and Leland 1987) and howler monkeys (Clarke and Glander 1984; Crockett and Eisenberg 1987; Jones 1980; Smith 1977). Efforts to assess the dominance rank of female chimpanzees (de Waal 1982; Goodall 1986; Hasegawa and Hiraiwa-Hasegawa 1983; Nishida 1979) and bonobos (Kuroda 1980) have generally been unsuccessful, although it may be possible to rank chimpanzee females in broad categories (Goodall 1986). Finally, agonistic interactions are quite rare among female gorillas (Harcourt 1979; Watts 1985), and there is some evidence that dominance interactions among females may have inconsistent outcomes (Stewart and Harcourt 1987).

The extent of consistency observed in groups that do form linear dominance hierarchies is difficult to explain. If females were readily differentiated by some attribute that was related to fighting ability, it would be easy to see how the outcomes of aggressive contests might produce a linear rank order. However, physical traits such as weight, stature, agility, and strength are not related to the outcome of dominance contests in any obvious way (Walters and Seyfarth 1987).

One factor that may influence a female's success in competitive or aggressive encounters—and remain relatively unchanged over long periods—is the number or power of her allies. Although dominance relationships are defined in terms of the outcomes of dyadic contests, many disputes among primate females involve more than two participants. Many observers have concluded that these polyadic interactions play an important role in the establishment and maintenance of dominance rank among females.

The role of female kin in the establishment and maintenance of dominance rank is well-documented among macaques (Berman 1983; Bernstein and Ehardt 1985; Chapais 1983; Datta 1983a, 1983b; Kaplan 1977, 1978; Kurland 1977; Massey 1977; Silk 1982; Watanabe 1979), vervets (Cheney 1983; Horrocks and Hunte 1983; Lee 1983), baboons

(Cheney 1977, 1978; Walters 1980), and geladas (Dunbar 1984, 1986). When two juvenile females are involved in a dispute, the outcome depends partly upon the relative age and size of the opponents—older and larger individuals tend to defeat younger and smaller ones (Datta 1983a; Horrocks and Hunte 1983). However, if the dispute takes place in proximity to the juveniles' mothers, the outcome may be very different. Then, the relative dominance rank of the two mothers is more likely to determine the outcome of the dispute (Horrocks and Hunte 1983). The presence of the mother matters because mothers are likely to intervene when their offspring are harassed. Thus, juveniles may come to be able to defeat everyone that their mothers can defeat.

Females also support their adult female kin in aggressive encounters, and this may help them to maintain their dominance positions. In geladas, mothers and daughters form cooperative alliances, and old females who had coalition partners outranked old females who did not have coalition partners (Dunbar 1986).

Support of kin leads to the formation of corporate dominance hierarchies in which maternal kin occupy adjacent ranks. Matrilineal dominance hierarchies are characteristic of baboons (Hausfater et al. 1982; Lee and Oliver 1979; Moore 1978; Samuels et al. 1987; Smuts 1985; Walters 1980), geladas (Dunbar 1984, 1986), vervets (Cheney 1983; Fairbanks and McGuire 1986; Horrocks and Hunte 1983; Lee 1983), and macaques (Angst 1975; Chikazawa et al. 1979; Datta 1983a, 1983b; Kawai 1958; Massey 1977; Missakian 1972; Silk et al. 1981a). It is suggestive that members of large lineages often outrank members of small lineages (Silk and Boyd 1983), although the causal connection between lineage size and lineage rank has not been established.

It is interesting to note that stable linear dominance hierarchies are found only in species that form female-bonded groups. These include the macaques, vervets, baboons, and capuchins. Moreover, several of the species for which there is relatively strong evidence that female dominance hierarchies are absent, unstable, or nonlinear—such as chimpanzees, howler monkeys, red colobus, and gorillas—live in groups from which females disperse. Information that has recently become available from field studies of two closely related species of squirrel monkeys is particularly instructive (Mitchell et al. 1991).

Among *S. oerstedi* in Costa Rica, females migrate. These females do not form long-term affiliative bonds nor do they establish dominance hierarchies. In *S. sciureus* at Manu Park in Peru, females do not disperse, but they do form stable affiliative alliances with kin and can be ranked in linear dominance hierarchies.

However, we cannot conclude that the female is consistently associated with the formation of linear dominance hierarchies because there are also several species—such as langur species, and the black-and-white colobus—in which females are philopatric, but do not establish linear dominance hierarchies. In at least one of these species—Hanuman langurs—females do not often form alliances with one another (Hrdy 1977). Female chimpanzees do form alliances in support of their daughters (Goodall 1986; Wrangham 1987), but female philopatry is relatively uncommon among chimpanzees (Pusey and Packer 1987). This suggests that both female philopatry and the support of stable related females may be necessary for the formation of linear dominance hierarchies among females.

Conflict over group membership. If the extent of competition over resources is a function of group size, it might sometimes be advantageous for females to reduce the number of competitors within their groups. There are several ways in which this might be accomplished.

In female-bonded groups, females might attempt to reduce the number of animals who are born or mature within their groups. Gelada females selectively harass cycling females (Dunbar 1980), and Hanuman langur females interrupt copulations (Hrdy 1977). Intense aggression directed by female baboons toward cycling females can cause their sexual swellings to abruptly detumesce (Wasser 1983b). Female infants and juveniles are permanent residents of their natal groups, and will become important sources of competition over resources when they mature. This may be why selective harassment of infant and juvenile females is observed in some groups of macaques (Dittus 1977, 1979; Silk et al. 1981b) and vervet monkeys (Horrocks and Hunte 1983). Hostile reactions to unfamiliar immigrant females may also represent attempts to limit group size.

It is important to keep in mind that females are expected to attempt to limit recruitment into their groups only when group size exceeds the optimal threshold. Cheney and Seyfarth (personal communication) note the absence of selective aggression by vervet females toward juvenile females in Amboseli. They note that this is a population in which many groups are very small, some groups have fused, and others have become extinct (Hauser et al. 1986). Cheney and Seyfarth suggest that when groups are too small to be viable, females may encourage recruitment by accepting migrants and tolerating immature females.

In non-female-bonded groups, females may compete over group membership. Intense aggression among females sometimes precedes the emigration of female mantled and red howler monkeys (Crockett 1984; Crockett and Eisenberg 1987; Jones 1980). Female chimpanzees are sometimes hostile to immigrant females (Hasegawa and Hiraiwa-Hasegawa 1983; Pusey 1979).

Although most of the preceding discussion and examples pertain to females who live in multifemale groups, conflict among females also appears to occur in monogamous and polyandrous species. In species in which males participate actively in raising young, the reproductive success of resident females may be limited by access to food and the availability of males to care for their offspring. Thus, the presence of other mature females may be doubly disadvantageous. Other females are likely to provide competition for food. If they reproduce successfully, their progeny will compete for paternal care.

This may be one of the reasons that intense aggression is directed by captive female marmosets and tamarins toward their maturing daughters and toward unrelated female residents (Kleiman 1979; McGrew and McLuckie 1986). In a captive colony of cotton-top tamarins, adult females chased, attacked, and wounded their oldest daughters, forcing observers to remove the latter from their natal groups. In this colony, nine of fourteen females born in four groups were the targets of severe aggression, while only one of fourteen males born in the same groups were the targets of severe aggression (McGrew and McLuckie 1986). Normally, in *Callitrichid* groups only one female mates and produces offspring, even if several mature females are present. The breeding female apparently retards the maturation of her female progeny (Tardif 1984) and suppresses the ovarian cycles of mature daughters and subordinate adult females (Abbott 1984; Epple and Katz 1984; Evans and Hodges 1984; Hearn 1978; Woodcock 1983). When noncycling females are removed from their natal groups and housed with males, they begin to cycle normally (Evans and Hodges 1984; French et al. 1984; Woodcock 1983). Females reared with unrelated adult males mature more rapidly than do females reared with their parents (Tardif 1984).

In solitary and monogamous species, females actively exclude potential competitors from their groups. Adult female gibbons are aggressive toward their mature female offspring (Leighton 1987), and *Callicebus* females are antagonistic toward strange members of their own sex (Robinson, 1981).

In several solitary *Galago* species, adult females exclude unfamiliar females from their territories (Bearder 1987), although mothers and daughters establish overlapping ranging areas and may share resources (Bearder 1987; Bearder and Doyle 1974). A pronounced male bias in the secondary sex ratios of *G. crassicadautus* may represent a means to reduce competition for such resources (Clark 1978).

Fitness Consequences of Competition

As a result of direct competition over resources and the maintenance of stable dominance relationships, food is sometimes distributed unequally among females. In Samburu National Park, Kenya, high-ranking female vervet monkeys monopolized access to clumped and desirable foods (Whitten 1983). In Amboseli, high-ranking female vervets also monopolize access to scarce resources (Wrangham 1981; Cheney et al. 1981). In Samburu, high-ranking females weighed more, and were presumably better nourished than were low-ranking females of the same stature. In a captive group of rhesus monkeys, high-ranking females also had more body fat than did low-ranking females (Small 1981).

Differential access to resources may contribute to observed differences in the reproductive performance of high- and low-ranking females. There are groups of baboons, macaques, and vervets in which dominance rank is positively correlated with age at menarche, age at first conception, age at first birth, probability of conceiving in a given year, number of sterile years, interbirth intervals, length of postpartum amenorrhea, number of offspring produced, and infant survival (table 3.1). There are also many *Cercopithecine* groups in which these patterns do not reach statistical significance and a few groups in which negative relationships between dominance rank and some parameters of reproductive performance have been found (table 3.1, reviewed in Fedigan 1983; Gray 1985; Silk 1987). It seems fair to emphasize that significant positive correlations between dominance rank and measures of reproductive success greatly outnumber significant negative correlations between these variables.

At the same time, however, it is important to acknowledge that dominance rank does not account for all of the variation in female reproductive performance. In some groups, variation in reproductive success may depend more on females' present age (Small 1984b), longevity (Cheney et al. 1988), associations with protective males (Smuts, 1985), and vulnerability to predation (Cheney et al. 1981) than on their rank. In species that do not form linear dominance hierarchies, there must be other sources of variation in female reproductive success. Thus, in mantled howler monkeys and Hanuman langurs, age seems to have an important impact upon reproductive success (Clarke and Glander 1984; Dolhinow et al. 1979). It seems quite likely that dominance rank is a more important component of variation in reproductive success in some species and in some groups than in others.

There is another reason to question the biological significance of the relationship between dominance rank and reproductive performance in *Cercopithecine* species that form linear dominance hierarchies. That is, we do not know the effects of dominance rank upon lifetime fitness. If females maintain the same rank positions throughout their lives, dominance rank may account for a significant fraction of the variation in fitness among females. On the other hand, if dominance rank fluctuates over time, dominance rank may account for considerably less of the variation in fitness among females. Long-term observations of several *Cercopithecine* groups demonstrate that dominance relationships can remain stable over long periods of time, but

Table 3.1 Components of Female Reproductive Success

Component of Reproductive Success	Species	Studies Finding a Positive Correlation (Author)	Studies Finding No Correlation (Author)
MATURATION			
Age at menarche, first birth, or conception	*Macaca mulatta*	Drickamer 1974 Wilson et al. 1983	
	M. sylvanus	Paul and Thommen 1984	
	M. fuscata	Sugiyama and Ohsawa 1982	Wolfe 1983 Gouzoules et al. 1982
	M. arctoides		Nieuwenhuijsen et al. 1985
	M. radiata		Silk et al. 1981c
	Cercopithecus aethiops		Cheney et al. 1988
	Papio cynocephalus	Altmann et al. 1988	
CONCEPTION RATE			
Probability of conceiving in given year, number of sterile years	*M. mulatta*	Drickamer 1974 Meikle et al. 1984 Wilson et al. 1978	
	M. fuscata	Gouzoules et al. 1982 Sugiyama and Ohsawa 1982	Takahata 1980. Sugiyama and Ohsawa 1982
	M. arctoides	Nieuwenhuijsen et al. 1985	

Table 3.1 *Continued*

Component of Reproductive Success	Species	Studies Finding a Positive Correlation (Author)	Studies Finding No Correlation (Author)
CONCEPTION RATE			
	M. sylvanus		Paul and Thommen 1984
	C. aethiops	Whitten 1983 Fairbanks and McGuire 1984	Cheney et al. 1988
	P. cynocephalus		Altmann et al. 1988
	Theropithecus gelada	Dunbar 1984	
INTERBIRTH INTERVAL			
Length of interval between births, length of postpartum amenorrhea	M. fuscata	Sugiyama and Ohsawa 1982	Gouzoules et al. 1982
	M. mulatta	Drickamer 1974	
	M. arctoides	Nieuwenhuijsen et al. 1985	
	M. radiata		Silk et al. 1981c
	C. aethiops	Fairbanks and McGuire 1984	
FERTILITY			
Number of offspring produced	M. mulatta		Anderson and Simpson 1979
	M. fuscata		Gouzoules et al. 1982 Wolfe 1984

Table 3.1 *Continued*

Component of Reproductive Success	Species	Studies Finding a Positive Correlation (Author)	Studies Finding No Correlation (Author)
FERTILITY			
	M. sinica	Dittus 1979	
	C. aethiops	Whitten 1983	
	C. patas		Loy 1981
	T. gelada	Dunbar 1984	
	Presbytis entellus		Dolhinow et al. 1979
MATURATION INFANT MORTALITY			
Fraction of infants that survive	*M. mulatta*	Drickamer 1974 Wilson et al. 1978	Anderson and Simpson 1979
	M. fasicularis	van Noordwijk and van Schaik 1987	
	M. fuscata	Sugiyama and Ohsawa 1982 Mori 1979	Wolfe 1984 *Gouzoules et al. 1982
	M. radiata	Silk et al. 1981c	
	M. sinica	Dittus 1979	
	M. sylvanus	Paul and Thommen 1984	
	C. aethiops	Whitten 1983	Cheney et al. 1981

Table 3.1 *Continued*

Component of Reproductive Success	Species	Studies Finding a Positive Correlation (Author)	Studies Finding No Correlation (Author)
MATURATION INFANT MORTALITY			
	P. cynocephalus		Altmann et al. 1988
	P. ursinus	Busse 1982	
	P. entellus		Dolhinow et al. 1979
	Alouatta palliata		Glander 1980

*In this study, a negative correlation between rank and infant mortality was found.

can also change substantially. For example, in one group of baboons in Amboseli, a stable female dominance hierarchy was maintained for more than a decade (Hausfater et al. 1982). Shortly after a report on the stability of their dominance relationships was published, rapid and substantial changes in female dominance were observed in the same group (Samuels et al. 1987). Major changes in dominance rank have also been observed in other groups (Chance et al. 1977; Koyama 1970). We cannot yet estimate the rate and magnitude of changes in dominance rank among females in species that form linear dominance hierarchies, but it would be surprising if such changes entirely eliminated the effects of high rank on lifetime female fitness.

SUMMARY AND CONCLUSIONS

The data reviewed here suggest that conflict plays an integral role in the behavioral strategies of female primates. Conflicts over access to resources needed to survive and reproduce sometimes arise among females who live in solitary territories, females who form pair bonds

Table 3.2 Studies Cited in Table 3.1

Authors	Species	Study Site	Ecological Condition	Duration of Study
Altmann et al. 1988	*Papio cynocephlaus*	Amboseli, Kenya	Free-ranging	13 years
Anderson and Simpson 1979	*Macaca mulatta*	Madingley, England	Captive	18 years
Busse 1982	*Papio ursinus*	Botswana	Free-ranging	
Dittus 1979	*Macaca sinica*	Polowonanura, Sri Lanka	Free-ranging	
Dolhinow et al. 1979	*Presbytis entellus*	Berkeley, California	Captive	
Drickamer 1974	*Macaca mulatta*	La Parguera, Puerto Rico	Semi-free-ranging, provisioned	10 years
Dunbar 1984	*Theropithecus gelada*	Sankaber, Ethiopia	Free-ranging	2 years
Fairbanks and McGuire 1984	*Cercopithecus aethiops*	Los Angeles, California	captive	
Glander 1980	*Alouatta palliata*	Costa Rica	free-ranging	
Gouzoules et al. 1982	*Macaca fuscata*	Texas	captive, provisioned	8 years
Loy 1981	*Cercopithecus patas*		captive	3 years
Meikle et al. 1984	*Macaca mulatta*	La Parguera, Puerto Rico	semi-free-ranging, provisioned	
Mori 1979	*Macaca fuscata*	Koshima Island, Japan	free-ranging, provisioned	
Nieuwenhuijsen et al. 1985	*Macaca arctoides*	Oss, The Netherlands	captive	
Paul and Thommen 1984	*Macaca sylvanus*	Salem, Germany	captive	5 years
Silk et al. 1981c	*Macaca radiata*	Davis, California	captive	10 years

Table 3.2 *Continued*

Authors	Species	Study Site	Ecological Condition	Duration of Study
Sugiyama and Ohsawa 1982	*Macaca fuscata*	Takasakiyama, Japan	free-ranging, provisioned	
Takahata 1980	*Macaca fuscata*	Arashiyama, Japan	free-ranging, provisioned	2 years
van Noordwijk and van Schaik 1987	*Macaca fasicularis*	Ketambe, Sumatra	free-ranging	
Whitten 1983	*Cercopithecus aethiops*	Samburu, Kenya	free-ranging	2 years
Wilson et al. 1978	*Macaca mulatta*	Atlanta, Georgia	captive	6 years
Wilson et al. 1983	*Macaca mulatta*	Atlanta, Georgia	captive	
Wolfe 1983	*Macaca fuscata*	Arashiyama, Japan	free-ranging, provisioned	3 years

Ecological Conditions:
Free-ranging = unconstrained dispersal, not provisioned, and not protected from predators;
free-ranging, provisioned = unconstrained dispersal, not protected from predators, but provisioned regularly;
Semi-free-ranging, provisioned = dispersal limited, protected from predators, and provisioned; and
Captive = no dispersal, protected from predators, and provisioned.

and live in small family groups, females who live among maternal kin in large social groups, and among females who disperse from their natal groups and join groups of unfamiliar females. In most cases, conflict can be ultimately traced to shortages of resources that limit females' abilities to reproduce successfully.

Ecological pressures appear to shape the context, frequency, and structure of conflict among primate females. In the presence of such pressures, natural selection is expected to favor the evolution of morphological and behavioral traits that enable females to compete successfully in contests over resources. Behavioral traits that enhance competitive ability may be quite diverse. They may include cognitive abilities to weigh the relative value of a resource against the costs of obtaining an alternate resource; social skills needed to establish and maintain alliances with females who will cooperate in defense of resources; and aggressive traits that enable females to effectively harass, dominate, or intimidate competitors.

The argument proposed here—that shortages of resources needed for survival and reproduction are ultimately responsible for much conflict observed among female primates—is clearly incomplete. The empirical links among density-dependent demographic processes, social behavior, reproductive performance, and female fitness are quite tenuous. As we learn more about the factors that affect the life histories of female primates, we may be able to build a more comprehensive model.

REFERENCES

Abbott, D. H. (1984) Behavior and physiological suppression of fertility in subordinate marmoset monkeys. *American Journal of Primatology* 6:169–186.

Altmann, J.; Hausfater, G.; Altmann, S. A. (1988) Determinants of reproductive success in savannah baboons, *Papio cynocephalus*. "*Reproductive Success*," Clutton-Brock, T. H. (ed.). Chicago: University of Chicago Press. 403–418.

Altmann, J.; Hausfater, G; Altmann, S. A. (1985) Demography of Amboseli baboons, 1963–1983. *American Journal of Primatology* 8:113–125.

Altmann, S. A.; Altmann, J. (1970) "Baboon Ecology." Chicago: University of Chicago Press.

Andelman, S. J. (1986) Ecological and social determinants of cercopithecine mating patterns. In Rubenstein, D. I., Wrangham, R. W. (eds.), "Ecology and Social Evolution: Birds and Mammals." Princeton, N.J.: Princeton University Press. 201–216.

Anderson, D. H.; Simpson, M. J. A. (1979) Breeding performance of a captive colony of rhesus macaques *(Macaca mulatta)*. *Laboratory Animal* 13:275–281.

Angst, W. (1975) Basic data and concepts in the social evolution of *Macaca fasicularis*. In Rosenblum, L. A. (ed.), "Primate Behavior," vol. 4. New York: Academic Press. 325–388.

Baldwin, J. D.; Baldwin, J. I. (1981) The squirrel monkeys, genus *Saimiri*. In Coimbra-Filho, A. F., Mittermeier, R. A. (eds.), "Ecology and Behavior of Neotropical Primates," vol. 1. Rio de Janeiro: *Academia Brasileira de Ciencias.* 277–330.

Bearder, S. K. (1987) Lorises, bushbabies, and tarsiers: Diverse societies in solitary foragers. In Smuts, B. B., Cheney, D. L., Seyfarth, R. M., Wrangham, R W., Struhsaker, T. T. (eds.), "Primate Societies." Chicago: University of Chicago Press. 11–24.

Bearder, S. K., Doyle, G. A. (1974) Ecology of bushbabies *Galago senegalensis* and *Galago crassicaudatus*, with some notes on their behavior in the field. In Martin, R. D., Doyle, G. A., Walder, A. C. (eds.), "Prosimian Biology." London: Duckworth. 109–130.

Berman, C. (1983) Early differences in relationships between infants and other group members based on the mothers' status: Their possible relationship to peer-peer rank acquisition. In Hinde, R. A. (ed.), "Primate Social Relationships: An Integrated Approach." Oxford: Blackwell. 154–156.

Bernstein, I. S.; Ehardt, C. L. (1985) Agonistic aiding: Kinship, rank, age and sex influences. *American Journal of Primatology* 8:37–52.

Boyd, R. (1982) Density dependent mortality and the evolution of social behaviour. *Animal Behaviour* 30:972–982.

Brown, J. L. (1964) The evolution of diversity in avian territorial systems. *Wilson Bulletin* 76:160–169.

Busse, C. D. (1982) Social dominance and offspring mortality among female chacma baboons. *International Journal of Primatology* 3:267.

Chance, M. R. A.; Emory, G. R.; Payne, R G.. (1977) Status referents in long-tailed macaques *(Macaca fasicularis):* Precursors and effects of a female rebellion. *Primates* 18:611–632.

Chapais, B. (1983) Dominance, relatedness, and the structure of female relationships in rhesus monkeys. In Hinde, R. A. (ed.), "Primate Social Relationships: An Integrated Approach." Oxford: Blackwell. 209–217.

Chase, I. D. (1980) Social process and hierarchy formation in small groups: A comparative perspective. *American Sociological Review* 45:905–924.

Cheney, D. L. (1977) The acquisition of rank and the development of reciprocal alliances among free-ranging immature baboons. *Behavioral Ecological and Sociobiology* 2:303–318.

Cheney, D. L. (1978) Interactions of immature male and female baboons with adult females. *Animal Behaviour* 26:389–408.

Cheney, D. L. (1983) Extra-familial alliances among vervet monkeys. In Hinde, R. A. (ed.), "Primate Social Relationships: An Integrated Approach." Oxford: Blackwell. 278–286.

Cheney, D. L. (1987) Interactions and relationships between groups. In Smuts, B. B., Cheney, D. L., Seyfarth, R. M., Wrangham, R. W., Struhsaker, T. T. (eds.), "Primate Societies." Chicago: University of Chicago Press. 267–281.

Cheney, D. L.; Wrangham, R. W. (1987) Predation. In Smuts, B. B., Cheney, D. L., Seyfarth, R. M., Wrangham, R. W., Struhsaker, T. T. (eds.), "Primate Societies." Chicago: University of Chicago Press. 267–281.

Cheney, D. L.; Lee, P. C.; Seyfarth, R. M. (1981) Behavioral correlates of non-random mortality among free-ranging adult female vervet monkeys. *Behavioral Ecological and Sociobiology* 9:153–161.

Cheney, D. L.; Seyfarth, R. M.; Andelman, S. J.; Lee, P. C. (1988) Reproductive success in vervet monkeys. In Clutton-Brock TH (ed.), "Reproductive Success." Chicago: University of Chicago Press. 384–402.

Chepko-Sade, B. D.; Sade, D. S. (1976) Patterns of group splitting within matrilineal kinship groups: A study of social group structure in *Macaca mulatta (Cercopithecidae: Primates)*. *Behavioral Ecological and Sociobiology* 5:67–87.

Chikazawa, D.; Gordon, T.; Bean, C.; Bernstein, I. (1979) Mother-daughter dominance reversals in rhesus monkeys *(Macaca mulatta)*. *Primates* 20:301–305.

Chivers, D. J. (1974) "The siamang in Malaya." In: "Contributions to Primatology," vol. 4. Basel: Karger.

Clark, A. B. (1978) Sex ratio and local resource competition in a prosimian primate. *Science* 201:163–165.

Clarke, M. R.; Glander, K. E. (1984). Female reproductive success in a group of free-ranging howler monkeys *(Alouatta palliata)* in Costa Rica. In Small, M. F. (ed.), "Female Primates: Studies by Women Primatologists." New York: Alan R. Liss. 111–126.

Clutton-Brock, T. H.; Harvey, P. H. (1976) Evolutionary rules and primate societies. In Bateson, P. P. G., Hinde, R. A. (eds.), "Growing Points in Ethology." Cambridge: Cambridge University Press. 195–238.

Clutton-Brock, T. H.; Harvey, P. H.; Rudder, B. (1977) Sexual dimorphism, secondary sex ratio, and body weight in primates. *Nature* 269:797–799.

Coelho, A. M.; Bramblett, C. A.; Quick, L. B. (1977) Social organization and food resource availability in primates: A socio-bioenergetic analysis of diet and disease hypotheses. *American Journal of Physical Anthropology* 46:253–264.

Cords, M. (1987) Forest guenons and patas monkeys: Male-male competition in one-male groups. In Smuts, B. B., Cheney, D. L., Seyfarth, R. M., Wrangham, R. W., Struhsaker, T. T. (eds.), "Primate Societies." Chicago: University of Chicago Press. 98–111.

Crockett, C. M. (1984) Emigration by female red howler monkeys and the case for female competition. In Small, M. F. (ed.), "Female Primates: Studies by Women Primatologists." New York: Alan R. Liss, 159–173.

Crockett, C. M.; Eisenberg, J. F. (1987) Howlers: Variations in group size and demography. In Smuts, B. B., Cheney, D. L., Seyfarth, R. M., Wrangham, R. W., Struhsaker, T. T. (eds.), "Primate Societies." Chicago: University of Chicago Press. 54–68.

Darwin, C. (1859) "The Origin of Species by Means of Natural Selection." London: John Murray.

Datta, S. J. (1983a) Relative power and the acquisition of rank. In Hinde, R. A. (ed.), "Primate Social Relationships: An Integrated Approach." Oxford: Blackwell. 93–102.

Datta, S. J. (1983b) Relative power and the maintenance of dominance. In Hinde, R. A. (ed.), "Primate Social Relationships: An Integrated Approach." Oxford: Blackwell. 103–111.

de Waal, F. M. B. (1982) "Chimpanzee Politics." New York: Harper and Row.

Dittus, W. P. J. (1977) The social regulation of population density and age-sex distribution in the toque monkey. *Behaviour* 63:281–322.

Dittus, W. P. J. (1979) The evolution of behavior regulating density and age-specific sex ratios in a primate population. *Behaviour* 69:265–302.

Dittus, W. P. J. (1980) The social regulation of primate populations: A synthesis. In Lindburg, D. G. (ed.), "The Macaques: Studies in Ecology, Behaviour, and Evolution." New York: Van Nostrand Reinhold. 263–286.

Dittus, W. P. J. (1986) Sex differences in fitness following a group takeover among Toque macaques: Testing models of social evolution. *Behavioral Ecology and Sociobiology* 19:257–266.

Dolhinow, P.; McKenna, J. J.; Vonder Haar Laws, J. (1979) Rank and reproduction among female langur monkeys: Aging and improvements (They're not just getting older, they're getting better). *Journal of Aggressive Behavior* 5:19–30.

Drickamer, L. C. (1974) A ten-year summary of reproductive data for free-ranging *Macaca mulatta*. *Folia Primatologica* 21:61–80.

Dunbar, R. I. M. (1980) Determinants and evolutionary consequences of dominance among female gelada baboons. *Behavioral Ecological and Sociobiology* 7:253–265.

Dunbar, R. I. M. (1984) "Reproductive decisions." Princeton, N.J.: Princeton University Press.

Dunbar, R. I. M. (1986) The social ecology of gelada baboons. In Rubenstein, D. I., Wrangham, R. W. (eds.), "Ecological Aspects of Social Evolution: Birds and Mammals." Princeton, N.J.: Princeton University Press. 332–351.

Dunbar, R. I. M.; Sharman, M. (1983) Female competition for access to males affects birth rate in baboons. *Behavioral Ecological and Sociobiology* 13:157–159.

Epple, G.; Katz, Y. (1984) Social influences on estrogen secretion and ovarian cyclicity in saddle-back tamarins *(Saguinus fuscicolis)*. *American Journal of Primatology* 6:215–227.

Evans, S.; Hodges, J. K. (1984) Reproductive status of adult daughters in family groups of common marmosets *(Callithrix jacchus jacchus)*. *Folia Primatologica* 42:127–133.

Fairbanks, L. A., McGuire, M. T. (1984) Determinants of fecundity and reproductive success in captive vervet monkeys. *American Journal of Primatology* 7:27–38.

Fairbanks, L. A., McGuire, M. T. (1986) Age, reproductive value, and dominance-related behaviour in vervet monkey females: Cross-generational influences on social relationships and reproduction. *Animal Behaviour* 34:1710–1721.

Fedigan, L. M. (1983) Dominance and reproductive success in primates. *Yearbook of Physical Anthropology* 26:91–129.

French, J. A.; Abbott, D. H.; Snowdon, C. T. (1984) The effect of social environment on estrogen excretion, scent marking, and sociosexual behavior in tamarins *(Saguinus oedipus)*. *American Journal of Primatology* 6:155–167.

Fujii, H. (1975) A psychological study of the social structure of a free-ranging group of Japanese monkeys in Katsuyama. In Kondo, S., Kawai, M., Ehara, A. (eds.), "Contemporary Primatology: Proceedings of the Fifth Congress of the International Primatological Society." Basel: S. Karger. 428–436.

Furiuchi, T. (1983) Interindividual distance and influence of dominance on feeding in a natural Japanese macaque troop. *Primates* 24:445–455.

Furuya, Y. (1969) On the fission of troops of Japanese monkeys, 2: General view of troop fission of Japanese monkeys. *Primates* 10:47–69.

Gaulin, S. J. C.; Sailer, L. (1984) Sexual dimorphism in weight among the primates: The relative impact of allometric and sexual selection. *International Journal of Primatology* 5:515–535.

Glander, K. (1980) Reproduction and population growth in free-ranging mantled howling monkeys. *American Journal of Physical Anthropology* 53:25–36.

Goldizen, A. Wilson (1987) Tamarins and marmosets: Communal care of offspring: In Smuts, B. B., Cheney, D. L., Seyfarth, R. M., Wrangham, R. W., Struhsaker, T. T. (eds.), "Primate Societies." Chicago: University of Chicago Press. 34–43.

Goodall, J. (1986) "The Chimpanzees of Gombe: Patterns of Behavior." Cambridge: Belknap Press.

Gouzoules, H.; Gouzoules, S.; Fedigan, L. M. (1982) Behavioural dominance and reproductive success in female Japanese monkeys *(Macaca fuscata)*. *Animal Behaviour* 30:1138–1151.

Gray, J. P. (1985) "Primate Sociobiology." New Haven, Conn.: HRAF Press.

Harcourt, A. H. (1979) Social relationships among adult female mountain gorillas. *Animal Behaviour* 27:251–264.

Hasegawa, T.; Hiraiwa-Hasegawa, M. (1983) Opportunistic and restrictive matings among wild chimpanzees in the Mahale Mountains, Tanzania. *Journal of Ethology* 1:75–85.

Hauser, M. D.; Cheney, D. L., Seyfarth, R. M. (1986) Group extinction and fusion in free-ranging vervet monkeys. *American Journal of Primatology* 11:63–77.

Hausfater, G . (1975) Dominance and reproduction in baboons *(Papio cynocephalus)*. In "Contributions to Primatology," vol. 7. Basel: S. Karger.

Hausfater, G.; Altmann, J.; Altmann, S. (1982) Long-term consistency of dominance relations among female baboons *(Papio cynocephalus)*. *Science* 217:752–755.

Hearn, J. P. (1978) The endocrinology of reproduction in the common marmoset, *Callithrix jacchus*. In Kleiman, D. (ed.), "The Biology and Conservation of the *Callitrichidae*." Washington, D.C.: Smithsonian Institution Press. 163–171.

Horrocks, J.; Hunte, W. (1983) Maternal rank and offspring rank in vervet monkeys: An appraisal of the mechanisms of rank acquisition. *Animal Behaviour* 31:772–782.

Hrdy, S. B. (1977) "The Langurs of Abu." Cambridge: Harvard University Press.

Hrdy, S. B.; Williams, G. C. (1983) Behavioral biology and the double standard. In Wasser, S. K. (ed.), "Social Behavior of Female Vertebrates." New York: Academic Press. 3–17.

Jolly, A. (1985) "The evolution of primate behavior," 2d ed. New York: Macmillan.

Jones, C. B. (1980) The functions of status in the mantled howler monkey, *Alouatta palliata* Gray: Intraspecific competition for group membership in a folivorous neotropical primate. *Primates* 21:389–405.

Kaplan, J. R. (1977) Patterns of fight interference in free-ranging rhesus monkeys. *American Journal of Physical Anthropology* 47:279–288.

Kaplan, J. R. (1978) Fight interference and altruism in rhesus monkeys. *American Journal of Physical Anthropology* 47:241–249.

Kawai, M. (1958) On the system of social ranks in a natural group of Japanese monkeys. *Primates* 1:11–48.

Kirkwood, J. K. (1983) The effects of diet on health, weight, and litter size in captive cotton-top tamarins *Saguinus oedipus oedipus*. *Primates* 24:515–520.

Kleiman, D. G. (1979) Parent-offspring conflict and sibling competition in a monogamous primate. *American Naturalist* 194:753–760.

Koford, C. B. (1965) Population dynamics of rhesus monkeys on Cayo Santiago. In DeVore, I. (ed.), "Primate Behavior: Field Studies of Monkeys and Apes." New York: Holt, Rinehart, and Winston. 160–174.

Koford, C. B. (1966) Population changes in rhesus monkeys: Cayo Santiago, 1960–1964. *Tulane Studies in Zoology* 13:1–7.

Koyama, N. (1970) Changes in dominance rank and division of a wild Japanese monkey troop in Arashiyama. *Primates* 11:335–390.

Koyama, N.; Norikoshi, K.; Mano, T. (1975) Population dynamics of Japanese monkeys at Arashiyama. In Kondo, S., Kawai, M., Ehara, A. (eds.), "Contemporary Primatology: Proceedings of the Fifth Congress of the International Primatological Society." Basel: S. Karger. 407–410.

Kurland, J. A. (1977) Kin selection in the Japanese monkeys. In "Contributions to Primatology," vol. 12. Basel: S. Karger.

Kuroda, S. (1980) Social behavior of the pygmy chimpanzees. *Primates* 21:181–197.

Lee, P. C. (1983) Context specific unpredictability in dominance interactions. In Hinde, R. A. (ed.), "Primate Social Relationships: An Integrated Approach." Oxford: Blackwell. 35–44.

Lee, P. C.; Oliver, J. I. (1979) Competition, dominance, and the acquisition of rank in juvenile yellow baboons *(Papio cynocephalus)*. *Animal Behaviour* 27:576–585.

Leighton, D. (1987) Gibbons: Territoriality and monogamy. In Smuts, B. B., Cheney, D. L., Seyfarth, R. M., Wrangham, R. W., Struhsaker, T. T. (eds.), "Primate Societies." Chicago: University of Chicago Press. 135–145.

Leutenegger, W.; Cheverud, J. (1982) Correlates of sexual dimorphism in primates: Ecological and size variables. *International Journal of Primatology* 3:387–402.

Loy, J. (1985) The descent of dominance in *Macaca:* Insights into the structure of human societies. In Tuttle, R. H. (ed.), "Socioecology and Psychology of Primates." The Hague: Mouton. 153–180.

Loy, J. (1981) The reproductive and heterosexual behavior of adult patas monkeys in captivity. *Animal Behaviour* 29:714–726.

Malik, I.; Seth, P. K.; Southwick, C. H. (1985) Group fission in free-ranging rhesus monkeys of Tughlaqabad, Northern India. *International Journal of Primatology* 6:411–422.

Massey, A. (1977) Agonistic aids and kinship in a group of pig-tailed macaques. *Behavioral Ecological and Sociobiology* 6:81–83.

Masui, K.; Sugiyama, Y.; Nishimura, A.; Ohsawa, H. (1975) The life table of Japanese monkeys at Takasakiyama. In Kondo, S., Kawai, M., Ehara, A. (eds.), "Contemporary Primatology: Proceedings of the Fifth Congress of the International Primatological Society." Basel: S. Karger. 401–406.

Maynard Smith, J. (1982) "Evolution and the Theory of Games." Cambridge: Cambridge University Press.

McGrew, W. C.; McLuckie, E. C. (1986) Philopatry and dispersion in the cotton-top tamarin, *Saguinus (o.) oedipus:* An attempted laboratory simulation. *International Journal of Primatology* 7:401–422.

Meikle, D. B.; Tilford, D. B.; Vessey, S. H. (1984) Dominance rank, secondary sex ratios, and reproduction of offspring in polygynous primates. *American Naturalist* 124:173–187.

Melnick, D. J.; Pearl, M. C. (1987) Cercopithecines in multi-male groups: Genetic diversity and population structure. In Smuts, B. B., Cheney, D. L., Seyfarth, R. M., Wrangham, R. W., Struhsaker, T. T. (eds.), "Primate Societies." Chicago: University of Chicago Press. 121–134.

Milton, K. (1982) Dietary quality and demographic regulation in a howler monkey population. In Leigh, E. G., Jr, Rand, A. S., Windsor, D. M. (eds.), "The Ecology of a Tropical Forest: Seasonal Rhythms and Long-Term Changes." Washington, D.C.: Smithsonian Institution Press. 273–289.

Missakian, E. M. (1972) Genealogical and cross-genealogical dominance relations in a group of free-ranging rhesus monkeys *(Macaca mulatta)* on Cayo Santiago. *Primates* 13:169–180.

Missakian, E. M. (1973) The timing of fission among free-ranging rhesus monkeys. *American Journal of Physical Anthropology* 38:621–624.

Mitchell, C. L.; Boinski, S.; van Schaik, C. P. (1991) Competitive regimes and female bonding in two species of squirrel monkeys (*Saimiri oerstedi* and *S. sciureus*). *Behavioral Ecological and Sociobiology* 28:55–60.

Moore, J. (1978) Dominance relations among free-ranging female baboons in Gombe National Park. In Chivers, D. J., Herbert, J. (eds.), "Recent Advances in Primatology," vol. 1. London: Academic Press. 67–70.

Mori, A. (1979) Analysis of population changes by measurement of body weight in the Koshima troop of Japanese monkeys. *Primates* 20:371–397.

Nash, L. T. (1976) Troop fission in free-ranging baboons in the Gombe Stream National Park, Tanzania. *American Journal of Physical Anthropology* 44:63–78.

Nieuwenhuijsen, A. H.; Lammers, J. C.; de Neek, K. J.; Koos Slob, A. (1985) Reproduction and social rank in female stumptail macaques *(Macaca arctoides). International Journal of Primatology* 7:77–79.

Nishida, T. (1979) The social structure of chimpanzees of the Mahali Mountains. In Hamburg, D. A., McCown, E. R. (eds.), "The Great Apes." Menlo Park, Calif.: Benjamin/Cummings. 73–122.

Ohsawa, H.; Dunbar, R. I. M. (1984) Variations in the demographic structure and dynamics of gelada baboon populations. *Behavioral Ecological and Sociobiology* 15:231–240.

Paul, A.; Thommen, D. (1984) Timing of birth, female reproductive success and infant sex ratio in semi-free-ranging barbary macaques *(Macaca sylvana). Folia Primatologica* 42:2–16.

Portman, O. (1970) Nutritional requirements (NRC) of non-human primates. In Harris, R. S. (ed.), "Feeding and Nutrition of Nonhuman Primates." New York: Academic Press. 87–116.

Pusey, A. E. (1979) Intercommunity transfer of chimpanzees in Gombe National Park. In Hamburg, D. A., McCown, E. R. (eds.), "The Great Apes." Menlo Park, Calif.: Benjamin/Cummings. 465–479.

Pusey, A. E.; Packer, C. (1987) Dispersal and philopatry. In Smuts, B. B., Cheney, D. L., Seyfarth, R. M., Wrangham, R. W., Struhsaker, T. T. (eds.), "Primate Societies." Chicago: University of Chicago Press. 250–266.

Rawlins, R. G.; Kessler, M. J. (1986) Demography of the free-ranging Cayo Santiago macaques (1976–1983). In Rawlins, R. G., Kessler, M. J. (eds.), "The Cayo Santiago Macaques: History, Behavior, and Biology." Albany, N.Y.: State University of New York Press. 47–72.

Ridley, M. (1986) The number of males in a primate troop. *Animal Behaviour* 34:1848–1858.

Robinson, J. G. (1981) Vocal regulation of inter- and intragroup spacing during boundary encounters in the titi monkey, *Callicebus moloch. Primates* 22:161–172.

Robinson, J. G.; Janson, C. H. (1987) Capuchins, squirrel monkeys, and atelines: Socioecological convergence with Old World primates. In Smuts, B. B., Cheney, D. L., Seyfarth, R. M., Wrangham, R. W., Struhsaker, T. T. (eds.), "Primate Societies." Chicago: University of Chicago Press. 69–82.

Robinson, J. G.; Wright, P. C.; Kinzey, W. G. (1987) Monogamous cebids and their relatives: Intergroup calls and spacing. In Smuts, B. B., Cheney, D. L., Seyfarth, R. M., Wrangham, R. W., Struhsaker, T. T. (eds.), "Primate Societies." Chicago: University of Chicago Press. 44–53.

Samuels, A.; Silk, J. B.; Altmann, J. (1987) Long-term changes in dominance relationships among free-ranging female baboons. *Animal Behaviour* 35:785–793.

Shopland, J. M. (1982) An intergroup encounter with fatal consequences in yellow baboons *(Papio cynocephalus)*. *American Journal of Primatology* 3:263–266.

Silk, J. B. (1982) Altruism and female *Macaca radiata:* Explanations and analysis of patterns of grooming and coalition formation. *Behaviour* 79:162–188.

Silk, J. B. (1987) Social behavior in evolutionary perspective. In Smuts, B. B., Cheney, D. L., Seyfarth, R. M., Wrangham, R. W., Struhsaker, T. T. (eds.), "Primate Societies." Chicago: University of Chicago Press. 318–329.

Silk, J. B., Boyd, R. (1983) Cooperation, competition, and mate choice in matrilineal macaque groups. In Wasser, S. K. (ed.), "Social Behavior of Female Vertebrates." New York: Academic Press. 315–347.

Silk, J. B.; Samuels, A.; Rodman, P. S. (1981a) Hierarchical organization of female *Macaca radiata. Primates* 22:84–95.

Silk, J. B.; Samuels, A.; Rodman, P. S. (1981b) The influence of kinship, rank, and sex on affiliation and aggression between adult female and immature bonnet macaques *(Macaca radiata). Behaviour* 78:111–137.

Silk, J. B.; Clark-Wheatley, C. B.; Rodman, P. S.; Samuels, A. (1981c) Differential reproductive success and facultative adjustment of sex ratios among captive female bonnet macaques *(Macaca radiata). Animal Behaviour* 29:1106–1120.

Small, M. F. (1981) Body fat, rank, and nutritional status in a captive group of rhesus macaques. *International Journal of Primatology* 2:91–96.

Small, M. F. (1984a) "Female Primates: Studies by Women Primatologists." New York: Alan R. Liss.

Small, M. F. (1984b) Aging and reproductive success in female *Macaca mulatta*. In Small, M. F. (ed.), "Female Primates: Studies by Women Primatologists." New York: Alan R. Liss. 249–259.

Smith, C. C. (1977) Feeding behaviour and social organization in howling monkeys. In Clutton-Brock, T. H. (ed.), "Primate Ecology: Studies of Feeding and Ranging Behaviour in Lemurs, Monkeys, and Apes." London: Academic Press. 97–126.

Smuts, B. B. (1985) "Sex and Friendship in Baboons." Hawthorne, N.Y.: Aldine.

Southwick, C. H. (1967) An experimental study of intragroup agonistic behavior in rhesus monkeys (Macaca mulatta). Behaviour 28:182–209.

Stewart, K. J.; Harcourt, A. H. (1987) Gorillas: Variation in female relationships. In Smuts, B. B., Cheney, D. L., Seyfarth, R. M., Wrangham, R. W., Struhsaker, T. T. (eds.), "Primate Societies." Chicago: University of Chicago Press. 155–164.

Struhsaker, T. T. (1973) A recensus of vervet monkeys in the Masai-Amboseli Game Reserve, Kenya. Ecology 54:930–932.

Struhsaker, T. T. (1976) A further decline in numbers of Amboseli vervet monkeys. Biotropica 8:211–214.

Struhsaker, T. T.; Leland, L. (1987) Colobines: Infanticide by adult males. In Smuts, B. B., Cheney, D. L., Seyfarth, R. M., Wrangham, R. W., Struhsaker, T. T. (eds.), "Primate Societies." Chicago: University of Chicago Press. 83–97.

Strum, S. C.; Western, D. (1982) Variations in fecundity with age and environment in olive baboons (Papio anubis). American Journal of Primatology 3:61–76.

Sugiyama, Y.; Ohsawa, H. (1975) Life history of male Japanese macaques at Ryozenyama. In Kondo, S., Kawai, M., Ehara, A. (eds.), "Contemporary Primatology: Proceedings of the Fifth Congress of the International Primatological Society." Basel: S. Karger. 401–406.

Sugiyama, Y.; Ohsawa, H. (1982) Population dynamics of Japanese monkeys with special reference to the effect of artificial feeding. Folia Primatologica 39:238–263.

Takahata, Y. (1980) The reproductive biology of a free-ranging troop of Japanese monkeys. Primates 21:303–329.

Tardif, S. D. (1984) Social influences on sexual maturation of female Saguinus oedipus oedipus. American Journal of Primatology 6:199–209.

Teas, J.; Richie, T. L.; Taylor, H. G.; Siddiqi, M. F.; Southwick, C. H. (1981) Natural regulation of rhesus monkey populations in Kathmandu, Nepal. Folia Primatologica 35:117–123.

Terborgh, J. (1983) "Five New World Primates: A Study in Comparative Ecology." Princeton, N.J.: Princeton University Press.

Trivers, R. L. (1972) Parental investment and sexual selection. In Campbell, B. (ed.), "Sexual Selection and the Descent of Man: 1871–1971." Chicago: Aldine.

van Noordwijk, M. A.; van Schaik, C. P. (1987) Competition among female long-tailed macaques, Macaca fasicularis. Animal Behaviour 35:577–589.

van Schaik, C. P. (1983) Why are diurnal primates living in groups? Behaviour 87:120–144.

Walters, J. R. (1980) Interventions and the development of dominance relationships in female baboons. Folia Primatologica 34:61–89.

Walters, J. R.; Seyfarth, R. M. (1987) Conflict and cooperation. In Smuts, B. B., Cheney, D. L., Seyfarth, R. M., Wrangham, R. W., Struhsaker, T. T. (eds.), "Primate Societies." Chicago: University of Chicago Press. 306–317.

Wasser, S. K. (1983a) (ed.) "Social Behavior of Female Vertebrates." New York: Academic Press.

Wasser, S. K. (1983b) Reproductive competition and cooperation among yellow baboons. In Wasser, S. K. (ed.), "Social Behavior of Female Vertebrates." New York: Academic Press. 349–390.

Watanabe, K. (1979) Alliance formation in a free-ranging troop of Japanese macaques. *Primates* 20:459–474.

Watts, D. (1985) Relations between group size and composition and feeding competition in mountain gorilla groups. *Animal Behaviour* 33:72–85.

Whitten, P. L. (1983) Diet and dominance among female vervet monkeys *(Cercopithecus aethiops). American Journal of Primatology* 5:139–159.

Whitten, P. L. (1987) Infants and adult males. In Smuts, B. B., Cheney, D. L., Seyfarth, R. M., Wrangham, R. W., Struhsaker, T. T. (eds.), "Primate Societies." Chicago: University of Chicago Press. 343–357.

Wilson, M. E.; Gordon, T. P.; Bernstein, I. S. (1978) Timing of births and reproductive success in rhesus monkey social groups. *Journal of Medical Primatology* 7:202–212.

Wilson, M. E.; Walker, M. L.; Gordon, T. P. (1983) Consequences of first pregnancy in rhesus monkeys. *American Journal of Physical Anthropology* 61:103–110.

Wolfe, L. M. (1984) Female rank and reproductive success among Arashiyama B Japanese macaques *(Macaca fuscata). International Journal of Primatology* 5:133–143.

Woodcock, A. (1983) The social relationships of twin common marmosets *(Callithrix jaccchus)* and their breeding success. *Primates* 24:501–514.

Wrangham, R. W. (1980) An ecological model of female-bonding primate groups. *Behaviour* 75:262–300.

Wrangham, R. W. (1981) Drinking competition in vervet monkeys. *Animal Behaviour* 29:904–910.

Wrangham, R. W. (1987) Evolution of social structure. In Smuts, B. B., Cheney, D. L., Seyfarth, R. M., Wrangham, R. W., Struhsaker, T. T. (eds.), "Primate Societies." Chicago: University of Chicago Press. 282–296.

Wright, P. C. (1984) Biparental care in *Aotus trivirgatus* and *Callicebus moloch.* In Small, M. F. (ed.), "Female Primates: Studies by Women Primatologists." New York: Alan R. Liss. 59–75.

CHAPTER FOUR

Social Conflict on First Encounters

SALLY P. MENDOZA

First encounters between unfamiliar animals are generally assumed to elicit aggression. Review of the available data for nonhuman primates suggests that this may not always be the case. In fact, serious fights are apparently uncommon during initial encounters, and agonistic overtures are rarely responded to in kind. Even when individuals are closely matched as opponents, equally endowed as potential winners (or losers) in agonistic encounters, they quickly show asymmetries in behavior and physiology. Apparently, differences between individuals are created upon first encounters where they previously did not exist. The relative speed, regularity, and transindividual consistency of this process of differentiation suggest that social rules or conventions govern the initial stages in the formation of relationships, and function to mitigate social conflicts.

Primates are social in a highly structured manner. Individuals are differentially attracted to and tolerant of conspecifics of particular age/sex classes leading to a species-characteristic modal grouping pattern which is defined by the number and type of animals that maintain proximity with one another (Mason 1976). While such grouping tendencies define the broad outline of the social group, it is the formation of highly personalized interindividual relationships that seem to best characterize the social dynamics within the group for most primate

Sally P. Mendoza—Department of Psychology and California Regional Primate Research Center, University of California, Davis

Preparation of this manuscript was supported by Grant RR00169 from the National Institutes of Health, Division of Research Resources.

species (Hinde 1976). Affinitive social relationships are generally viewed as providing the cohesive infrastructure of the group. Antagonistic relationships are considered to be dispersive and limit participation in group activities.

In actuality, most relationships are more complex than what an affinitive-agonistic dichotomy suggests. Primate social relationships vary along multiple dimensions including the type, quality, and intensity of interactions which characterize the relationship; the degree of reciprocity or complementarity evident in the give-and-take between participants; and the importance of the relationship to the individuals involved (Hinde 1979). Relationships can contain seemingly contradictory attributes, such as those that are characterized by a high degree of mutual dependency along with high levels of antagonistic interactions.

Whatever their form, relationships imply a historical process. At some point the participants encounter one another for the first time. They interact in a way that facilitates or at least permits subsequent interactions. Over time, they establish a consistent pattern of interaction with one another that, in some ways, reflects the characteristics of each individual and, in other ways, reflects the mutual influence which they have on one another. Once established, departure from expected patterns of interaction can lead to attempts to reinstate the relationship (Coe and Rosenblum 1984; Miller 1971. See also de Waal, chapter 5 in this volume). If these attempts are unsuccessful, physiological changes indicative of severe distress can ensue which may place the participants at risk (see Sapolsky, chapter 7 in this volume). Thus, at some point during relationship formation, individuals establish conventions which define the qualities of the relationship. Deviations from conventional modes of interaction are recognized and lead to behavioral and physiological adjustments which tend to maintain the relationship under most circumstances.

This chapter is concerned with how qualities of relationships are established. Given the consistency of social structure for a particular species, there are undoubtedly some constraints on the kinds of relationships formed (Mendoza and Mason 1989). Upon encountering a stranger, the individual primate must evaluate qualities of the stranger in relation to itself. Depending on the outcome of the evaluation, a course of interaction is selected which is consistent with maintaining the species-typical social structure.

Sustaining a monogamous life-style, for example, depends upon each individual differentially interacting with same-sex and opposite-sex strangers. The behavior expressed during initial encounters ensures the probability that abiding relationships between like-sex individuals will be formed is near zero, whereas, the probability that an abiding relationship will be formed between a lone adult male and a lone adult female is extremely high.

The stereotypic nature of interactions when animals initially meet makes it reasonable, in some cases, to describe first encounters as being rule-bound. The rules governing initial interactions for a monogamous species may be responding to a like-sex stranger antagonistically and discouraging further interaction or responding to an opposite-sex stranger affinitively and encouraging further interaction. Rules of interaction for monogamous animals may be altered if either the male or female is mated, reducing the likelihood that one adult will establish relationships with two heterosexual partners (see Anzenberger, chapter 11 in this volume). Such simple rules may dictate with whom relationships will be established, as well as broadly defining the qualities of those relationships.

As in the preceding example, the most commonplace rules during first encounters are based on the ability to make categorical distinctions—such as male versus female, or adult versus young—and to adjust behavioral responses accordingly (Mendoza and Mason 1989). When relationships are formed between individuals representing different age or sex categories, the role which each animal plays in the relationship can also be largely defined by whether it is the adult or the infant, the male or female. For many primate species, stable relationships are formed regularly between individuals of the same age and sex categories. Despite the lack of obvious distinctions between the participants prior to relationship formation, role differentiation—dominant and subordinate—is generally a prominent feature of within-sex relationships as is adherence to very clear rules of interaction. The questions of primary concern in this chapter are: During formation of relationships, when does role differentiation occur? and Do rules of interaction contribute to this process? Two possibilities will be considered.

The first possibility is that interactions between strangers are not rule-bound. Conventions regarding how individuals will engage with one another develop gradually as relationships are established. Interactions during first encounters are likely to be varied and unpredictable. As animals become increasingly familiar with one another, each shapes the behavior of the other, and the range of interactions becomes more limited. Thus, through a process of mutual shaping, conventions or rules of interaction are established as the animals learn what can and cannot be done with one another. Participants in the relationship develop a set of shared expectancies and respond to deviations from their own customary modes of interaction. Similarities in rules governing different relationships would occur by chance or when similar overtures—for example, a biting attack—produce similar responses in different animals.

The second possibility that will be considered is that first encounters between strangers are structured from the outset by a code of conduct. Upon engaging with one another, each animal follows an etiquette of what may be done when and with whom. Interactions during first encounters with strangers are likely to be stereotyped and predictable in form and sequence. The rules of interaction serve as a guide for the types of relationships to be formed, and they should be the same for all normal members of the species. Increased familiarity would require less formal adherence to the rules. Therefore, as relationships are established, interactions will become more varied, personal, and idiosyncratic.

DOMINANCE RELATIONSHIPS

Traditionally, it has been assumed that the formation of dominance relationships is a simple case of mutual shaping. During the early stages of relationship formation, animals fight frequently. These aggressive encounters are likened to contests of strength. The loser of the encounters learns that the strength and fighting ability of the other supersedes its own. The subordinate role is established when the loser responds at the outset of each agonistic interaction or competitive encounter with submissive responses, thus avoiding the predictable outcome of an actual fight. The dominant animal learns to pursue whatever course of action it chooses without regard to the subordinate. In this way, dominance relations function to ritualize aggressive tendencies between potential rivals (Bernstein 1981). For subordinate individuals, dominance relations may reduce the likelihood of injurious aggression but other disadvantages persist including reduced access to resources, social exclusion, impaired reproduction, and increased need for vigilance.

According to the traditional view, the clear advantages accrued by being dominant makes high status an objective worthy of competition. Status competition would be keenest when strangers meet for the first time and it is not clear to the animals which might win in a contest of strength. Acceptance of lower status is an expedient means to avoid injury for the moment. Subordinate animals can be viewed as awaiting their opportunity to usurp the dominant role.

An alternate view of dominance is that role differentiation is a necessary prerequisite for the formation of multifaceted relationships between similar animals (Kummer 1984; de Waal 1986a, 1986b. Also see de Waal, chapter 5 in this volume). Society requires mechanisms that inhibit some members and disinhibit others in situations that require different courses of action and where individual differences are not enough to ensure that divergent responses occur (Kummer 1973). Dominance creates differences in otherwise similar animals. Both dominant and subordinate animals benefit from their association together. Both dominant and subordinate animals actively maintain the relationship (Mendoza 1984). Status differentiation may be more important to the participants than high or low status per se, and may be achieved with minimal trauma to either party. Competition for status would not occur even during initial stages of relationship formation.

The studies reviewed here examine behavioral and physiological sequelae of formation of dominance relationships. The primary concern is with studies investigating initial encounters between individuals that are all unfamiliar with one another at the time of introduction. Thus, studies examining the introduction of individuals to established groups will be considered only as they provide clarifications of other data.

BEHAVIOR DURING FIRST ENCOUNTERS

One of the first attempts to examine the behavioral responses to formation of a new social group with animals previously unfamiliar with
one another was conducted by Bernstein and Mason (1963). The
expressed intent of this study was to examine the factors involved in
the establishment of relationships among a group of previously unfamiliar rhesus monkeys. Subjects were selected to approximate the
age/sex distribution of a natural group and contained two adult
males, three adult females, two subadults (one male and one female),
and four juveniles (one male and three females). All animals were
released simultaneously into a large enclosure.

As might be anticipated, agonistic interactions occurred frequently
during the initial hour following introduction of the animals to one
another and accounted for more than 80% of the total social interactions.
Most agonistic interactions occurred between adults and, hence, between
the largest and presumably most powerful members of the new social
unit. The frequency of agonistic interactions dropped precipitously following the initial hour of group formation. Comparable periods during
the next seventy-four days indicated that agonistic interactions were initially some twenty times higher than at any subsequent time. Affiliative
interactions—such as grooming and huddling—occurred only rarely
during the first hour following group formation and subsequently
accounted for a considerable portion of social activities.

Despite the high rate of agonistic interactions observed during
the initial phase of group formation, the expression of aggressive inclinations was surprisingly orderly. Ritualized agonistic behavior—
facial threats, vocalizations, or postures—accounted for the majority of
agonistic acts. Physical aggression was largely restricted to one adult
male and one adult female. Fighting—attack followed by counterattack—was never observed among adults and only during the first few
minutes following release in juveniles. Moreover, the direction of agonistic interactions among adults suggested that status relations were
established rapidly and were discernable even within the first hour.
These initial status relationships were not subsequently reversed. The
authors conclude that "...status was determined principally by the

perception of physical attributes, posture and bearing of companions, rather than by a contest of strength" (Bernstein and Mason 1963, 31).

The results of this study suggest that relationship formation does not proceed gradually. Moreover, initial interactions were quite orderly. Physical aggression was minimal, fighting was not observed, and directionality in agonistic interactions was evident during the first hour of group formation. Apparently, within minutes of the first encounter between unfamiliar animals, the broad outlines of interindividual relationships are established.

These findings were replicated in a study that compared the response to group formation in four species of macaques (Hawkes 1970). All four species studied showed higher levels of agonistic behavior during the first hour following group formation than they engaged in subsequently. Sexual behavior followed a similar trend. Species differed, however, in the extent to which these trends were expressed. Rhesus monkeys interacted most frequently and, except for a few sexual overtures, all interactions were agonistic. Stumptail monkeys, on the other hand, interacted sexually more frequently than agonistically during the first hour. Crabeater and bonnet monkeys interacted less than either stumptails or rhesus and had fewer agonistic interactions than did the rhesus and fewer sexual interactions than did the stumptails. An important finding in this study was that aggressive overtures were not responded to with counteraggression in any species. Only one such occurrence was observed, and this was in the bonnet macaques, the least aggressive species overall. Behaviors surrounding the aggressive episode also suggested that the attack-and-counterattack sequence was embedded within a sexual context.

The behavioral responses of the macaques to group formation in these early studies point to two dimensions of initial encounters that have received attention in subsequent studies. The first concerns the tendency for individuals to respond to conspecifics in a species-typical manner. The second dimension that the early group formation studies highlight is that, at least among macaques, the pattern of interactions viewed temporally is similar. Between-sex interactions, when they occur, are predominantly sexual during the initial periods following group formation. Group-formation induced breeding has been documented in a number of species including macaques, talapoins, and squirrel monkeys (Keverne 1979; Mendoza et al. 1979; Rose et al. 1972). It has been suggested that the sexual behavior evident during initial encounters may reduce the likelihood of aggressive interchange and allow cohabitation during initial stages of relationship formation. This possibility may be particularly pertinent in natural populations where individuals typically emigrate from their natal groups and enter new social units. The entry into new social groups is most likely to be successful during periods of high sexual activity.

Within-sex interactions during initial encounters are predominantly agonistic. The expression of aggressive tendencies, however, is orderly and suggestive of a general tendency for individuals of like age and sex to form status relationships rapidly. Inasmuch as the social groups in these studies were constituted to approximate the social configuration of natural groups, no attempt was made to match animals within age/sex categories for size or demeanor. In general, the outcome of aggressive contacts can be reliably predicted if the contestants are not closely matched for size (Rowell 1988). It may be that the lack of overt aggression in these studies reflected the animals' ability to evaluate one another's fighting potential relative to themselves. As Bernstein and Mason (1963) reasonably concluded, the animals were provided with multiple cues that could be used to predict the outcome of fights without having to engage in aggressive contests. If this conclusion is correct, then two further predictions regarding the process of status delineation can be made. First, if animals differ in observable characteristics which are predictive of outcome in aggressive contests, and if each individual is capable of ranking the others relative to themselves in the possession of these characteristics, then only a brief period of mutual assessment may be necessary for status determination. Interactions during initial encounters should reflect the results of the mutual assessment. Those animals possessing characteristics predictive of success in aggressive contests should readily exhibit behaviors consistent with high status and individuals with less pronounced characteristics predictive of success should exhibit behaviors consistent with low status.

A second, related prediction is that the degree to which individuals differ in characteristics that may be used as cues in status delineation should influence the complementarity and accuracy of the assessment process. For dyads composed of animals that are extremely different in status cues, the assessments made by each should be relatively unambiguous. Which animal would be the winner or the loser of an aggressive contest, should it occur, would be obvious to both participants without ever interacting aggressively. If, on the other hand, individuals were more similar, ambiguity regarding probable outcome would be increased. Predictions as to the probable winner may not be in agreement, and the likelihood of an actual contest of strength would be increased. Thus, when differences in status cues are extreme, the period of mutual assessment should be brief, and the instances of agonistic interactions relatively low and immediately directional. However, when differences in status cues are minimal the assessment period should be prolonged, the instances of agonistic interactions relatively high and directionality would not be immediately apparent.

Influence of Status Characteristics during First Encounters

A particularly useful data set for investigating the influence of differences in status characteristics on behavioral interactions during first encounters is provided by Maxim (1976). The data were originally collected in an effort to describe the communicative value of specific behavioral acts typically displayed by rhesus monkeys (Maxim 1976, 1978, 1985). Hence, the data were used in a different manner from the manner in which they will be viewed here. Rather than attempting to precisely delineate specific characteristics that contribute to status delineation, Maxim relied on the ability of caretakers to assess the potential of monkeys to achieve high status. On this basis, six index monkeys (three males and three females) were selected from a colony of 250 rhesus macaques to represent the most dominant, the most subordinate, and a typical middle-ranking individual of each sex. The index animals were subsequently observed in a series of paired tests with ten randomly selected, unfamiliar individuals of each sex from the remaining colony. Assuming that the caretakers were accurate in their evaluations of the index animals, these pairings provide an opportunity to compare initial interactions between animals when clear cues were available to the participants that could be used to predict the outcomes of aggressive contests (same-sex pairs involving the high-status animal or the low-status animal) with initial interactions when clear cues were unavailable to the participants (same-sex pairings involving the typical middle-ranking individual).

Maxim was not concerned with the incidence of aggression per se during first encounters. Unfortunately, for our purposes, he did not record sequences of interactions during the first two minutes following introduction or, if a fight occurred, data collection was postponed until one minute following the conclusion of the fight. His methods suggest that physical aggression was a frequent occurrence in the initial reactions of animals to one another. However, it is impossible to determine from his reports whether aggression typically included series of attacks and counterattacks, hence qualifying as a fight or an aggressive contest. It is also unknown whether aggression was differentially prevalent in pairings with the three types of index animals. We can presume, however, that the severity of aggression was moderate because pairs were not disbanded due to injury, and the delays in data collection accounted for only a few minutes. Following this initial period during which no data were collected, all interactions for the next thirty minutes were recorded. Data were organized into behavioral episodes, defined as interaction sequences preceded and followed by fifteen-second periods during which no interactions occurred.

The results indicate that the caretakers were accurate in their assessments of the potential dominance ranks of the index animals. The index male judged to be of high status directed considerably more threats to his male partner than he received; whereas, the reverse was true of the index animal judged to be of low status (see figure 4.1). Similarly, the index male judged to be of high status exhibited no fear responses and few submissive responses to the other males; whereas, when paired with the same males, the index male judged to be of low status exhibited numerous fear and submissive behaviors, but elicited no such behaviors from them. The index male judged to be of middle status was intermediate on each measure. Although the data are less clear for female-female pairings, the same patterns were evident (see figure 4.2).

MALE-MALE DYADS

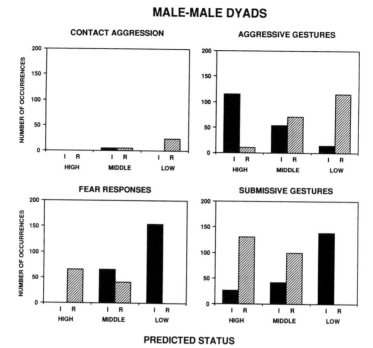

Figure 4.1 The frequency of interactions initiated (I) and received (R) by rhesus monkeys judged to be a high ranking male, a middle ranking male or a low ranking male during the first 30 minutes following dyad formation. Data taken from Maxim, 1976.

Examination of the content of behavioral episodes reported by Maxim reveal a striking degree of constraints in the interaction patterns expressed in dyads of any combination of animals. A notable finding was that contact aggression was an infrequent behavior (see figures 4.1 and 4.2). Moreover, agonistic behaviors of any type were not responded to in kind. For two-move behavioral episodes, physical aggression—attack, hit, or bite—was never responded to with counteraggression of any type. Similarly, noncontact antagonistic gestures—threat, or stare—were responded to with a variety of submissive gestures, but the reactions to these agonistic overtures never consisted of aggressive acts or gestures. There were also remarkably few behavioral episodes of three or more moves that included aggres-

FEMALE-FEMALE DYADS

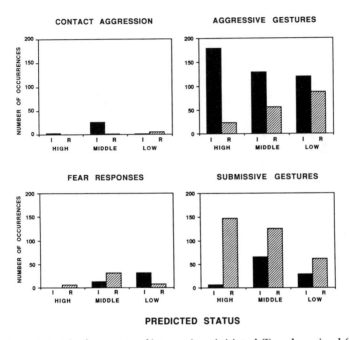

Figure 4.2 The frequency of interactions initiated (I) and received (R) by rhesus monkeys judged to be a high ranking female, a middle ranking female or a low ranking female during the first 30 minutes following dyad formation. Data taken from Maxim, 1976.

sive acts or gestures by both interactants (70 of 2,829, or 2.5%). Furthermore, most of the behavioral episodes (60 of 70) that included aggressive acts by both individuals were limited to low-intensity agonistic gestures—stare—by both animals.

The results of Maxim's study indicate that individual characteristics can and do influence status determinations. Clearly, the characteristics utilized by the animal handlers in Maxim's study correctly identified animals that would emerge from dyadic encounters as high- or low-status animals. The characteristics were salient to animals and handlers alike.

The degree of disparity in status characteristics did not influence the frequency of aggressive encounters. In all types of dyads, agonistic interactions involving physical aggression were rare. Noncontact agonistic interactions were prevalent, but even the dyadic encounters with the index animals judged to be a typical middle-ranking animal were composed of asymmetric interactions.

These findings are consistent with the idea that rules of interaction guide relationship formation rather than emerge gradually from relationship formation. However, other possibilities should be considered. For example, it is possible that, even though the middle-ranking index animal fell within the range of characteristics important for status delineation of the ten randomly selected individuals, for each dyad formed there may have been sufficient differences between individuals that the participants were able to predict winners or losers of an agonistic contest and, therefore, could avoid engaging in the contest. If this is the case, then one would predict that, if handlers were instructed to carefully match animals for any attributes that may influence status delineation, the incidence of agonistic contests should be quite high and, correspondingly, the incidence of symmetrical aggressive interchanges should be frequent. Another possibility is that aggressive contests did occur, but were restricted to the first few minutes following initial encounters. If this is the case, then we can assume that aggressive contests are rapidly resolved. Thus, we must direct our observation to the period immediately following introduction of animals to one another.

First Encounters between Individuals Judged to be Equivalent

These possibilities were tested in another study of rhesus macaques which examined interactions during the first forty-five minutes following formation of three member groups (Barchas and Mendoza 1984; Mendoza and Barchas 1983). The subjects for this study were twenty-seven female and twelve male rhesus macaques. All subjects were fully mature, newly imported monkeys and thus judged to have been fully socialized. Prior to testing, subjects were assigned to within-sex triads (nine female triads and four male triads). Members of a triad were carefully matched for size, probable age, reproductive state, and general activity levels by an experienced veterinarian. So far as was known, members of a triad had no prior experience with one another.

The monkeys were introduced to one another by simultaneously raising guillotine doors of three holding cages thereby allowing the animals to enter a larger test room. No attempt was made to force the monkeys to leave their holding cages nor to prevent their return. All monkeys readily exited. Thus, we may conclude that the opportunity to interact with conspecifics or the lure of greater freedom of movement provided sufficient incentive for the monkeys to emerge into the larger test environment.

Upon entering the larger test environment, the monkeys engaged predominantly in exploratory behaviors and spent considerable time investigating the room and their social partners. The frequency of social exploration—such as close visual, manual, or olfactory investigation of another animal—decreased across the observation period, although it remained the most frequent form of interaction even during the final fifteen minutes of observation. The high frequency of this behavior suggests that the animals were engaged in mutual assessment, perhaps necessitated by the attempt to minimize the cues available to permit differentiation of animals. Affiliative behaviors, exemplified by the frequency of grooming episodes, tended to increase throughout the observation session.

Attack by biting, hitting, or grabbing another animal was most frequently observed during the first fifteen minutes of observations, although frequencies were never very high. As in the previous studies, attack was never followed by counterattack. Noncontact behaviors expressive of assertiveness were much more common than were behaviors involving physical interaction. However, even these behaviors were only infrequently followed by a retaliatory act or gesture. Indeed, only 3.8% of occurrences of any behavior indicative of assertiveness were followed by an assertive response by the initial target directed toward the initial actor.

Analysis of behavioral sequences in this study suggests that triadic, as well as dyadic, interactions follow particular sequences and not others. Patterns of interaction that are logically possible and intuitively probable generally do not occur. Following the methods described by Chase (1980, 1984) sequences of interactions were categorized according to whether they followed:

1. *Repeat,* in which animal A initiated two successive acts of the same type to the same recipient B.

2. *Reversal,* in which animal A directed an act to animal B, and the next interaction of that type was initiated by B and directed to A.

3. *Double Initiate,* in which animal A directed a behavior toward animal B, and the next interaction of the same type was initiated by A and directed to the third animal, C.

4. *Double Receive,* in which an act initiated by A and directed to B was followed by C directing a similar behavior to B as well.

5. *Initiator Receives,* an act initiated by A and received by B is followed by animal C directing a similar behavior to A.

6. *Pass-On,* in which animal B, the recipient of the first act initiated by A, initiates the second act of the same type directed toward C.

The results of this analysis are presented in table 4.1. Interactions involving only two animals tended to be repeated rather than reversed. This is most clearly the case with dominance behaviors—namely, attack, threat, stare, displace—and submissive behaviors—described as grimace, present, and avoid. That investigation and grooming tend to follow the same pattern suggests that animals are not likely to respond to any act by expressing the same type of behavior in return.

When sequential acts involved all three animals, some patterns were frequently utilized, whereas others were infrequent. For all types of interactions, the Double Initiate and/or the Double Receive patterns occurred considerably more often than anticipated by chance (p's <.001). For aggressive interactions, Chase argues that these two patterns of interaction establish a temporal sequencing of behaviors which will facilitate the formation of linear dominance hierarchies (Chase 1984). These data suggest that interactions between animals are constrained during initial encounters. The rules of interaction are relatively simple and may be stated as follows: (1) Do not do unto others as they have done unto you; (2) Treat others as you have seen them treated; and (3) Treat all other animals in the same way.

Table 4.1 Patterns of Sequential Interactions Following
Formation of Rhesus Triads

	Repeat A→B*	Reverse B→A	Double Initiate A→C	Double Receive C→B	Initiate to Initiator C→A	Pass-On B→C
Dominance	46.8	3.2	5.8	20.5	3.8	4.3
Submission	67.1	1.6	32.7	10.1	3.9	3.2
Investigation	46.2	28.3	44.9	49.5	29.2	24.0
Grooming	19.7	2.1	0.8	4.7	1.3	1.7

Source: Reanalyses of data originally presented in Barchas and Mendoza 1984.
*Sequential interactions were categorized by determining the initator and
recipient of the second act given an initial act of the same type initiated by animal A and directed to animal B. Data were collected during the first forty-five
minutes following introduction.

The infrequent occurrence of the Initiator Receives and Pass-On
patterns of triadic interaction is counterintuitive. Evidence is strong that
these patterns occur in primate interactions. For example, redirected
aggression, which relies on the Pass-On pattern of interaction is held to
be a frequent occurrence in primate groups. (See Mason, chapter 2, and
de Waal, chapter 5 in this volume.) Similarly, agonistic interference or
recruitment relies on interaction sequences involving the Initiator
Receives pattern of interaction. This suggests that, once relationships
are established, the rules may not be as closely followed as they are during initial encounters or that new rules of interaction are established.

Given the lack of data on comparative frequency of patterns of
interaction, it is also possible that the Pass-On (redirected aggression)
and Initiator Receives (agonistic interference) patterns are equally
improbable occurrences, regardless of the time from the initial group
formation. Even infrequent expression of these particular patterns of
interactions, however, may signify an important departure from the
norm and, thereby, communicate significant information to group
members.

It should be noted that the patterning of successive acts accord-ing to the Double Initiate or Double Receive patterns need not imply that dyadic interactions are unidirectional. Investigative behaviors were equally likely to be initiated and received by all triad members. Directionality was apparent only when analysis focused on successive acts. For dominance and submissive behaviors, directionality was not immediately apparent, but emerged in most dyads by the end of the observation period.

In order to estimate when role differentiation occurred, arbitrary criteria were employed. A dyad was considered to have settled which was dominant if two successive dominance acts—attack, threat, stare, or displace—were initiated by one individual and received by the other within a two-minute period, followed by five minutes during which the recipient of the initial interactions did not direct any domi-nance behaviors to the initiator nor did the initiator direct any submis-sive gestures—grimace, present, or avoid—to the recipient. Similar cri-teria were used to identify status delineation by way of submissive gestures. The number of minutes from the start of observation to achievement of the criteria was considered to be the time to status delineation. The minimum time by definition was six minutes.

Of the thirty-nine dyads, thirty-four achieved the criteria within the forty-five minute observation period, indicating that the relative status of the dyad members was determined. The average time to meeting the criteria was only 13.9 minutes.

The data were further examined for reversals. None were found despite the fact that there was ample time for reversals to have occurred. Males and females did not differ in time to status delin-eation. Following achievement of the criteria—but not before—domi-nance and submissive interactions within each dyad were unidirec-tional. That is, all dominance behaviors were initiated by the dominant members of the pairs and directed to the subordinate members, while submissive behaviors were initiated by the subordinate and directed to the dominant.

In summary, the data from this study indicate that initial encounters among rhesus monkeys do not entail prolonged series of dominance struggles nor contests in which group members vie for high status even when subjects are closely matched. Moreover, to the extent that it is possible to compare these data with those obtained in other studies, it appears that reducing the ability of animals to predict outcomes of fights leads to less, not more, instances of physical aggression. Interactions between animals are constrained according to some relatively simple, but apparently potent, rules of interaction. Adherence to the rules does not, in and of itself, imply that behaviors tend to be initiated by some individuals and received by others. Investigative and affiliative behaviors follow the rules of interaction, but all individuals within the triads are equally likely to initiate these interactions. Initially, dominance and submissive acts are likewise initiated equally often by all group members. Within a short period, however, individuals limit themselves to behaviors consistent with only the dominant or the subordinate role within each dyad. It may seem obvious, but it is important to note that interactions of the middle-ranking animal in each triad must differentiate between the group member to whom it is subordinate and the group member to whom it is dominant. These distinctions seem to not pose any difficulties for the animals.

PHYSIOLOGICAL DIFFERENTIATION DURING
FIRST ENCOUNTERS

Work with macaques suggests that, when animals meet with defeat, physiological changes ensue which persist well beyond the immediate response to the incidents which produce the response (Rose et al. 1972, 1975). Specifically, defeat has been associated with increased adrenocortical activity and suppression of gonadal activity. Since subordination and defeat are often considered to be synonymous, these results have been used to suggest that the process of subordination is a stressful event that engenders a variety of physiological changes which lead to impaired reproductive function. The behavioral data just reviewed, however, indicates that the process of subordination need not include defeat. It is likely, therefore, that the physiological changes associated with formation of dominance relationships are also different.

In a series of studies using squirrel monkeys as experimental subjects, we have monitored physiological activity prior to and following creation of new social groups in order to evaluate the effects, if any, of relationship formation on neuroendocrine function. With respect to the pituitary-gonadal system, the results are clear and consistent. The relative dominance status established between males is critical in determining each male's hormonal response to group formation. Alpha males have the highest levels of plasma testosterone within each social group (Coe et al. 1979; Mendoza et al. 1979; Saltzman 1991). It is not uncommon, however, for subordinate males to exhibit higher testosterone levels than the dominant animal of another group. Moreover, testosterone levels prior to group formation cannot be used to predict subsequent status of the males. Therefore, formation of social groups differentially alters each male's physiological state and the differentiation between males is particular to the group (Everitt et al. 1981).

Heterosexual stimulation also alters testosterone levels in male squirrel monkeys. Males introduced to novel females exhibit increased testosterone levels, except when a more dominant male is present. Similar findings have been reported for male rhesus macaques (Rose et al. 1972, 1975) and talapoins (Everitt et al. 1981).

The influence of dominance on pituitary-adrenal activity is neither clear nor consistent. In our studies, formation of unisexual units has resulted in increased cortisol (Mendoza et al. 1979) and decreased cortisol (Mendoza and Mason 1991; Mendoza et al. 1991b). Dominance can influence cortisol in male squirrel monkeys but, in some cases, the alpha male exhibits the higher cortisol (Mendoza et al. 1979). In other cases, the subordinate male exhibits the highest cortisol (Mendoza et al. 1991b). It may be that whether dominant or subordinate males show lower cortisol activity depends on the subspecies of squirrel monkeys (Coe et al. 1983; Mendoza et al. 1991a). Among monkeys imported from Bolivia, dominant animals tend to exhibit higher cortisol than do subordinates, whereas monkeys imported from Colombia, Peru, or Guyana exhibit the reverse pattern with subordinates showing higher cortisol levels than their more dominant cagemates. The pattern of adrenocortical activity exhibited by Bolivian squirrel monkeys is relatively rare in primates and has been noted in only one other species—vervet monkeys (McGuire et al. 1986). The more usual pattern—found in some squirrel monkeys, and several species of macaques and baboons—in which subordinate monkeys exhibit higher cortisol levels than their more dominant counterparts has been interpreted as evidence for increased social stress in subordinate monkeys. Other physiological changes which occur as a result of relationship formation lead to altered interpretation of these data.

Cardiovascular activity was monitored before, during, and after formation of unisexual relationships in fourteen adult male squirrel monkeys. Pair formation was accomplished by introducing two males simultaneously into a familiar test cage for one hour. Their behavior and heart rate during pair formation was compared to responses to the test cage while living and being tested alone, and to their response to the same cage after the males had been living continuously together for two to three weeks.

The initial encounters between males were notably devoid of any social interactions aside from furtive glances at one another from opposite sides of the cage. Males were slow to approach one another but, by the end of one hour, most (six of seven pairs) had achieved proximity. Contact between the males was restricted to the last twenty minutes of the observation period and occurred in only three pairs. No aggressive interactions were observed, and only one instance of behavior even remotely indicative of agonism—a genital display—was observed. After the pairs had lived continuously together for two to three weeks, the frequency of interactions increased. The males were in proximity to one another approximately 50% of the time, and most (six of seven pairs) spent at least some portion of the observation period sitting in passive physical contact. Aggressive interactions were notably absent, and only two genital displays were observed.

In each exposure to the test cage, heart rate was initially high and rapidly declined. The rate and magnitude of decline was greater during pair formation in all males than previously when the monkeys were living and being tested alone. Following a two to three week period of cohabitation, a further decline in heart rate was evident. Although reduction in heart rate was evident in both dominant and subordinate males, the magnitude of the effect was greater for the dominant. Thus, both animals were altered physiologically by formation of dominance relationships. Since cardiovascular ailments associated with high sympathetic activity are relatively common in this species, we may presume that the physiological changes were beneficial for both animals.

The behavioral results in this study parallel those of the rhesus macaques, in that fights were not observed in squirrel monkey males coming together for the first time. Unlike the rhesus, squirrel monkeys seem very inhibited during their initial encounters—to the point where interactions generally do not occur until the animals have been together for some time. Despite this reluctance to interact, the social situation was clearly salient to the animals as evident by the altered cardiac activity in this situation and as compared to these measures when animals were being tested and living alone.

How two squirrel monkeys encountering one another for the first time determine which is dominant and which is subordinate cannot be determined from these data. Unlike rhesus monkeys, squirrel monkey males do not engage in agonistic interactions during the initial stages of relationship formation. Their rules of interaction also appear to be different than those which macaques employ. For squirrel monkeys, extreme caution in interacting seems to be the predominant theme in their first encounters. Overt interaction is avoided. The animals even observe one another furtively and tend to be active only when the other is inactive. It is as if the animals simply decided which ones would be dominant and which would be subordinate. Often, the smaller animal emerges as dominant. However dominance is determined among male squirrel monkeys, we usually cannot discern status through behavior prior to the time when it has clear physiological effects for the animals.

CONCLUSIONS

Taken together, the data regarding first encounters are contrary to common assumptions regarding the likelihood of social conflict and its physiological consequences. According to Bernstein, "When two animals [upon initial encounter] do something different, then we must find different explanations for the behavior of each; we cannot explain

both responses by the same situational variable. We assume something different in the two that accounts for their distinct responses"
(1981, 419). The data we have reviewed in this chapter suggest that,
even when animals are closely matched for characteristics that may
influence dominance, behavioral differences rapidly emerge. Physiological differences also emerge from the formation of new social relationships. Physiological differentiation does not apparently result
from the differential experiences of defeat or victory. These behavioral
elements are not evident during first encounters.

Mutual shaping does not account for the formation of dominance
relationships. There are no fights. There are no winners or losers.
Whether or not agonistic interactions occur depends, in part, on the
species. In all primates examined to date, however, agonistic overtures
are not responded to in kind. From the outset interactions between
unfamiliar animals are rule-bound. The rules vary from species to
species and appear to set broad limits on the types of relationships to
be formed.

The rapidity with which status delineations are made suggests
that who attains high status and who attains low status is not all that
important. Dominance rank is not worthy of competition. Creating
differences between otherwise similar animals appears to be the primary goal. The behavioral and physiological differentiation probably
functions to reduce the potential for destructive conflict during the initial stages of relationship formation, thereby permitting the development of more elaborate and personal relational qualities.

The consistency of patterns of interaction during first encounters—particularly among individuals with no current relationships—
suggests that animals are prepared to engage with one another in
ways which lead to predictable outcomes at the dyadic or group level.
Rules of interaction guide relationship formation. One can envision a
social niche available, and that unfamiliar animals of the appropriate
age/sex class are viewed as candidates for filling the niche.

Status delineation and adherence to the rules of interaction do
not necessarily imply that relationship formation is complete. Vandenbergh (1967) observed rhesus monkeys released onto a spacious island
habitat off Puerto Rico and found that stabilization of social structure

was a gradual process requiring two or more years for groups to achieve stable membership. Moreover, attempts to form bands by grouping monkeys under captive conditions prior to release was generally unsuccessful and, in no instance, did the preformed group remain entirely together following release. In fact, in two of the four attempts, the preformed groups disbanded entirely upon release. Similar observations have been made of Japanese macaques (Kawai 1960).

If, as Kummer (1984) suggests, status differentiation is an essential first step in the formation of affinitive relationships, it may be that we too readily assume that relationship formation is completed as soon as we can discern dominant and subordinate. It is quite likely that other dimensions of relationship formation proceed by quite different mechanisms than the relatively simple process of deciding which role each animal will assume.

In conclusion, there is every indication that, even in their initial encounters with one another, social primates are guided by rules of interaction. The types of rules which primates follow during first encounters appear to be species-specific and to vary according to the types of relationships that typify the social organization. It is clear that role differentiation proceeds with minimal overt conflict between the participants. Precisely how role differentiation is accomplished by the animals is unknown. It is as if the animals create the behavioral and physiological differences between themselves for the sole purpose of being different.

REFERENCES

Barchas, P.; Mendoza, S. P. (1984) Emergent hierarchical relationships in rhesus macaques: An application of Chase's model. In Barchas, P. R. (ed.), "Social Hierarchies: Essays toward a Sociophysiological Perspective." Westport, Conn.: Greenwood Press. 81–95.

Bernstein, I. S. (1981) Dominance: The baby and the bathwater. *Behavioral and Brain Sciences* 4:419–457.

Bernstein, I. S.; Mason, W. A. (1963) Group formation by rhesus monkeys. *Animal Behaviour* 11:28–31.

Chase, I. D. (1980) Social processes and hierarchy formation in small groups. *American Sociological Review* 45:905–924.

Chase, I.D. (1984) Social process and hierarchy formation in small groups: A comparative perspective. In Barchas, P. R. (ed.), "Social Hierarchies: Essays toward a Sociophysiological Perspective." Westport, Conn.: Greenwood Press. 45–80.

Coe, C. L.; Mendoza, S. P.; Levine, S. (1979) Social status constrains the stress response in the squirrel monkey. *Physiology and Behavior* 23:633–638.

Coe, C. L.; Rosenblum, L. A. (1984) Male dominance in the bonnet macaque: A malleable relationship. In Barchas, P. R., Mendoza, S. P. (eds.), "Social Cohesion: Essays toward a Sociophysiological Perspective." Westport, Conn.: Greenwood Press. 31–63.

Coe, C. L.; Smith, E. R.; Mendoza, S. P.; Levine, S. (1983) Varying influence of social status on hormone levels in male squirrel monkeys. In Steklis, H. D., Kling, A. S. (eds.), "Hormones, Drugs, and Social Behavior in Primates." New York: Spectrum Publications. 7–32.

de Waal, F. B. M. (1986a) Class structure in a rhesus monkey group: The interplay between dominance and tolerance. *Animal Behaviour* 34:1033–1040.

de Waal, F. B. M. (1986b) Dynamics of social relationships. In Smuts, B. B., Cheney, D. L., Seyfarth, R. M., Wrangham, R. W., Struhsaker, T. T. (eds.), "Primate Societies." Chicago: University of Chicago Press. 421–429.

Everitt, B. J.; Herbert, J; Keverne, E. B.; Martensz, N. D.; Hansen, S. (1981) Hormones and sexual behaviour in rhesus and talapoin monkeys. In Fuxe, K., Gustafsson, J.-A., Wetterberg, L. (eds.), "Steroid Hormone Regulation of the Brain." London: Pergamon Press. 317–330.

Hawkes, P. N. (1970) Group formation in four species of macaques. Unpublished doctoral dissertation. University of California at Davis.

Hinde, R. A. (1976) On describing relationships. *Journal of Child Psychology and Psychiatry* 17:1–19.

Hinde, R. A. (1979) "Toward Understanding Relationships." London: Academic Press.

Kawai, M. (1960) A field experiment on the process of group formation in the Japanese monkey (*Macaca fuscata*), and the releasing of the group at Ohirayama. *Primates* 2:181–255.

Keverne, E. B. (1979) Sexual and aggressive behaviour in social groups of talapoin monkeys. In Porter, R., Whelan, J. (eds.), "Sex, Hormones and Behaviour." Amsterdam: Excerpta Medica. 271–316.

Kummer, H. (1973) Dominance versus possession: An experiment on hamadryas baboons. In Menzel, E. W., Jr. (ed.), "Symposium Fourth International Congress of Primatology," vol. 1, "Precultural Primate Behavior." New York: Karger. 226–231.

Kummer, H. (1984) From laboratory to desert and back: A social system of hamadryas baboons. *Animal Behaviour* 32:965–971.

Mason, W. A. (1976) Primate social behavior: Pattern and process. In Masterton, R. B., Bitterman, M. E., Campbell, C. B. G., Hotton, N. (eds.), "Evolution of Brain and Behavior in Vertebrates." Hillsdale, N.J.: Lawrence Erlbaum Associates. 425–455.

Maxim, P. E. (1976) An interval scale for studying and quantifying social relations in pairs of rhesus monkeys. *Journal of Experimental Psychology: General* 105:123–147.

Maxim, P. E. (1978) Quantitative analysis of small group interaction in rhesus monkeys. *American Journal of Physical Anthropology* 48:283–296..

Maxim, P. E. (1985) Multidimensional scaling of macaque social interaction. *American Journal of Primatology* 8:279–288.

McGuire, M. T.; Brammer, G. L.; Raleigh, M. J. (1986) Resting cortisol levels and the emergence of dominant status among male vervet monkeys. *Hormones and Behavior* 20:106–117.

Mendoza, S. P. (1984) The psychobiology of social relationships. In Barchas, P. R., Mendoza, S. P. (eds.), "Social Cohesion: Essays toward a Sociophysiological Perspective." Westport, Conn.: Greenwood Press. 3–29.

Mendoza, S. P.; Barchas, P. R. (1983) Behavioral processes leading to linear status hierarchies following group formation in rhesus monkeys. *Journal of Human Evolution* 12:185–192.

Mendoza, S. P.; Mason, W. A. (1989) Primate relationships: Social dispositions and physiological responses. In Seth, P. K., Seth, S. (eds.), "Perspectives in Primate Biology," vol. 2, "Neurobiology." New Delhi: Today & Tomorrow's Printers and Publishers. 129–143.

Mendoza, S. P.; Mason, W. A. (1991) Breeding readiness in squirrel monkeys: Female-primed females are triggered by males. *Physiology and Behavior* 49:471–479.

Mendoza, S. P.; Coe, C. L.; Lowe, E. L.; Levine, S. (1979) The physiological response to group formation in adult male squirrel monkeys. *Psychoneuroendocrinology* 3:221–229.

Mendoza, S. P.; Lyons, D. M.; Saltzman, W. (1991a) Sociophysiology of squirrel monkeys. *American Journal of Primatology* 23:37–54.

Mendoza, S. P.; Saltzman, W.; Lyons, D. M.; Schiml, P. A.; Mason, W. A. (1991b) Within-sex relationships in squirrel monkeys regulate pituitary-adrenal activity. In Ehara, A., Kimura, T., Takenaka, O. (eds.), "Primatology Today: Proceedings of the Thirteenth Congress of the International Primatological Society." Amsterdam: Elsevier. 443–446.

Miller, R. E. (1971) Experimental studies of communication in the monkey. In Rosenblum, L. A. (ed.), "Primate Behavior: Developments in Field and Laboratory Research," vol 2. New York: Academic Press. 139–175.

Rose, R. M.; Bernstein, I. S.; Gordon, T. P. (1975) Consequences of social conflict on plasma testosterone levels in rhesus monkeys. *Psychomatic Medicine* 37:50–61.

Rose, R. M.; Gordon, T. P.; Bernstein, I. S. (1972) Plasma testosterone levels in the male rhesus: Influences of sexual and social stimuli. *Science* 178:643–645.

Rowell, T. E. (1967) A quantitative comparison of the behavior of a wild and caged baboon group. *Animal Behaviour* 15:499–509.

Rowell, T. E. (1988) Beyond the one-male group. *Behaviour* 104:189–201.

Saltzman, W. (1991) Social relationships among squirrel monkeys (*Saimiri sciureus*): Context, dynamics, and physiological consequences. Unpublished doctoral dissertation. University of California at Davis.

Vandenbergh, J. G. (1967) The development of social structure in free-ranging rhesus monkeys. *Behaviour* 29:179–194.

CHAPTER FIVE

Reconciliation among Primates: A Review of Empirical Evidence and Unresolved Issues

FRANS B. M. DE WAAL

Primates invest considerable energy in the maintenance of social relationships. Their agonistic behavior should be studied, therefore, not only from the standpoint of immediate gains (for example, access to resources) and costs (such as injury), but also in the light of long-term social commitments and the benefits derived from these. Agonistic incidents can have, undoubtedly, an undermining effect on beneficial relationships. Globally, there are two ways in which primates may avoid this. One is through inhibitions on aggression, as reflected in a tolerant attitude toward the partner. The other method is by conciliatory postconflict reunions with adversaries. Both social mechanisms depend on reassuring and appeasing forms of body contact. This chapter reviews detailed quantitative studies on reconciliation behavior and social tolerance in several primate species. Although the interspecies differences are dramatic, the resolution of conflict by peaceful means is a common primate strategy.

Frans B. M. de Waal—Yerkes Regional Primate Research Center and Psychology Department, Emory University, Atlanta, Ga.

I thank Mary Schatz and Jackie Kinney for typing the manuscript, and Linda Endlich for drawing the figures. I am grateful to Tine Griede and Gerard Willemsen for data collected in Arnhem. Writing was supported by grant no. RR00167 of the National Institutes of Health to the Wisconsin Regional Primate Research Center. This is publication no. 27–004 of the WRPRC.

The little creature, which I had punished for the first time,
shrank back, uttered one or two heart-broken wails, as she
stared at me horror-struck, while her lips were pouted
more than ever. The next moment she had flung her arms
round my neck, quite beside herself, and was only com-
forted by degrees, when I stroked her.

Wolfgang Köhler (1925)

Köhler saw a need for forgiveness reflected in the above behavior by a
juvenile chimpanzee. Similar stories can be heard from people who
have raised apes at home. For example, Kellogg and Kellogg experienced emotional reunions after reprimanding their infant chimpanzee,
and noted that she would express relief during the embrace by "heav-
ing her great sigh, audible a meter or more away" (1933, 172).

Apes seem to perceive a conflict with their caretakers as a threat to the
relationship, and try to control the damage by means of affectionate
behavior. Although the environment of hand-reared apes is obviously
unnatural, human influence does not appear to explain the phenomenon. Reconciliations also occur after fights in naturalistic settings in
which nonhuman primates interact amongst themselves. This chapter
reviews the evidence collected over the past decade by our team and
adds recent work by others. It also pays attention to the methodology
of the various studies, and presents a framework for further inquiry.

In doing so, the data will be placed before the theory. Conflict
resolution is still an area of inductive research—one in which each
new finding stimulates our thinking, and in which economic models
of explanation are as yet lacking. Interest in the subject did not arise
from existing theories, which generally assume that animals are ruth-
less competitors. Instead, the picture emerging from the research
reported here supports Smith's correction of this assumption. He
wrote, "Surely, no matter how complexly social their lives become,
individuals continue to have divergent needs. But these diverging
needs result in their competing with each other *in addition* to cooperating, not...instead of cooperating" (1986, 74).

The study of conflict resolution essentially concerns the value
that animals attach to their social relationships, and the balance they

strike between short-term competitive gains and peaceful coexistence. Because competition over resources is constrained by the need to conserve cooperative relationships, victory is rarely absolute. Mechanisms of social repair following aggression are to be expected in species such as primates with highly differentiated, long-term social relationships (de Waal 1989a, 1989b).

TWO BASIC HYPOTHESES

Thus far, the principal aim of reconciliation research has been to test two alternative expectations concerning the effect of aggressive encounters on social relations.

Dispersal hypothesis. Losers of aggressive incidents tend to avoid winners. The traditional notion of aggression as a spacing mechanism (Lorenz 1963; Scott 1958) was based on experience with territorial species, and on Hediger's (1941) influential concept of individual distance. It predicts a decreased probability of contact between individuals following aggressive behavior.

Reconciliation hypothesis. Individuals try to undo the damage that aggression inflicts upon valuable social relationships. This hypothesis predicts *(a)* an increased contact probability following aggression, and *(b)* the use of special reassuring and appeasing behavior patterns during these contacts.

The reconciliation hypothesis is supported, therefore, if individuals contact each other more frequently and with more calming gestures after aggression than in the absence of preceding aggression. Yet, it is good to realize that, strictly speaking, such a result would not demonstrate the specific function suggested by the term *reconciliation,* which is to repair a disturbed relationship. Rather, the term is a heuristic label from which further predictions can be derived. One such prediction is that postconflict contacts are selectively made—that is, it is not just a matter of everyone contacting everyone. Whereas calming body contact is, in principle, possible with any social partner, reconciliation requires interaction between the antagonists themselves, because only they can mend their relationship.

The primatological literature contains numerous descriptions of reassuring and calming behavior. While initially stressing this behavior's arousal-reducing effects on the internal state of aggressive, excited or frightened individuals, investigators have also begun considering the social implications of such behavior (Blurton-Jones and Trollope 1968; Ehrlich and Musicant 1977; Ellefson 1968; Lindburg 1973; Mason 1964; McKenna 1978; Poirier 1968; Seyfarth 1976; van Lawick-Goodall 1968). The first study to deal explicitly with these questions concerned the large chimpanzee (*Pan troglodytes*) colony of the Arnhem Zoo in the Netherlands. The dispersal hypothesis was rejected because aggressive incidents in this colony were associated with an average decrease in interindividual distance (de Waal and van Roosmalen 1979). Continuous video recording revealed that the chimpanzees were more often within two meters from one another after aggression had occurred between them than before. Obviously, aggressors were avoided during the fight itself, but rapprochement often began as soon as the hostilities ceased. This was not due to a lack of space, as the group lived on an island of nearly one hectare with plenty of room for adversaries to stay out of each other's way.

Participants in aggressive episodes were observed for forty-five minutes afterward to document their reunions. Although control procedures were not followed, the data strongly indicated that reassuring body contacts occurred relatively shortly after the conflict, were preferentially made between former opponents, and involved special behavior patterns, which were rarely seen outside this context. Typically, contact between former opponents was initiated by an invitational hand gesture, with outstretched arm and open-handed palm, often followed by mouth-to-mouth kissing (figure 5.1). This intensive contact pattern was less common toward third parties. So-called *consolations*, in which a participant in an agonistic conflict afterward sought contact with an uninvolved bystander, more often involved embracing than kissing (table 5.1).

Table 5.1 Frequencies of kissing and embracing among the chimpanzees in two postconflict contexts

Context	Kiss	Embrace
Reconciliation[a]	23	8
Consolation[b]	19	57

Source: A study at Arnhem Zoo in the Netherlands by de Waal and van Roosmalen (1979).
[a]Contact between two former opponents.
[b]Contact by participants in a fight with uninvolved individuals.

Figure 5.1a An adult male (left) and female chimpanzee engage in a mouth-to-mouth kiss after an aggressive incident between them (reconciliation; from de Waal, 1989b).

Figure 5.1b A juvenile male (left) embraces a screaming adult male who is withdrawing from a confrontation with a rival (consolation; from de Waal, 1982).

CONTROLLED OBSERVATIONS

Contact Rates

After the Arnhem Zoo exploratory study, carefully controlled investigations of reconciliation behavior were needed. They have taken two forms: *(a)* observational studies comparing behavior following aggression with behavior during control periods, and *(b)* studies of experimentally provoked aggression.

A number of observational studies followed a paradigm developed by de Waal and Yoshihara (1983), which consists of a focal observation on an individual immediately following an aggressive incident in which this individual had participated (the postconflict observation or PC), and a control observation of the same duration on the same individual during the next possible observation day, starting at exactly the same time of day as the PC observation (the matched-control observation or MC). This paradigm has been applied to captive groups of rhesus macaques (*Macaca mulatta*; de Waal and Yoshihara 1983); patas monkeys (*Erythrocebus patas*; York and Rowell 1988); stumptail macaques (*M. arctoides*; de Waal and Ren 1988); and longtail macaques (*M. fascicularis*; Aureli et al. 1989; Cords 1989). Slightly different control procedures were followed in studies of free-ranging vervet monkeys (*Cercopithecus aethiops*; Cheney and Seyfarth 1989); captive pigtail macaques (*M. nemestrina*; Judge, 1991); and captive bonobos (*Pan paniscus*; de Waal 1987). The study by Cheney and Seyfarth compared postconflict and preconflict observations of the same individuals. The study by Judge compared postconflict observations with general baseline data, and the study by de Waal included both of these controls.

Figure 5.2 shows the results of the stumptail study by de Waal and Ren (1988). A greater proportion of opponent pairs established contact after aggression than during control observations. One way to express this difference in a single measure is the so-called *conciliatory tendency*—that is, the percentage of attracted opponent pairs. A pair is said to show attraction if the two individuals make contact during PC observation only, or at least earlier in the PC than in the MC observation. The advantage of this measure is its built-in correction for the normal contact rate. Figure 5.3 compares the conciliatory tendencies of three species based on a ten-minute time window, while taking nonagonistic body contact as a criterion for reconciliation.

Studies of longtail macaques indicate a conciliatory tendency in the range of that of rhesus and patas monkeys (Aureli et al. 1989; Cords 1989). This means that the probability of reconciliation among stumptail monkeys is remarkably high compared to the other monkey species investigated with controlled procedures. Thierry, however, completed an uncontrolled study of another highly conciliatory monkey, the tonkeana macaque (*Macaca tonkeana*) (1984, 1985).

Because in all studies the contact level was found to be significantly higher during postconflict as compared to control observations, the dispersal hypothesis can be safely rejected. Before favoring the reconciliation hypothesis, however, we need to exclude two obvious alternative explanations. The first one is that, because of having just interacted, animals are frequently at short range from one another after an aggressive incident. This would increase the probability of

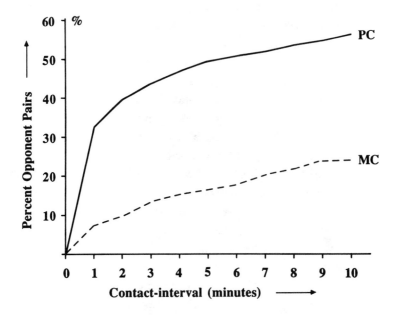

Figure 5.2 The cumulative percentage of opponent pairs that established nonagonistic body contact within a given time interval following aggression (Postconflict, or PC), or during control observations (Matched-Control, or MC), in a captive group of stumptail macaques. From de Waal and Ren (1988).

contact between them if postconflict contacts were blindly initiated. One way of excluding this explanation is to define aggressive incidents in such a way that initial spacing is likely. Thus, many of the studies required a chase or pursuit of more than two meters by the aggressor. In addition, York and Rowell (1988) required proximity between individuals before taking a control observation on them, thus excluding the possibility that the higher contact rate during PC observations was due merely to differences in spatial distribution.

This precaution was not taken in the control procedure of other studies, but de Waal and Ren (1988) did keep a record of interindividual distances among stumptail monkeys, both immediately following the aggressive incident and at the beginning of the scheduled control observation. They found no significant difference. Moreover, even individuals that were distant from one another by the end of their conflict com-

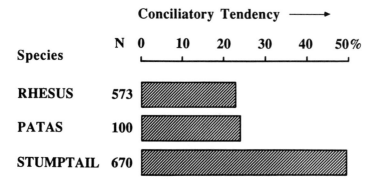

Figure 5.3 Conciliatory tendency is defined as the percentage of attracted opponent pairs, i.e. pairs making contact within a shorter time interval after an aggressive incident than during a matched control observation. The data concern captive groups of three monkey species, based on ten minute observations by de Waal and Yoshihara (1983), York and Rowell (1988), and de Waal and Ren (1988). N is the total sample of opponent pairs.

pared to the beginning of the control observation, contacted one another more often following the conflict than during control periods.

The second alternative explanation for frequent contact between former opponents is that this reflects a general surge in contact following aggression, which does include participants in the incident but not on a selective basis. Yet, several studies found that the *proportion* of contacts involving former adversaries was greater during postconflict periods as compared to control periods (Aureli et al. 1989; de Waal and Ren 1988; de Waal and Yoshihara 1983; York and Rowell 1988). This result includes a correction for changes in overall social activity. The conclusion is that *selective attraction* exists between individuals which previously opposed one another in a fight.

The most rigorous control procedures, with results similar to those already reported, were followed in an observational study of the rare bonobos, or pygmy chimpanzees (*Pan paniscus*), of the San Diego Zoo (de Waal 1987). The apes were observed continuously, allowing a comparison of the following conditions:

1. The fifteen-minute block preceding aggression (pre);
2. The first fifteen-minute block following aggression (post 1);

3. The second fifteen-minute block following aggression (post 2); and

4. Other times of the day, when there was no aggression and no food competition in the group (base level).

Data were analyzed separately per dyadic combination of individuals so that overall differences in interaction frequencies could not be due to the behavior of only a few individuals.

After aggression within a dyad, an increase occurred in the rate of affiliative and sociosexual interaction between the same individuals compared to both their base level and their preconflict rate (figure 5.4). Since the preconflict rate of contact was close to the base level, a common cause for aggression and contact behavior was ruled out. The order of events—first aggression, then an increase in contact—rather suggests a causal link between the two, as predicted by the reconciliation hypothesis.

Table 5.2 compares the rate of postconflict contact for the chimpanzees of Arnhem and the bonobos of San Diego, demonstrating a higher rate in the bonobos. The ten bonobos in San Diego were divided into an all-juvenile group in a spacious enclosure, and two adult groups in a smaller grotto enclosure. The data are presented separately for the adult and juvenile groups in San Diego, and for the indoor and outdoor conditions in Arnhem. These results are not directly comparable to the conciliatory tendency in figure 5.3, as only the latter measure corrects for control contact rates. Moreover, the ten-minute interval used in both analyses may not mean the same for pongids, with their longer memory span and slower pace of interaction, than for monkeys. For example, Aureli and colleagues defend an even shorter interval for longtail macaques (Aureli et al. 1989).

Contact-Initiative and Aggression-Intensity

The initiative for reconciliation following clear-cut unidirectional conflict is presented in table 5.3. For three macaque species, contact initiative following an agonistic encounter is compared to the initiative during matched-control observations. Such control data are not available

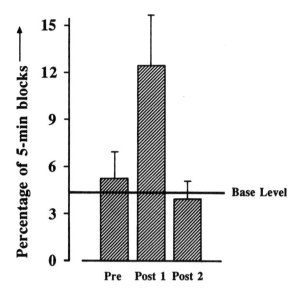

Figure 5.4 Average (± SEM) dyadic rate of interaction among captive bonobos. Rates are expressed as the percentage of 5-min time blocks in which a dyad engaged at least once in sociosexual or affiliative behavior, excluding grooming. Four conditions are compared, i.e. the three 5-min blocks preceding aggression within the dyad (pre), the three blocks immediately following aggression (post 1), the three blocks after that (post 2), and the base level of interaction. From de Waal (1987).

for the other species. Instead, table 5.3 shows the initiative between dominant and subordinate chimpanzees as recorded by Willemsen (1981) for contacts not preceded by aggression during one outdoor period in the Arnhem colony. For bonobos, the table presents the initiative for a large sample of specific contact forms (such as grooming, sexual mounts and mating, embracing, kissing, and patting) observed in the San Diego colony. In this case, as well, the dominance relationship is taken as a reference point. It should be noted that all species show strong agreement between dominance and the performance of unidirectional aggression.

In three of the seven species, reconciliations are mostly initiated by aggressors—that is, in rhesus and patas monkeys, and in bonobos. Yet, there are reasons to interpret this bias differently for bonobos.

Table 5.2 Percentage of reconciled conflicts among the chimpanzees and bonobos

Study		N[1]	Percentage Reconciled[2]
CHIMPANZEES	Indoors 1975–1976[3]	150	34.7
	Outdoors, 1976[3]	200	29.5
	Outdoors, 1980[4]	395	26.6
BONOBOS	Adult subgroups[5]	333	43.8
	Juvenile subgroup[5]	179	55.9

Sources: Studies in Arnhem by [3]de Waall and van Roosmalen (1979) and [4]Griede (1981). A study in San Diego by [5]de Waal (1987).

[1]N = the number of opponent pairs observed.

[2]Reconciliation is defined here as nonagonistic body contact between former opponents within ten minutes of their aggressive interaction.

Bonobos have a high rate of reconciliation (table 5.2), and much more elaborate patterns of reassurance behavior than do rhesus and patas monkeys. Although the bonobos' control data are not directly comparable with the reconciliation data, they do suggest that the contribution of dominants to the establishment of contact increases after aggressive incidents. Such an increase is not evident in the two other species.

It has been suggested by de Waal and Ren (1988) that the relatively low conciliatory tendency of rhesus monkeys is due to reluctance in subordinates to approach dominants. De Waal and Luttrell (1989) have expanded on this issue by characterizing the dominance style of rhesus monkeys as rather intolerant and strict. A similar argument may apply to patas monkeys, as postconflict interactions seem particularly tense in this species, as reflected in frequent approach/avoidance interactions noted by York and Rowell (1988). The bias toward aggressor-initiated reconciliations, therefore, appears to have a different origin in the two monkey species when compared to bonobos. In patas and rhesus monkeys, this bias may be due to a fearful passivity of subordinates, while in bonobos, it may be due to an increased activity of dominants.

Table 5.3 Initiative for the first contact following an agonistic encounter (reconciliation) or during control periods

Species	RECONCILIATIONS		CONTROL CONTACTS	
	N	Percentage of Aggressors	N	Percentage of Aggressors (or Dominants)
Patas monkey[1]	31	67.7	—	—
Rhesus macaque[2]	142	67.6	109	65.1
Stumptail macaque[3]	263	38.4	94	47.9
Longtail macaque[4]	88	36.4	24	58.3
Pigtail macaque[5]	162	34.6	—	—
Bonobo[6]	246	61.4	1795	47.4
Chimpanzee[7]	379	44.3	528	38.1

Sources: N = number of contacts. % Aggressor = percentage of N in which the previous aggressor initiated the contact. Data are from York and Rowell (1988)[1], de Waal and Yoshihara (1983)[2], de Waal and Ren (1988)[3], Aureli et al., (1989)[4], Judge (1991)[5], de Waal (1987)[6], and from a combination of studies on the Arnhem chimpanzees, including de Waal and van Roosmalen (1979)[7]. The control level for rhesus, stumptail and longtail monkeys concerns matched-control observations on the same individuals; see text for an explanation of the other control levels.

The view that the behavior of bonobo aggressors is exceptional is supported by an analysis of the effect of aggression-intensity on the probability of reconciliation. De Waal (1987) presents evidence that physical attacks among captive bonobos are more often reconciled than are incidents of lower intensity, and that this increase is entirely due to the increased contact rate of aggressors. For example, after biting another individual, aggressors would soon return to the victim to inspect the spot where they had put their teeth, and lick and clean it if an injury had resulted. The opposite relation between aggression intensity and contact initiative was found in a study of the Arnhem chimpanzees (Griede 1981). While the tendency of victims to initiate reconciliation was unaffected by the intensity of the previous conflict, aggressors were significantly *less* likely to initiate contact after high-intensity encounters than after low-intensity ones.

Such interspecific differences are quite puzzling, and may be resolved only after a further standardization of methods, and a careful differentiation as to the age and sex combination of the individuals involved in the aggression. As a further complication, Aureli and colleagues (1989) have demonstrated that, in longtail macaques, contact initiative is different in the first three minutes than in the remaining time of the ten-minute postconflict interval. The initiative changes from mostly by-the-victim to mostly by-the-aggressor. (Table 5.3 lumps the data over the entire interval.) The authors view this reversal as an expression of the victim's greater need for stress reduction immediately following the receipt of aggression.

Behavior Patterns

The greatest interspecific variability concerns the form of postconflict reunions. Even closely related species, such as chimpanzees and bonobos, may use totally different behavior patterns. According to studies by de Waal and van Roosmalen (1979), Griede (1981), and Willemsen (1981) on the Arnhem chimpanzee colony, and Goodall's (1986) descriptions of wild chimpanzees, reconciliations in this species typically involve kissing, embracing, hold-out-hand invitations, and gentle touching. Mounts and matings do occur in this context, but do not rank among the most common behavior patterns (de Waal, 1992). In contrast, bonobos typically reconcile by means of genital stimulation, and invitations for postconflict contact are often of a sexual nature. For example, females present their genital swellings either dorsally or while lying on their backs; males present their erect penises sitting upright with legs spread apart. The reconciliation patterns described by de Waal (1987) for bonobos include mutual penis thrusting between males; genito-genital rubbing between females (Kuroda 1980); ventro-ventral and ventro-dorsal matings between the sexes; manual genital massage; and a variety of nonsexual contact forms such as ventro-ventral embracing and gentle touching. By and large, it seems that what chimpanzees do by kissing and embracing, bonobos do by means of sexual and erotic behavior.

Rhesus monkeys show a significant increase in lipsmacking and embracing during postconflict reunions as compared to control contacts (de Waal and Yoshihara 1983). Yet, while this result does support the prediction of behavioral distinctness of such reunions, less than ten percent of the reunions involve these particular behavior patterns. This means that, most of the time, reconciliations among rhesus monkeys are relatively inconspicuous. To contrast this to the situation in stumptail macaques, de Waal and Ren (1988) have labeled the rhesus style of peacemaking as *implicit* reconciliation and the stumptail style as *explicit* reconciliation. That is, stumptails explicitly refer to the previous aggressive incident by means of conspicuous behavior patterns that they rarely show outside this context (de Waal 1989b).

The behavior most characteristic of reconciliations among stumptail monkeys is the so-called hold-bottom ritual, in which one individual presents its hindquarters, and the other clasps the presenter's haunches (figure 5.5). Whereas 34.3% of the postconflict contacts between opponents are preceded by genital presentations, and 20.5% result in a hold-bottom, these respective behavior patterns are observed during only 1.3% and 0.9% of the control contacts. In addition, several other reassurance behaviors significantly increase in frequency after aggression among stumptails—that is, gentle touching, mouth-to-mouth contact, genital inspection, grooming, and teeth-chattering (de Waal and Ren 1988. See also Blurton-Jones and Trollope 1968). So stumptails combine a high conciliatory tendency (figure 5.3) with a remarkably rich repertoire of reassurance gestures.

Patas monkeys do not exhibit systematic behavioral differences between reconciliations and contacts in other contexts. The failure to find such differences might be due to the relatively small sample size of the study by York and Rowell (1988). It might also reflect a general characteristic of the species, which has been said to rely on monitor/adjust behavior for social organization rather than on communication gestures and displays. Another difference is that the social hierarchy of patas monkeys is not as consistent and clear-cut as that of macaques, and is rarely expressed in ritualized status signals (Kaplan and Zucker 1980; Rowell and Olson 1983). De Waal's (1986a) model of the facilitating effect of dominance relationships on conflict resolution predicts that reconciliation behavior will be less developed in species,

Figure 5.5 After an aggressive incident three adult female stumptail monkeys engage in a clasping contact. The female in the middle squeals while holding the hips of her presenting opponent (right). The female on the left holds on to the center female, who protected her during the fight. From de Waal (1989b).

such as the patas monkey, with a weakly formalized hierarchy. In agreement with this prediction, postconflict approaches in this species often lead to avoidance by the subordinate and renewed aggression by the dominant. These monkeys clearly do follow up on their conflicts, but only less than half of their postconflict approaches lead to friendly contact due to an apparent inability to overcome the resulting tensions (York and Rowell 1988).

In rhesus monkeys, similar tensions are frequently observed but, in this species, adversaries solve the problem by threatening low-ranking bystanders (de Waal and Yoshihara 1983. See also Cords 1988, for longtail macaques). Such redirection is relatively rare in the stumptail macaque. Possibly, all species have a need to underline status relationships after a fight but, instead of doing this in an aggressive manner—and thus endangering the peace process—some species have turned the reconciliation process itself into a status ritual. Thus, during hold-bottom interactions among stumptail monkeys the dominant party does by far most of the clasping and the subordinate performs most of the presenting (de Waal and Ren 1988). In this respect, stumptails seem to resemble male chimpanzees in which successful reconciliation requires status communication before or during the approach (de Waal 1986a).

Conclusions

Whereas the initial effect of aggression in primate groups is often dispersal, this effect is overcome within minutes by the tendency of adversaries to seek contact with one another. The increase in contact between former adversaries is both absolute and relative—that is, they show selective attraction toward each other. The strength of the attraction, and its predominant direction between aggressor and victim, differs among species. It is especially difficult to generalize as even closely related species may differ considerably with respect to who will make up when.

Also, the behavior by means of which former adversaries resume contact is highly variable, ranging from mostly inconspicuous reconciliations among patas, rhesus, and longtail monkeys, to intensive sexual contacts among bonobos, kissing among chimpanzees, or hold-bottom rituals among stumptail monkeys. Behavioral differences between postconflict contacts and contacts in other contexts—together with the demonstrated selective attraction—provide strong support for the reconciliation hypothesis.

EXPERIMENTAL STUDIES

Provoked Aggression

The tendency of monkeys to compete over food provides an opportunity to experimentally induce aggression, and to determine the effects

on subsequent relationships. This method was first applied by de
Waal (1984) to small isosexual groups of juvenile rhesus monkeys.
Three all-male and three all-female groups received a single piece of
apple and were observed for half an hour afterward. The behavior
during these tests was compared with that during control tests without provision of extra food. In a previous study, it had been found that
male rhesus monkeys in a mixed group reconciled more than did
females, but interpretation of this result was ambiguous as males also
have a higher average dominance rank. The purpose of the experimental study was to tease apart the effects of rank and sex.

Following aggressive competition over the apple piece—and the
consumption of the piece by one of the monkeys—grooming behavior
and social cohesiveness increased significantly in the male groups. The
female groups, in contrast, showed a nonsignificant decrease in the
same behavior (figure 5.6). This confirmed the previously found sex
difference in reconciliation behavior, but this time it was uncon-
founded by effects of social dominance. However, it is doubtful if the
apple tests measured exactly the same phenomenon. First, grooming
among the males was *not* related to the amount and direction of
aggression during the initial food competition. Individuals who
opposed one another in a particular apple test were not necessarily the
ones who did most of the subsequent grooming. Second, in another
series of tests a handful of small apple pieces was spread over the cage
floor. While the monkeys hurried around to fill their cheek pouches,
they performed the same amount of aggression as in the tests with a
single apple piece. Yet, subsequent affiliative behavior decreased in
groups of both sexes. This means that it was the method of food distribution rather than the occurrence of aggression that determined subsequent affiliation patterns.

For this reason, de Waal (1984) avoids the term *reconciliation*,
speaking instead of *restorative behavior*. Such calming contact, it is
argued, occurs in response to social tensions in the group, and tensions are related to the absence of food-sharing. More specifically, tensions are caused by envy and frustration in nonpossessors of food. Figure 5.7 summarizes the model. According to this model the observed
sex difference in the response of juvenile rhesus monkeys to a monop-

olizable food source can mean two things. Either female relationships are not strained by unequal access to food, whereas male relationships are; or females are less active than males in trying to reduce the social tensions caused by unequal access.

Another experiment, involving juvenile male longtail macaques, was conducted by Cords (1988). Conflicts were provoked by giving a small tidbit to a lower-ranking male in the presence of a higher-ranking one. This was done in selected dyads at moments when they were partially isolated from the rest of the captive group to which they belonged. Three dyads consisted of matrilineally related males, and three dyads of unrelated males. Behavior was followed for fifteen minutes after the provoked aggression, and compared with control observations taken on a different occasion after the same two males had interacted in a friendly or neutral manner.

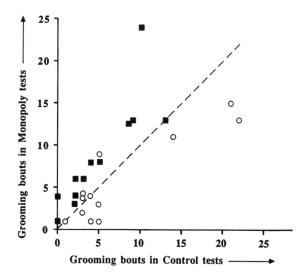

Figure 5.6 Comparison of grooming bouts performed per individual rhesus monkey during so-called Monopoly tests, in which the group received a single piece of apple, and Control tests. Data for each individual are from ten 30-min tests of both types. Solid squares represent males, open circles represent females. Eleven of the 12 males showed a grooming increase following the food competition induced in the Monopoly tests, whereas only 3 of the 12 females did. From de Waal (1984).

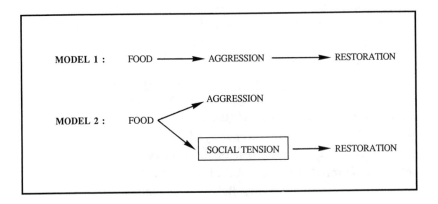

Figure 5.7 According to our initial model, restorative behavior, such as grooming, occurs in response to the disruptive effects of food-induced aggression (Model 1). The new model incorporates a hypothetical component, social tension, which is not linked to aggression but to the type of food provision (Model 2). According to Model 2 restorative behavior occurs in response to "frustrations" caused by unequal access to food regardless of the amount and direction of previous aggression. From de Waal (1984).

Unrelated males showed attraction similar to that found in the observational studies previously discussed—that is, former opponents interacted again sooner and more often after aggression than during control tests, and this did not reflect a general increase in social activity. In the dyads of related males, postconflict attraction was less pronounced or absent. To explain the difference between kin and nonkin, Cords (1988) introduces the concept of the *security* of a social relationship—that is, its predictability and resilience. According to this view, relationships among kin are not only of greater value but also more secure than are relationships among nonkin and may, for this reason and under certain circumstances, require less repair after aggressive disturbance.

Although the concept of security seems useful, the problem is that the large majority of observational studies on monkeys indicate more reconciliation among kin than among nonkin (Aureli et al. 1989; de Waal and Ren 1988; de Waal and Yoshihara 1983; Judge 1991; York and Rowell 1988. See also Cords 1989). An alternative explanation for

Cords's (1988) experimental results is given by the model in figure 5.7—namely, that affiliative behavior after food competition depends on the mediating mechanism of social tension. Unequal access to food may create less social tension among kin than among nonkin. Although both dominant kin and nonkin make aggressive attempts to obtain the food item, there is perhaps less frustration in the loser if the winner is a relative. After all, related monkeys also show greater tolerance in competitive situations, as reflected in the amount of codrinking and cofeeding among kin (de Waal 1986b; Yamada, 1963).

Celebrations and Food Sharing

A major difference between macaques and chimpanzees is that chimpanzees respond with reassurance behaviors before or during competition for food rather than afterward, as did the macaques in the already described studies. Thus, if an animal caretaker arrives with a bucket full of fruits and vegetables the apes rush towards each other—embracing, kissing, and patting each other on the back—before coming forward to obtain their shares of food. De Waal (1989c) recorded more than an one-hundred fold increase in reassurance behavior in chimpanzees upon the sight of attractive food. In macaques, by contrast, the same situation induces a tense, competitive atmosphere, and the monkeys immediately seek positions as close to the source of food as their dominance rank permits. In view of this difference, it is not surprising that noncompetitive mechanisms of food allocation are more developed in chimpanzees (de Waal 1989c; Goodall 1963; Nishida 1970; Teleki 1973).

Speculating that the celebrations of chimpanzees reduce the probability of aggressive interaction over food, de Waal (1989c) designed an experiment with two types of food delivery. One test allowed one to two minutes to elapse between the visibility of the food and its provisioning to the waiting colony. The second test allowed no such delay. The second type of provisioning gave the chimpanzees hardly any time to go through the characteristic appeasement rituals prior to food distribution. As predicted, aggressive competition over food was more common during food trials that were not preceded by a celebration.

Similarly, in order to investigate the role of reassurance mechanisms during interactions over food, de Waal (1987) recorded the behavior of captive bonobos around feeding time, comparing it with base levels throughout the rest of the day. Upon seeing the keeper arrive with foliage, the apes would show penile erections, sexual invitations, and mounts in virtually all possible positions and combinations. A similar sociosexual response to food has been observed in other captive bonobo colonies (Jordan 1977; Tratz and Heck 1954), and in the wild (Kano 1980; Kuroda 1980, 1984; Thompson-Handler et al. 1984). Observations at the provisioning site in the forest of Wamba, Zaire, have led to a specific hypothesis concerning the functions of food-related sex—namely that it "works to ease anxiety or tension and to calm excitement" and "thus to increase tolerance, which makes food-sharing smooth" (Kuroda 1980, 190).

This hypothesis was supported by the captive study of de Waal (1987). The fact that the apes responded to food with the same sociosexual behavior as shown during reconciliations following aggression unrelated to food, suggests a shared function of the behavior—that is, tension reduction. Moreover, some subordinates behaved considerably more assertively following sociosexual contact with dominant food-possessors than without such prior contact (figure 5.8). This change in self-confidence was particularly marked in one adult female vis-à-vis the dominant male. After, or sometimes even during a mating with him, this female would claim all his food. The possibility of trade between sexual favors and access to food, as first suggested by Yerkes (1941) for chimpanzees and echoed by Kuroda (1984) for bonobos, is of considerable theoretical importance in connection with current scenarios of the evolution of the human family and the task-division between the sexes (Fisher 1983; Lovejoy 1981).

Conclusions

Experimental studies confirm the paradoxical effect of aggression on affiliative behavior; that is, an increase in affiliative behavior after and, in some species, during food-induced competition. The morphological similarity of this behavior to reconciliation behavior suggests a shared function. In a few species, social tensions created by unequal access to food may be buffered so effectively as to make food-sharing possible.

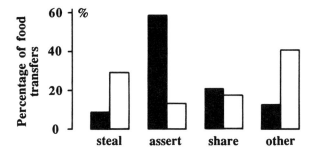

Figure 5.8 The relative frequencies of four methods of food transfer from dominant to subordinate bonobos in the San Diego Zoo colony. The four types are stealing (the subordinate grabs food and runs away), assertive claiming (the subordinate calmly appropriates the dominant's food), sharing (feeding side-by-side on the same pile), and other types of food transfer (e.g., the subordinate waits for discarded pieces of food). Interactions over food which are preceded by affiliative or sociosexual behavior (black) more often involve assertive claims by the subordinate than do interactions unpreceded by such behavior (white). From de Waal (1987).

THEORETICAL ISSUES

Constraints on Competition

If primates respond with reassurance and appeasement behavior to social tensions, and seek reunions with former adversaries, this can hardly mean anything other than that they value peaceful coexistence. As Kummer (1979) points out, long-term social relationships are an investment worth maintaining and defending. This insight has profound consequences for the way in which we view intragroup competition. Competition involves not only the risk of injury, but it also endangers established relationships. Primates are constantly faced with the dilemma—familiar among people as well—that one sometimes cannot win a fight without losing a friend. This means that, when two individuals compete over a particular resource, they must take into account not only the value of the resource itself and the risk of bodily harm, but also the value of the relationship with the competitor. Sometimes, the resource may not be worth the straining of a cooperative relationship, even if an individual could easily win the fight.

This possibility considerably complicates models of competition (de Waal 1989a).

Traditionally, research on competitive relationships and social dominance has focused on the benefits derived from winning, particularly in terms of reproductive success. This has been reviewed for primates by Fedigan (1983) and Shively (1985). Besides this focus on the outcome of competition, researchers should also compare the number of opportunities for competition with the number of actual instances. Social tolerance, defined as a low competitive tendency, can be expected to correlate with a high rate of reconciliation following conflicts. The reason is that both traits—tolerance and reconciliation—supposedly serve the same function of preserving valuable relationships. The one does so by limiting aggression, while the other limits the damage caused by aggression. If such a correlation can be demonstrated, this will provide a basis for more comprehensive models of conflict resolution.

A promising approach to this issue is that of comparisons between closely related species on dimensions such as group cohesion, the strictness of the dominance hierarchy, the symmetry of aggressive encounters, social tolerance, conciliatory tendency, and so on. A number of comparative studies on members of the genus *Macaca* have been conducted, or are presently under way (de Waal and Luttrell 1989; Thierry 1985, 1987). They explore conflict resolution and the nature of dominance relationships from the perspective of the advantages of group living and cooperation. These advantages vary with species, environment, and social partner, providing both proximate and ultimate reasons for variation in dominance style (de Waal 1989a; van Schaik 1989).

Social Sophistication and Cognition

Even in species with a high reconciliation rate, a large proportion of conflicts go unreconciled. We know very little about the conditions of peacemaking. Variations across relationships have been studied by most of the investigators of reconciliation behavior. The results are often contradictory. For example, Cords (1988, 1989) found more reconciliations among nonkin than among kin, whereas all other monkey studies report the opposite. Moreover, de Waal and Yoshihara (1983) found that, in rhesus monkeys, the effect of kinship on conciliatory tendency disappears if social bond strength (in terms of time spent in association) is taken into account. However, de Waal and Ren (1988) failed to confirm an effect of bond strength on the conciliatory tendency among stumptail monkeys. Instead, they found that the kinship effect persists after correcting for differences in bond strength. (See also Aureli et al. 1989.)

Similarly, if we ask how the nature of previous aggression affects the probability of reconciliation, contradictory answers may be obtained. For example, there is the already reported difference between chimpanzees and bonobos. In bonobos, high-intensity aggression is more often reconciled by the aggressor, but in chimpanzees this occurs less often. It seems, then, that the basis on which primates decide to reconcile or not is a wide-open area of investigation. To understand the underlying process, we may need much more subtle psychological concepts for animals than hitherto accepted. The literature suggests the following tentative lists of new concepts: rank acceptance, envy, the value and security of the relationship, the need to reduce uncertainty about the other's intentions, a distinction between reasonable and unreasonable aggression, and the withholding of reconciliation as a form of blackmail (Aureli et al. 1989; Cords 1988; de Waal 1982, 1984, 1986a, 1989b; de Waal and Ren 1988).

Several studies have systematically addressed triadic aspects of reconciliation in monkeys. De Waal and Yoshihara (1983) noted an increase in grooming activity following violent aggression. Most of this grooming was aimed by the aggressor at an uninvolved third individual. This was explained as a substitute reconciliation resulting from two incompatible tendencies—attraction and hostility toward the opponent. In agreement with this explanation, it was found that the grooming occurred especially after fights between individuals who normally had a close relationship. The phenomenon was dubbed *redirected affection*.

Judge (1991), de Waal and Ren (1988), and Cheney and Seyfarth (1989) investigated the possibility that postconflict contact with outsiders concerns individuals with a special relationship to the opponent. Thus, Cheney and Seyfarth (1989) distinguished *direct* reconciliation (postconflict reunion between two opponents), *simple* reconciliation (postconflict affiliation of one opponent with the kin of another), and *complex* reconciliation (postconflict affiliation between kin of both opponents). Judge (1991) as well as Cheney and Seyfarth (1989) found that such triadic patterns do occur more often than expected by chance, supporting the idea that both conflict and its resolution generalize beyond individual opponents to their entire families. (See also Cheney and Seyfarth 1986; de Waal 1989b.) However, when de Waal and Ren (1988) expressed contact with the opponent's kin as a percentage of all contact partners—thus correcting for overall contact tendencies—they failed to find the same difference between postconflict and control observations.

Cheney and Seyfarth (1989) and Kummer and colleagues (1990) are correct in pointing out that only experimental studies can provide the necessary controls and conditions to understand how individuals perceive the relationships among others, and how this perception influences triadic patterns of interaction. It should be added, however, that such experiments can only complement, not replace, the observation of spontaneous social interactions. Such observation remains necessary to provide insight into the function of particular cognitive capacities in daily social life, and to generate new hypotheses (de Waal, 1991). As an illustration, consider the following anecdotal indications of social sophistication in relation to peacemaking in the Arnhem chimpanzee colony:

Deception. On six occasions, a dominant female who had been unable to catch a fleeing opponent was observed to approach this individual some time afterward with a friendly appearance, holding out her hand, only to change her behavior when the other came within reach. Reasons to regard the subsequent attack as the female's real intention are its timing (very sudden, without warning signals), the fact that all the instances concerned victims capable of outrunning the aggressor, and the intensity of the punishment (de Waal 1986c).

Opportunistic reconciliation. Reconciliation may occur in a hurried fashion if continuation of the fight would harm the interests of both individuals. For example, in the years that the Arnhem colony was ruled by a coalition of Nikkie and Yeroen, the alpha male, Nikkie, could get in serious trouble during prolonged conflicts with his partner. The third male would begin an intimidation display, initially terrorizing the females and juveniles but later displaying closer and closer to the two quarreling males themselves. Nikkie was never observed to control the third male on his own. He would first approach his opponent, Yeroen, with a large grin, seeking an embrace. Only after reestablishment of contact with his partner would Nikkie go over to the third male to subdue him. (For a photograph of one such scene see, de Waal 1982, 140–141).

Third-party mediation. If, after a fight between them, two male rivals stayed in prolonged proximity without engaging in an actual physical reunion (an apparent dead-lock situation), an adult female might initiate a grooming contact with one of the two. After several minutes of grooming, she would slowly walk to the other male, often followed by her grooming partner. If he failed to follow she might return to tug at his arm. After the three individuals had been together for a while, with the female in the middle, she would then get up and stroll away leaving the males alone.

Functional and Evolutionary Perspectives

Thus far, research on reconciliation behavior has aimed at demonstrating that the phenomenon exists—that is, that a causal connection exists between aggression and subsequent reassurance behavior. To further support the concept of reconciliation, we now need studies on its *effects*. A first attempt is an unpublished study of the Arnhem chimpanzee colony by Gerard Willemsen and de Waal (Willemsen 1981). Comparing 127 agonistic incidents which were not followed by a reconciliation within half an hour with 39 reconciled incidents, we found that the probability of revival of the conflict was higher in the first category of relationships—that is, 26.0% versus 5.1% of the opponent pairs engaged in renewed antagonism within half an hour ($X^2 = 6.6$, p $= 0.01$). This is an inconclusive result, however, as there are at least two possible explanations: (1) a friendly reunion between former opponents may cause a reduction in antagonism, or (2) a friendly reunion may reflect a low level of residual antagonism. The second

explanation needs to be excluded before a conciliatory effect of reconciliation can be accepted.

One possible approach to the functional question is an experiment in which aggression-affiliation sequences among primates are
interrupted in order to prevent postconflict reunions. It has been suggested that such a procedure—if followed systematically with newly
introduced monkeys—will prevent them from forming a close social
bond even if they are given plenty of time for peaceful association. The
idea is that a social bond requires a formalized dominance relationship, and that aggression and subsequent reconciliation are a necessary part of the formation process (de Waal 1986a).

Recently, Cords (1990) reported an experiment in which affiliative contact following food-induced aggression between two longtail
macaques was either prevented or allowed. After this, the two monkeys were presented with resources that could be jointly exploited.
Pairs that had engaged in affiliative contact following their aggression
showed greater compatibility, as measured by proximity during the
resource test, than did pairs that had been distracted by the experi-
menter in order to prevent reconciliation. This experimental result
strongly supports a conciliatory function of postconflict reunion—that
is, the restoration of a tolerant attitude in the dominant and a corresponding reduction of fear in the subordinate.

Another angle from which the function of reconciliation can be
studied is through its effects on physiological and behavioral indicators
of stress. Aureli and colleagues (1989) conducted a detailed analysis in
longtail macaques of how the rate of self-scratching—a correlate of
sympathetic arousal—varies with the presence or absence of postconflict reunion between former antagonists. As predicted by the reconciliation hypothesis, they found a fast reduction in scratching rate and a
decrease in the reoccurrence of aggression following reconciliation.
This interesting approach—which must be further developed, prefer-
ably complemented with direct physiological measures—may provide
important information about the cost, in terms of stress-level, associated with unreconciled aggression. Hence, the adaptive significance of
reconciliation may be illuminated from a physiological perspective.

Since reconciliation behavior has hardly been studied in the natural habitat, we do not know how the present findings extrapolate to the context in which this behavior evolved. In the field, Seyfarth (1976) observed postconflict contacts among female baboons (*Papio cyno-cephalus*); Goodall (1986) studied reassurance behavior after aggression among chimpanzees; and Kuroda (1984) investigated food-related sexual interactions among bonobos. Cheney and Seyfarth (1989) applied a controlled observation paradigm to postconflict behavior in wild vervet monkeys. Although the observed rate of reconciliation was rather low, their work confirmed many of the findings of captive studies. In view of these studies, there is no reason to assume that reconciliation behavior is limited to animals in confinement. Yet, it would surprising if the dynamics were identical under captive and free-ranging conditions.

However, certain global patterns observed in captivity—such as dramatic interspecies or sex differences—probably reflect fundamental traits that are expressed under a wide variety of conditions. This has been confirmed for at least one pattern. Goodall (1986) and de Waal (1986a) independently report a higher conciliatory tendency in male chimpanzees as compared to females. The first study concerned the frequency of reassurance behavior following aggression within an unspecified time interval among wild chimpanzees. The second study concerned contact between former opponents within half an hour after aggression in the Arnhem colony. Both investigators attribute the low level of peacemaking among females to the virtual absence of a clear-cut hierarchy. This is only the proximate view, however. From an evolutionary perspective we may speculate that unifying social mechanisms are more important for the reproductive success of male chimpanzees than for females. Males need to stick together for territorial defense. In addition, their system of ever-changing coalitions requires them to stay on good terms even with rivals (de Waal 1982; Goodall 1986; Nishida 1983; Wrangham 1979).

Environmental pressures, such as predation and food distribution, shape the reconciliation behavior of each species to create an equilibrium between the centrifugal force of competition and the social cohesiveness required in that particular environment. On the basis of the behavior of captive stumptail monkeys, for example, we can expect this species to live in close-knit groups in its natural habitat. We do not actually know how these monkeys live, but it would not make sense for a species living in loosely organized groups in which members are widely dispersed to possess the observed powerful mechanisms of physical reassurance and conflict resolution. Similarly, we can expect female chimpanzees to have trouble coping with competition and, as a result, to disperse, which we know to be the case in the wild (Goodall 1986; Nishida 1979).

Conclusions

At the proximate level, we need data on the conditions under which primates reconcile or do not. Probably, the decision-making process cannot be properly understood without dramatically reconceptualizing animal psychology, including the way in which animals evaluate their own and others' social relationships. A broad approach to these issues would include all facets of competition, especially the conditions under which competition is suppressed or mitigated. At the functional level, we need to study the effects of reconciliation behavior on long-term relationships, and to relate mechanisms of tension regulation and social cohesiveness to the requirements of the natural environment.

REFERENCES

Aureli, F.; van Schaik, C. P.; van Hooff, J. A. R. A. M. (1989) Functional aspects of reconciliation among captive longtailed macaques (*Macaca fascicularis*). *American Journal of Primatology* 19:39–52.

Blurton Jones, N. G.; Trollope, J. (1968) Social behaviour of stumptailed macaques in captivity. *Primates* 9:365–394.

Cheney, D. L.; Seyfarth, R. M. (1986) The recognition of social alliances by vervet monkeys. *Animal Behaviour* 34:1722–1731.

Cheney, D. L.; Seyfarth, R. M. (1989) Redirected aggression and reconciliation among vervet monkeys (*Cercopithecus aethiops*). *Behaviour* 110:258–275.

Cords, M. (1988) Resolution of aggressive conflicts by immature long-tailed macaques (*Macaca fascicularis*). *Animal Behaviour* 36:1124–1135.

Cords, M. (1989) What happens after social animals fight? Variation in long-tailed macaques. Poster presentation at the 21st International Ethological Conference. Utrecht.

Cords, M. (1990) How immature long-tailed macaques cope with aggressive conflcit. Presented at the 13th Congress of the International Primatrological Society. Kyoto, Japan.

de Waal, F. B. M. (1982) "Chimpanzee Politics." London: Jonathan Cape.

de Waal, F. B. M. (1984) Coping with social tension: Sex differences in the effect of food provision to small rhesus monkey groups. *Animal Behaviour* 32:765–773.

de Waal, F. B. M. (1986a) Integration of dominance and social bonding in primates. *Quarterly Review of Biology* 61:459–479.

de Waal, F. B. M. (1986b) Class structure in a rhesus monkey group: The interplay between dominance and tolerance. *Animal Behaviour* 34:1033–1040.

de Waal, F. B. M. (1986c) Deception in the natural communication of chimpanzees. In Mitchell, R. W., Thompson, N. S. (eds.), "Deception: Perspectives on Human and Nonhuman Deceit." Albany, N.Y.: State University of New York Press. 221–224.

de Waal, F. B. M. (1987) Tension regulation and nonreproductive functions of sex among captive bonobos *(Pan paniscus). National Geographic Research* 3:318–335.

de Waal, F. B. M. (1989a) Dominance "style" and primate social organization. In Standen, V., Foley, R. (eds.), "Comparative Socioecology: The Behavioural Ecology of Humans and Other Mammals." Oxford: Blackwell. 243–264.

de Waal, F. B. M. (1989b) "Peacemaking among Primates." Cambridge, Mass.: Harvard University Press.

de Waal, F. B. M. (1989c) Food sharing and reciprocal obligations among chimpanzees. *Journal of Human Evolution* 18:433–459.

de Waal, F. B. M. (1991) Complementary methods and convergent evidence in the study of primate social cognition. *Behaviour* 118:297–320.

de Waal, F. B. M. (1992) Appeasement, celebration, and food sharing in the two *Pan* species. In Nishida, T., McGrew, W. C., Marler, P., Pickford, M., de Waal, F. B. M. (eds.), "Topics in Primatology:" vol. 1, "Human Origins." Tokyo: University of Tokyo Press. 37–50

de Waal, F. B. M.; Luttrell, L. M. (1989) Toward a comparative socioecology of the genus *Macaca:* Different dominance styles in rhesus and stumptail monkeys. *American Journal of Primatology* 19:83–109.

de Waal, F. B. M.; Ren, R. M. (1988) Comparison of the reconciliation behavior of stumptail and rhesus macaques. *Ethology* 78:129–142.

de Waal, F. B. M.; van Roosmalen, A. (1979) Reconciliation and consolation among chimpanzees. *Behavioral Ecological and Sociobiology* 5:55–66.

de Waal, F. B. M.; Yoshihara, D. (1983) Reconciliation and redirected affection in rhesus monkeys. *Behaviour* 85:224–241.

Ehrlich, A.; Musicant, A. (1977) Social and individual behaviors in captive slow lorises. *Behaviour* 60:195–220.

Ellefson, J. O. (1968) Territorial behavior in the common white-handed gibbon, *Hylobates Lar Linn.* In Jay, P. C. (ed.), "Primates: Studies in Adaptation and Variability." New York: Holt. 180–199.

Fedigan, L. (1983) Dominance and reproductive success in primates. *Yearbook of Physical Anthropology* 26:91–129.

Fisher, H. (1983) "The Sex Contract." New York: Quill.

Goodall, J. (1963) My life among wild chimpanzees. *National Geographic Magazine* 124:272–308.

Goodall, J., van Lawick- (1968) The behaviour of free-living chimpanzees in the Gombe Stream Reserve. *Animal Behaviour Monographs* 1(3):161–311.

Goodall, J. (1986) "The Chimpanzees of Gombe: Patterns of Behavior." Cambridge, Mass.: Belknap.

Griede, T. (1981). *Invloed op verzoening bij chimpansees.* Unpublished research report. University of Utrecht.

Hediger, H. (1941) *Biologische Gesetzmaszigkeiten im Verhalten von Wirbeltieren. Mitt Naturforsch Gesellschaft.* Bern, 1940. 37–55.

Jordan, C. (1977) *Das verhalten zoolebender zwergschimpansen.* Unpublished dissertation. Frankfurt: Goethe University.

Judge, P. G. (1991) Dyadic and triadic reconciliation in pigtail macaques *(Macaca nemestrina). American Journal of Primatology* 23:225–237.

Kano, T. (1980) Social behavior of wild pygmy chimpanzees *(Pan paniscus)* of Wamba: A preliminary report. *Journal of Human Evolution* 9:243–260.

Kaplan, J. R.; Zucker, E. (1980) Social organization in a group of free-ranging patas monkeys. *Folia Primatologica* 34:196–213.

Kellogg, W. N.; Kellogg, L. A. (1933) "The Ape and the Child." New York: McGraw-Hill.

Köhler, W. (1925) "The Mentality of Apes." London: Routledge and Kegan Paul.

Kummer, H. (1979) On the value of social relationships to nonhuman primates: A heuristic scheme. In Von Cranach, M. et al. (eds.), "Human Ethology." Cambridge: Cambridge University Press. 381–395.

Kummer, H.; Dasser, V.; Hoyningen-Huene, P. (1990) Exploring primate social cognition: Some critical remarks. *Behaviour* 112:84–98.

Kuroda, S. (1980) Social behavior of the pygmy chimpanzees. *Primates* 21:181–197.

Kuroda, S. (1984) Interaction over food among pygmy chimpanzees. In Susman, R. (ed.), "The Pygmy Chimpanzee." New York: Plenum Press. 301–324.

Lindburg, D. G. (1973) Grooming behavior as a regulator of social interactions in rhesus monkeys. In Carpenter, C. R. (ed.), "Behavioral Regulators of Behavior in Primates." Lewisburg, Pa.: Bucknell University, 85–105.

Lorenz, K. (1963) *"Das sogenannte Bose."* Vienna: Borotha-Schoeler.

Lovejoy, C. (1981) The origin of man. *Science* 211:341–350.

Mason, W. A. (1964) Sociability and social organization in monkeys and apes. In Berkowitz, L. (ed.), "Advances in Experimental Social Psychology." New York: Academic Press. 277–305.

McKenna, J. J. (1978) Biosocial function of grooming behavior among the common langur monkey *(Presbytis entellus). American Journal of Physical Anthropology* 48:503–510.

Nishida, T. (1970) Social behavior and relationships among wild chimpanzees of the Mahale Mountains. *Primates* 11:47–87.

Nishida, T. (1979) The social structure of chimpanzees of the Mahale Mountains. In Hamburg, D. A., McCown, E. R. (eds.), "The Great Apes." Menlo Park: Benjamin Cummings. 73–121.

Nishida, T. (1983) Alpha status and agonistic alliance in wild chimpanzees. *Primates* 24:318–336.

Poirier, F. E. (1968) Dominance structure of the Nigiri langur *(Presbytis johnii)* of South India. *Folia Primatologica* 12:161–186.

Rowell, T. E.; Olson, D. (1983) Alternative mechanisms of social organization in monkeys. *Behaviour* 86:31–54.

Scott, J. P. (1958) "Animal Behaviour." Chicago: University Chicago Press.

Seyfarth, R. (1976) Social relationships among adult female baboons. *Animal Behaviour* 24:917–938.

Shively, C. (1985) The evolution of dominance hierarchies in nonhuman primates society. In Ellyson, S., Dovidio, I. (eds.), "Power, Dominance, and Nonverbal Behavior." Berlin: Springer. 67–87.

Smith, W. J. (1986) An "informational" perspective on manipulation. In Mitchell, R. W., Thompson, N. S. (eds.), "Perspectives on Human and Nonhuman Deceit." Albany, N.Y.: State University of New York Press. 71–86.

Teleki, G. (1973) "The Predatory Behavior of Wild Chimpanzees." Lewisburg, Pa.: Bucknell University Press.

Thierry, B. (1984) Clasping behavior in *Macaca tonkeana. Behaviour* 89:1–28.

Thierry, B. (1985) A comparative study of aggression and response to aggression in three species of macaque. In Else, I., Lee, P. (eds.), "Primate Ontogeny, Cognition and Social Behaviour." Cambridge: Cambridge University Press. 307–313.

Thierry, B. (1987) *Coadaptation des variables sociales: l'exemple des systemes sociaux des macaques.* Colloques de l'INRA 38:92–100.

Thompson-Handler, N.; Malenky, R. K.; Badrian, N. (1984) Sexual behavior of

Pan paniscus under natural conditions in the Lomako Forest, Equateur, Zaire. In Susman, R. L.(ed.), "The Pygmy Chimpanzee." New York: Plenum Press. 347–368.

Tratz, E.; Heck, H. (1954) *Der afrikanische Anthropoide "Bonobo," eine neue Menschenaffengattung.* Saugetierk, Mitt 2:97–101.

van Schaik, C. P. (1989) The ecology of social relationships amongst female primates. In Standen, V., Foley, R. (eds.), "Comparative Socioecology: The Behavioural Ecology of Humans and Other Mammals." Oxford: Blackwell. 195–218.

Willemsen, G. (1981) *Verzoeningsgedrag van de chimpansee.* Unpublished research report. University of Utrecht.

Wrangham, R. W. (1979) Sex differences in chimpanzee dispersion. In Hamburg, D. A., McCown, E. R. (eds.), "The Great Apes." Menlo Park, Calif.: Benjamin/Cummings. 481–490.

Yamada, M. (1963) A study of blood relationships in the natural society of the Japanese macaque. *Primates* 4:43–65.

Yerkes, R. M. (1941) Conjugal contrasts among chimpanzees. *Journal of Abnormal Psychology* 36:75–199.

York A. D.; Rowell ,T. E. (1988) Reconciliation following aggression in patas monkeys (*Erythrocebus patas*). *Animal Behaviour* 36:502–509.

CHAPTER SIX

Social Conflict in Adult Male Relationships in a Free-Ranging Group of Japanese Monkeys

NAOSUKE ITOIGAWA

Japanese primatologists were among the first to appreciate that understanding social order and social dynamics required life-span demographic data based on identifiable individuals of known genealogy, living in free-ranging groups. This chapter deals with one group of Japanese monkeys, the Katsuyama group, observed over a period spanning nearly thirty years.

The focus is on conflicts among adult males. Severe antagonism was infrequent, owing largely to the presence and solidarity of a subgroup of high-ranking males. These animals were friendly, long-term associates of similar age, and they occupied the central part of the group. They remained close to each other in spite of large differences in their individual dominance ranks, and acted together against peripheral males (usually younger) who approached the central part of the troop. Radical changes in the composition and age-structure of this central core of high-ranking males had disruptive consequences which spread throughout the group, leading to increased con-

Naosuke Itoigawa—Department of Psychology, Osaka University, Japan
I thank colleagues of Osaka University for sharing their data on monkeys, the people of Katsuyama for their assistance, and Dr. Charles R. Menzel for his valuable comments and English language corrections on the manuscript. I also thank Dr. William A. Mason of the California Regional Primate Research Center for his helpful instruction. This study was supported in part by Japanese government grants including a grant-in-aid for special project research on Biological Aspects of Optimal Strategy and Social Structure.

145

flict and probably contributing to the eventual fissioning of the group. Thus, the integrity of the central core of high-ranking mature males plays an important role in the inhibition and resolution of conflict, and maintaining the stability of the group.

Free-ranging groups of Japanese monkeys (*Macaca fuscata*) are characterized by a multimale social organization in which more than one adult male can contribute to the maintenance of the group. The implicit benefits of the existence of multiple adult males for the group's maintenance—with regard to defense against enemies, foraging, reproduction, and the rearing of offspring—can only become explicit if severe aggression and social conflicts among adult males in the group are effectively inhibited or resolved.

Starting in 1958, long-term observations of a free-ranging group of Japanese monkeys at Katsuyama in Okayama Prefecture indicate that severe fighting causing serious injury and death has occurred only infrequently among adult males. The following two questions may be important in clarifying the inhibitory or resolutive mechanisms that prevent severe antagonism among adult male Japanese monkeys living in multimale groups: How do adult males maintain interindividual distances in the group so as to inhibit or resolve moment-to-moment occasions of social conflict and severe antagonism? How does the background of the adult males, in terms of their matriline origin, affect the stability and structure of the group?

This chapter will deal with the following four topics.

First, the chronology of the Katsuyama group will be described with regard to the group's fission, changes in dominance ranks among adult males, and the group's size and sex ratio. This will serve the purpose of clarifying long-term changes in social relationships among adult males in the group.

Second, changes in the matrilineal backgrounds and ages of the dominant adult males will be described for the purpose of clarifying how these factors contribute to the establishment of stable social patterns and to the inhibition or resolution of severe antagonism.

Third, the results of an analysis of spatial proximity among adult males in the group will be examined for the purpose of clarifying multifarious, long-term changes in their social relationships by a uniform method of analysis.

Fourth, instances of severe antagonism among adult males, which occurred only infrequently, will be reported to clarify the causes and consequences of severe antagonism among them.

SUBJECTS

Members of the Katsuyama group were the subjects of the present study. The group lives in a mountainous area in the northern part of Katsuyama. Its home range size averages about 3 kilometers, with seasonal and long-term chronological variation. The vegetation within the group's range includes deciduous and evergreen forests, planted coniferous forests, grass-bush fields, and cultivated land.

Artificial feeding of the Katsuyama group started in January 1958 and has continued until the present time. Food is provided in a section of the Kamba Waterfall valley, which is located approximately in the center of the group's range. Artificial feeding normally occurs twice a day, provided that the group has come to the feeding area.

The group's size was 104 individuals in March 1958. Group members were individually identified during the initial course of the artificial feeding. As of March 1958, mothers could not be specified for twenty-two male members (sixteen adults, estimated age, five years or more; six subadults and juveniles, estimated age, two to four years) and thirty-eight female members (thirty-five adults and subadults, estimated age, four years or more, and three juveniles, estimated age, three years). Each of the thirty-eight females whose mothers could not be specified was designated as the matriarch of an independent lineage starting in March 1958. The mothers of one subadult female and all remaining juveniles and infants of both sexes were specified, and these immature animals were classified within one of the thirty-eight matrilineages. A large majority of offspring born to each matriach in and after 1958 have been individually identified by researchers of Osaka University.

The twenty-two adult and six subadult males whose membership in a matrilineage was not specified—along with males born after 1958 whose matrilineages were not identified—and nonnatal males

were each given a name that was different from the names of the original thirty-eight matrilineages.

Table 6.1 shows the chronological changes in the number, identity,
and dominance ranks of the matrilineages in the years in which significant changes occurred (see also table 6.2). The determination of dominance ranks among matrilineages and dynamic aspects of the changes in
dominance ranks in Japanese monkey groups have been reported elsewhere (Itoigawa 1973; Kawai 1958; Kawamura 1958). One of the original
thirty-eight matrilineages was extinguished in April 1958, because of the
sudden disappearance of the matriarch and her infant. Perhaps, they
died. Furthermore, a total of seventeen matrilineages became extinguished between 1958 and 1973, because of death, disappearance, and
capture of matriline members, and the group's fission in April 1973.

CHRONOLOGY OF THE GROUP

Table 6.2 shows the chronology of the Katsuyama group. The alpha
male was M43*Romeo (identification:Rm) when the group was
observed in February 1958. His tenure ended with his sudden disappearance in June 1964. Perhaps he was shot by a farmer when the
group was foraging across cultivated land. The beta male—M42*Gabo
(identification: Gb)—became the alpha male following the disappearance of Rm. Gb's tenure ended in February 1970 when he became old
and subordinate to the beta male, M62Rika'58' (identification: Ri). Ri
became the new alpha male.

The group split into two groups in April 1973, chiefly because of
social conflicts and aggression between adult females of high-ranking
and middle-ranking matrilineages. In August 1976, Ri's tenure ended
with his disappearance. Perhaps he died due to unaccountable premature senility at the age of fourteen years. The beta male, M65Rika'60'
(identification: Rn), a cousin of Ri, became the new alpha male following Ri's disappearance. Rn's tenure has continued since August 1976.

The group's population census has been conducted each year
since 1958 on 1 September. As shown in table 6.2, the group's size has
increased approximately twofold during the past twenty-eight years.
The group's size decreased in 1973 because of the group's fission. The

Table 6.1 Chronological changes in the matriline composition of the Katsuyama group of Japanese monkeys

| 1958 | | 1964 | | 1970 | | 1973 | | 1976 | | 1986 | |
Rank	Matri	Rank	Matri	Rank	Matri	Rank	Matri	Rank	Matri	Rank	Matri
1	Dera	1	Rika	1	Rika	1	Bera	1	Bera	1	Bera
2	Rika	2	Mara	2	Mara	2	Mara	2	Mara	2	Mara
3	Mara	3	Dera	3	Dera	3	Masa	3	Masa	3	Masa
4	Masa	4	Masa	4	Masa	4	Tana	4	Tana	4	Elza
5	Bera	5	Bera	5	Bera	5	Fera	5	Fera	5	Kera
6	Mora	6	Mora	6	Dana	6	Elza	6	Elza	6	Tana
7	Bara	7	Bara	7	Tana	7	Bara	7	Bara	7	Tera
8	Malta	8	Malta	8	Fera	8	Tera	8	Tera	8	Bara
9	Dana	9	Dama	9	Elza	9	Kera	9	Kera	9	Fera
10	Tera	10	Tera	10	Bara	10	Lira	10	Lira	10	Lira
11	Tana	11	Tana	11	Tera	11	Fena	11	Fena	11	Fena
12	Rura	12	Rura	12	Kera	12	Mora	12	Mora	12	Mora
13	Mama	13	Mama	13	Lira	13	Pipa	13	Pipa	13	Pipa
14	Nana	14	Hilda	14	Fena	14	Rola	14	Rola	14	Jura
15	Hilda	15	Mona	15	Mora	15	Lipka	15	Lipka	15	Lisa
16	Mona	16	Rena	16	Pipa	16	Mona	16	Mona	16	Lipka
17	Rena	17	Pipa	17	Rola	17	Jura	17	Jura	17	Rola
18	Malga	18	Lipka	18	Lipka	18	Viva	18	Viva	18	Mona
19	Pola	19	Lira	19	Mona	19	Lisa	19	Lisa	19	Viva
20	Pipa	20	Rita	20	Jura	20	Cera	20	Cera	20	Cera
21	Sara	21	Fera	21	Viva						
22	Santa	22	Rola	22	Lisa						
23	Lipka	23	Elza	23	Cera						
24	Lira	24	Jura								
25	Rita	25	Viva								
26	Fera	26	Gilda								
27	Rola	27	Fena								
28	Elza	28	Lisa								
29	Moka	29	Cera								
30	Jura	30	Gina								
31	Viva	31	Kera								
32	Gilda										
33	Fena										
34	Lisa										
35	Cera										
36	Gina										
37	Kera										

Matri = Matrilineage

Table 6.2 Chronology of the Katsuyama group of Japanese monkeys

	1958 February Start of artificial feeding	1964	1970	1973 April Fission of group	1976	1986
Alpha male Tenure	---Rm------ Jun, '64	----Gb------	-------Ri------- Feb, '70		------Rn----- Aug, '76	
Number of animals	112	195	217	185	198	235
Adult male(A) 5 > yrs old	15	21	16	20	19	26
Adult female(B) 5 > yrs old	35	59	59	54	57	77
Sex ratio (B) / (A)	2.33	2.81	3.69	2.70	3.00	2. 96

sex ratio was determined by calculating the number of adult females per adult male, with adult age defined as five or more years for both sexes. It should be noted that the sex ratio was the lowest in the early stage of artificial feeding in February 1958. The relatively larger population of adult males—characteristic of wild, unprovisioned groups—may reflect the effects of having to forage under more difficult ecological conditions in competition with other groups.

COMPOSITION OF ADULT MALES BY AGE AND MATRILINE BACKGROUND

The age and matrilineal background of adult males is an important determinant of social relationships among individuals in a free-ranging group of Japanese monkeys, and constitutes a basic structure for the inhibition and resolution of social conflicts and aggression. This section will describe changes in the age-matriline composition of adult males of the Katsuyama group since 1958, for the purpose of clarifying the inhibitory and resolutive mechanisms of severe antagonism among the adult males. Table 6.3 shows the age-matriline composition of the eight most dominant adult males in the group during each period of the chronology of the Katsuyama group.

February 1958. There were four higher ranking, fully mature adult males, estimated to be twelve or more years old, and four lower ranking, younger adult males in February 1958. The difference in age between the eight adult males was estimated to be small. Each male differed by about one or two years, or a maximum of three years, from the males who were adjacent to it in rank. This social composition, in which higher ranking adult males were older and more fully mature than were lower ranking ones, and in which age differences were small between dominant adult males, might be the typical age composition of adult males in a wild, unprovisioned group of Japanese monkeys. This composition seemed to contribute to stable social relationships among a relatively larger number of adult males. The relatively larger population of adult males during this period before the start of artificial feeding may reflect a means of reducing the risks to survival that were greater in the natural habitat than under later conditions in which food was provided.

Table 6.3 Chronological changes in the age-matriline composition of the eight highest-ranking adult males of the Katsuyama group of Japanese monkeys

1958 February				1964 July				1970 July				1986 April			
Rank	Id	A	MatR	Rank	Id	A	MatR	Rank	Id	A	MatR	Rank	Id	A	MatR
1	Rm	14*	H	1	Gb	22*	M	1	Ri	8	1	1	Rn	21	(1)
2	Tt	13*	H	2	Ns	9*	H	2	Rn	5	1	2	Fr	16	9
3	Gb	15*	M	3	Hi	13*	M	3	Yn	13	M	3	Ms	19	3
4	Zn	12*	M	4	Br	10*	5	4	Ke	12	M	4	Fn	16	11
5	Ek	10*	M	5	Yn	7	M	5	Ln	13	L	5	Ki	15	5
6	Ro	11*	?	6	Ma	7	M	6	Tr	9	11	6	Mn	17	18
7	Wa	9*	L	7	Wo	9	L	7	Dn	7	6	7	Bk	12	8
8	Dd	7*	M	8	Ke	6	M	8	Mr	5	2	8	Ka	11	5

Id = Identification. A = Age in years. MatR = Matrilineal dominance rank. * = Estimation. ? = Inestimable. H = High-ranking. M = Middle-ranking. L = Low-ranking. ( ) = Matrilineal dominance rank preceding the group's fission.

With regard to the matriline composition, two of the eight most dominant males were estimated from physical appearance and social interactions to be from high-ranking matrilineages, four from middle-ranking matrilineages, one from a low-ranking matrilineage. The remaining one was inestimable. Thus, it may be reasonable to say that the eight most dominant males in the group were from matrilineages which varied widely in their dominance ranks. However, this is necessarily a tentative statement, owing to the crude basis for estimating the males' blood relations.

July 1964. In July 1964, Gb, at the estimated age of 22 years, became the new alpha male following the disappearance of Rm. The membership of the eight most dominant males of the group had greatly changed since February 1958. Gb was much older than other adult males. However, the age composition of the eight adult males was similar to that in February 1958 in one respect—the four higher ranking males were more fully mature and older, in general, than were the four lower ranking males.

The fourth-ranked male (Br) was identified as a member of the fifth-ranking matrilineage. The second-ranked male was estimated to be from a high-ranking matrilineage. Five males were estimated to be from middle-ranking matrilineages, and one male was estimated to be from a low-ranking matrilineage. Again, it may be reasonable to say that the matriline composition of adult males varied widely. Nevertheless, this statement must be qualified, because the estimate of blood relationships was crude.

July 1970. Gb disappeared in July 1970. The new alpha male (Ri) and beta male (Rn) were both young. Ri was eight years old, and Rn was only five years old. They were not fully mature in terms of physical growth and behavioral characteristics. By contrast, the third-, fourth-, and fifth-ranking males were fully mature and much older than the alpha and beta males. Ri and Rn were both from the first-ranking matrilineage, the Rika, and acquired their dominance ranks chiefly due to support from their mother and their grandmother, the matriarch of the Rika matriline.

The existence in the group of these two young dominant males—

who formed a contrast with the fully mature subordinate males—seemed to create unstable social relationships and conflicts among the adult males. Furthermore, this situation created even more distinctive social conflicts among the adult females than among the adult males. As a part of this instability, some adult females of middle-ranking matrilineages—such as the Bera and the Kera matrilines—began to mate with the young alpha male (Ri). This resulted in antagonism directed toward the adult female kin members of Ri—especially toward the mother and grandmother of Ri—and the emergence of social conflict between the adult females of high-ranking matrilineages and those of middle-ranking matrilineages. Adult females of high-ranking matrilineages were effectively and repeatedly chased off by adult females of middle-ranking matrilineages and formed a separate group. This was a major cause of the group's fission in April 1973. The population of the group had also reached a high level of more than 230 individuals before fission occurred.

The eight highest ranking males in 1970 were from matrilineages which varied widely in dominance rank. The two highest ranking males and the bottom-ranking male were from high-ranking matrilineages, the fifth-ranking male was from a low-ranking matrilineage, and the remaining four males were from middle-ranking matrilineages. Such wide variation in the age and matrilineal background of high-ranking adult males may have the effect of diversifying social relationships among the adult males and among the other individuals in the group, and perhaps this contributed to social stability. The opposite effect could occur, however, if social conflicts arising from variable social relationships are not resolved. This apparently is what happened in April 1973 when fission of the Katsuyama group occurred, seemingly because variability in the age-matrilineal composition of the high-ranking adult males was too large to allow the integrity of the group to be maintained.

April 1986. All eight of the highest ranking adult males were eleven or more years old and fully mature and, among them, the alpha male was the oldest. The six highest ranking males did not differ greatly in age. By contrast, the matrilineal composition of the eight males varied widely. The alpha male was from the former top-ranking matrilineage. The third-, fifth- and bottom-ranking males were from high-ranking matrilineages, the second-, fourth- and seventh-ranking males were from middle-ranking matrilineages, and the sixth-ranking male was from a low-ranking matrilineage. It should be noted that the sixth-ranking male—M69Mona'59' (identification: Mn)—was from the eighteenth-ranking matrilineage, nearly at the bottom of the total of twenty matrilines in the group.

The precise matrilineal composition of the eight highest ranking adult males could be completely determined for the first time in 1986. It was found that the males' individual dominance ranks did not correlate in a simple manner with the dominance ranks of their respective matrilines. As described previously, wide variation in age-matriline composition among adult males may normally create variable social relationships among the males and among the other individuals in the group. In particular, a wider range of likely mating possibilities exists when males from varied matrilineal backgrounds remain in the group than is the case when males from only one or two restricted matrilineages remain in the group.

Taken together, the changes described in the age-matriline composition of adult males in the Katsuyama group over the past 28 years suggest that the social organization of a *Macaca fuscata* group may depend largely upon the constituent structure of adult males in the group. The age-matriline composition of adult males may be an important determinant of the group's unity in connection with the inhibition and resolution of severe antagonism among the adult males. In particular—and in contrast to the situation in July 1970—age-matriline composition of the adult males in April 1986 seemed to contribute to stable social relationships among the adult males.

PROXIMITY AMONG ADULT MALES

Interindividual distance is a particularly useful measure for assessing multiphasic social relationships among the members of a primate group. Intragroup gregariousness and spacing mechanisms have been investigated in various groups of Japanese monkeys, including the

Katsuyama group (Fujii 1975; Koyama et al. 1981; Mori 1977). It was found that occurrences of proximity among adult and juvenile males reflected their social relationships in the group and seemed to be a determinant of which males left the group (Itoigawa 1975). The purpose of this section is to analyze occurrences of proximity among adult males in the group and to clarify the long-term changes in their social relationships and the spacing mechanisms which contribute to the inhibition and resolution of social conflicts.

Occurrences of proximity (within about ten meters) between adult males were recorded using a one-or-none method within an observation trial. A trial extended from about half an hour to one hour. During this time an observer travelled across the area where the group's members were dispersed and noted males in proximity. The total number of observational trials varied greatly across days and across months, owing to irregular observations in the field. Normally, more than six trials were performed each month. If any intense agonistic interaction such as chasing and biting occurred among adult males, the occurrences of proximity related to these intense agonistic interactions were excluded from the present analysis. Therefore, occurrences of proximity reported here presumably indicate the extent of affinitive social relationships among the adult males.

February–March 1958.   Figure 6.1 shows the results of a multidimensional scaling analysis of the occurrences of proximity among thirteen adult males in the group during a period from February to March 1958 based on Hayashi's type IV method (Hayashi 1954). Polyadic occurrences of proximity were broken down into their constituent dyads. The matrix of values provides an indication of the affinitive interactions between individuals. The method plots the minimal distance from the others on two dimensions with the intersection of the axes representing the center of the group. A total of fifty-six dyadic occurrences of proximity were analysed.

As shown in figure 6.1, the nine higher ranking males were located close to one another, forming a central grouping. The four lower ranking males were located separately from one another and each was located away from the central grouping. Among the nine

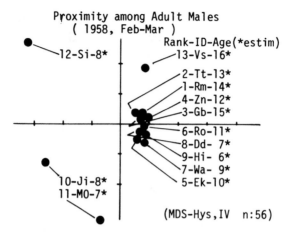

Figure 6.1 Proximity among 13 adult males of the Katsuyama group from February to March 1958, using Hayashi's type IV multidimensional scaling analysis method. Rank = individual dominance rank. ID = individual identification. Age = age in years. * = estimated age.

males of the central grouping, the four most dominant males of older ages seemed to form a subgrouping which was slightly separated from another subgrouping of five younger and more subordinate males. The oldest lowest ranking male, M41*Vaso (identification:Vs), was closer to the central grouping than were the other younger peripheral males. This may indicate that fully mature males tended to remain close to one another, irrespective of large differences in their dominance ranks. The results of figure 6.1 might reflect the typical social relationships among adult males in a wild unprovisioned group of Japanese monkeys, since the observations were made during the early stage of artificial feeding.

July–August 1964. Figure 6.2 shows the results of the analysis of proximity among eight adult males in the group from July to August 1964. At the start of these observations, Gb had just become the new alpha male. A small number of occurrences of proximity were recorded during this period, chiefly because the group came infrequently to the feeding area. This seemed to be due to the disappearance of the former alpha male (Rm), who was probably killed by a

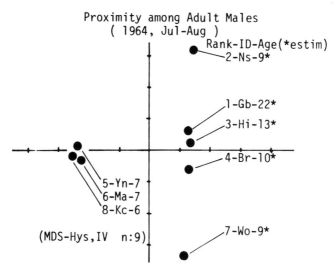

Figure 6.2 Proximity among eight adult males of the Katsuyama group from July to August 1964. See Figure 6.1 for analysis method and legends.

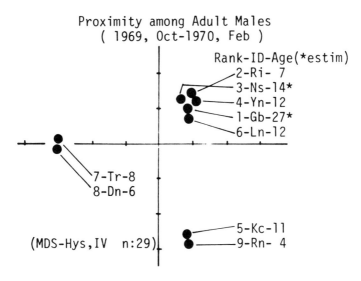

Figure 6.3 Proximity among nine adult males of the Katsuyama group from October 1969 to February 1970. See Figure 6.1 for analysis method and legends.

human. As the results show, a central grouping of dominant adult males, such as was seen in February 1958, was not observed during this period following the disappearance of Rm.

The results, however, indicate that the new alpha male Gb—the oldest among the eight males—was located closest to the second oldest third-ranking male, M51*Hiro (identification: Hi). Furthermore, the third oldest fourth-ranking male, M54*Bera' (identification: Br), was closer to Gb and Hr than to the other younger adult males. Therefore, the social relationships among adult males during this period had a structure similar to that in February 1958 in one respect. Fully mature males of similar age tended to be located close to one another, irrespective of differences in their dominance ranks.

October 1969–February 1970. Figure 6.3 shows the results of the analysis of proximity among nine adult males in the group during a period from late October 1969 to early February 1970. At the start of these observations, Gb had been the alpha male for more than five years. As shown in figure 6.3, a central grouping was composed of five high-ranking males. The sixth-ranking, twelve-year-old male, M57*Lino (identification: Ln), was included in the central grouping, whereas the fifth-ranking, eleven-year-old male, M58Kicho (identification: Kc), was not. It should be also noted that the second-ranking, seven-year-old male, Ri, was included in the central grouping.

This composition of the central grouping in which the young Ri was included was associated with unstable social relationships among the central adult males, particularly between Gb and Ri. The relationship between these two top-ranking adult males had been maintained on the basis of a balance between mutual alliance and competition until 24 February 1970 when fighting occurred between Gb and Ri. Ri became dominant over Gb following the fighting.

July–September 1970. Figure 6.4 shows the results of the analysis of proximity among nine adult males during the period from July to September 1970 when Ri was the new alpha male. A central grouping of dominant males was indistinctive in this early stage of Ri's tenure, a situation similar to the early stage of Gb's tenure in 1964 (see figure 6.2). Nevertheless, the five highest ranking males seemed to constitute a cen-

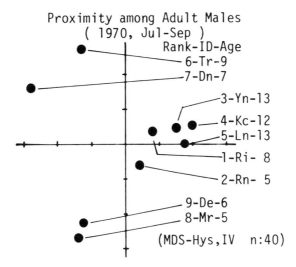

Figure 6.4 Proximity among nine adult males of the Katsuyama group from July to September 1970. See Figure 6.1 for analysis method and legends.

tral grouping. Among them, the three older males (Yn, Kc, and Ln) were located closer to one another than to the two younger males (Ri and Rn).

February 1973.   Figure 6.5 shows the results of the analysis of proximity among eleven adult males in February 1973. At this time, Ri had been the alpha male for about three years. A central grouping was composed of five dominant males. The sixth-ranking, nine-year-old male was included in the central grouping, whereas the fifth-ranking, nine-year-old male, M63 Dana' (identification: Dn), not only was not included but was also slightly separated from the central grouping. The fission of the Katsuyama group occurred in April 1973. When the fission occurred, Dn became the alpha male of the split group. The slight separation of Dn from the central grouping might have been predictive of this outcome.

As previously described, unstable social relationships were observed among the central males when Ri became the new alpha male in 1970. The instability was considerably reduced by 1973, since Ri had become fully mature and both Kc and Ln had died of severe injuries of an unknown cause in 1971. As described previously, Kc and Ln were older than Ri, but were subordinate to him. This created social conflicts among the adult males in the group. On the other hand, social conflicts

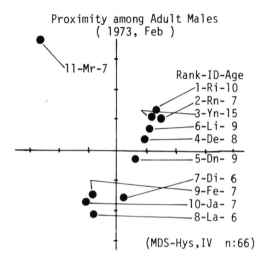

Figure 6.5 Proximity among eleven adult males of the Katsuyama group in February 1973. See Figure 6.1 for analysis method and legends.

were created and continued between the adult females of the high-ranking matrilineages—such as the Rika, the Mara, and the Dera matrilineages—and those of the middle-ranking matrilineages—such as the Bera and the Kera. The adult females of the Bera and the Kera matrilineages had mating relationships with Ri.

August 1976. Figure 6.6 shows the results of the analysis of proximity among ten adult males in August 1976. Rn became the new alpha male following Ri's disappearance on 7 August 1976. A central grouping was composed of six adult males in which two young, eight-year-old males—M68Bera'53*' (identification: Be), and M68Fera' (identification: Fe)—were included. These two young adult males were from high-ranking matrilineages and received support from their kin members, particularly from those adult females who had mated with the former alpha male (Ri). After Ri disappeared and Rn became the new alpha male, the two young males were unable to remain in the group. Fe left the group in July 1977 and Be left in early 1979. In contrast, two subcentral adult males remained in the group and later became the central males. They were seventh-ranking nine-year-old M67Masa'54* (identification: Ms) and the tenth-ranking seven-year-old M69Mona'59' (identification: Mn).

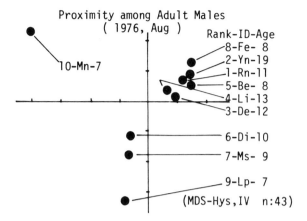

Figure 6.6   Proximity among ten adult males of the Katsuyama group in August 1976. See Figure 6.1 for analysis method and legends.

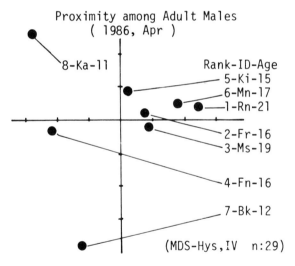

Figure 6.7   Proximity among eight adult males of the Katsuyama group in April 1986. See Figure 6.1 for analysis method and legends.

April 1986.   Figure 6.7 shows the results of the analysis of proximity among eight adult males in April 1986. The central grouping was composed of five males. The seventeen-year-old Mn was included in this group. He was the sixth-ranking male and was located closest to the alpha male (Rn). Ms was also a central male.

OCCURRENCES OF SEVERE ANTAGONISM

Severe antagonism has occurred among adult males in the group, although only infrequently. The following are the known instances of severe antagonism among adult males in the chronology of the Katsuyama group.

Antagonism in Early Stage of Artificial Feeding

In February 1958, the second-ranking male, M44*Toto (identification: Tt), attacked and severely bit the oldest lowest ranking male, Vs, who was located close to the central grouping (see figure 6.1). The aggression from Tt to Vs occurred as a redirected attack following mild aggression from the alpha male Rm to Tt. Although Vs was severely bitten, he continued to remain in the group and maintained relationships with the central males. Finally, however, he left the group in May 1958 and was found foraging alone. Therefore, it was possible that the aggression in February might have created social conflicts between Vs and Tt or other central males which induced to Vs to leave the group.

As previously described, there was a proportionally large number of adult males in the group in the early stage of artificial feeding. This was perhaps the characteristic age-sex composition of a wild, unprovisioned, free-ranging group of Japanese monkeys. This composition changed following the initiation of artificial feeding. Three of six fully mature adult males—those more than eleven years old—left the group within one year of February 1958. The departure of these males must have been caused by multiple factors, among which two factors seem to be of particular importance. First, there was presumably a sharp increase in the frequency of social conflicts among the adult males resulting from the spatially restricted food source and consequent reduction in interanimal distance. Second, decreasing environmental pressure allowed fewer adult males to remain in the group. The severe attack by Tt upon Vs and the departure of Vs from the group seemed to result from social conflicts among the adult males that were engendered by socioecological changes accompanying artificial feeding.

Antagonism Between the Alpha and Beta Males.

On 24 February 1970, Gb and Ri were found fighting with each other. Although the beginning of the fight was not observed, it was likely that Ri initiated the aggression. At the start of observations Gb appeared to dominate the fight, and he bit Ri. Immediately, Rn who was a cousin of Ri, rushed in and allied with Ri. The two of them bit Gb, and Gb was injured. Following the fighting, Gb declined from the first- to the sixth-ranking adult male. Ri became the new alpha male of the group.

This fighting seemed to occur chiefly because of social conflicts between young Ri and old Gb. Thus, on the one hand Ri was seven years old and rapidly achieved high rank with the support of his kin members (the first-ranking Rika matrilineage). On the other hand, Gb was estimated to be twenty-eight years old and had become weakened due to senility. In spite of the severe fighting, Gb continued to remain in the group and maintained proximity with Ri until his disappearance in August 1970. Gb probably died of old age.

Antagonism During the Group's Fission

As previously described, adult females of both the fifth-ranking Bera and the twelfth-ranking Kera matrilineages mated with the alpha male (Ri). These adult females initiated aggression toward the adult females of the first-ranking Rika matrilineage, including Ri's mother, grandmother, and sisters. This aggression also expanded to include the adult females of the other high-ranking matrilineages, such as the Mara, the Dera, the Masa, and the Dana matrilineages. Severe fighting continued intermittently for more than one year, until the original group split into two groups.

The alpha male Ri did not actively intercede in the fighting which occurred between his mating partners and his female kin. He even avoided those fights which involved his own mother and grandmother. During the approximately one-year period in which the group's fission was taking place, nine adult females were severely injured and two were killed. One of the two females killed was Ri's grandmother, the matriach of the first-ranking Rika matrilineage.

These severe fights started and continued predominantly among adult females and, over time, the fights came to involve the females' juvenile offspring. None of the adult males was injured or killed. Naturally, adult males were involved and interceded in the fighting among adult females. However, none of the adult males interceded effectively enough to inhibit or resolve the severe fighting among adult females which caused serious injury and death.

The adult males seemed to contribute to the termination of the severe antagonism chiefly in two ways. First, the adult males initiated the separation among themselves into two groups—a major group and a minor one—without the occurrence of any severe antagonism among the adult males. Second, the adult males maintained proper distance between the two groups, with each group being followed by particular adult females and their offspring. Thus, the two groups were established with independent membership and home ranges.

Conflict between the New Alpha Male and a Young Central Male

The alpha male Ri disappeared on 7 August 1976, presumably because of death due to unaccountable early senility at the age of fourteen years. The long-lasting severe antagonism and conflicts might have shortened his life. The beta male Rn became the new alpha male following Ri's disappearance. During the two days following the Ri's disappearance, Rn spent a considerable portion of his time in the peripheral part of the group, perhaps exploring the disappearance of Ri and observing the situation at the central part of the group from a distance. On the third day, he walked into the central part of the group, directly approached the alpha female (F53*Bera') and bit her. Apparently, he aimed to attack the most dominant of the adult females who had mated with the former alpha male Ri. Furthermore, the adult female whom he bit had been one of the most aggressive and active animals in expelling the adult female kin members of the Rika matrilineage during the group's fission. The attack from Rn on F53*Bera' was not severe, but it was definitely well-oriented and well-controlled. It seemed to be effective enough for Rn to establish his position as the alpha male in the group.

No apparent aggression occurred among the adult males in relation to the change of the alpha male. However, social conflicts were seen between the new alpha male Rn and the fifth-ranking young adult male Be, as previously described. Be was a son of F53*Bera' and was estimated to be a son of the former alpha male Ri. Following the attack by the new alpha male upon his mother, Be became gradually peripheralized and finally left the group in early 1979.

Aggression from Central Males to Peripheral Males

Two instances of severe aggression occurred from central or subcentral adult males to peripheral males. The first was in September 1985, and the second was in January 1986 (observations by Imakawa and colleagues, personal communication). In the first instance, the seventh-ranking, sixteen-year-old subcentral male Mn initiated the attack on the eleventh-ranking, eleven-year-old peripheral male, M74Elza' (identification: Ez). Following the initial attack by Mn, two low-ranking, eight-year-old, peripheral males allied with Mn, and these three males severely bit Ez. Ez was paralyzed, lying in a stream with half of his body sunk in water. He would have been killed, if researchers had not rescued him.

The aggression from Mn to Ez seemed to occur as a result of social conflicts between central and peripheral males. Ordinarily, Mn, who initiated the attack, was not an aggressive male. On the contrary, he appeared to react phlegmatically, especially when receiving aggression from dominant central adult males and females. This severe aggression, however, indicated that he was not tolerant of the peripheral adult male who had approached the central part of the group at the beginning of mating season. His reaction to the peripheral male seemed to be derived from experiences in his own life history. He was born in the eighteenth-ranking matrilineage, or at nearly the bottom-ranked among twenty matrilineages in the group. He became a subordinate, peripheral young adult male, left the group at age of eleven years and spent more than one year outside of the group. He returned to the peripheral part of the group at the age of twelve years and finally became one of the eight most dominant adult males in the group. It was a rare case that an adult male, such as Mn, returned to his natal group after spending more than one year outside of the group and became a central male.

On the other hand, Ez who received the attack from Mn, was born of the fourth-ranking matrilineage, which was much higher ranking than that of Mn. He became peripheralized at about five years of age, yet he began to approach the central part of the group in mid-September, the beginning of the mating season for the Katsuyama group. It was not too surprising that the two peripheral young adult males allied with Mn and attacked Ez. This was the typical manner for peripheral young adult males to cope with a fully mature peripheral adult male.

In the second instance, Mn again initiated a severe attack upon a peripheral, seven-year-old male in January 1986. In this instance, however, the alpha male Rn allied with Mn, and these two fully mature dominant adult males severely bit the young adult male. The victim was M78Bera'53*'67' (identification: Ba) who was born in the first-ranking Bera matrilineage and was the estimated son of the alpha male Rn. He was a high-ranking central juvenile male and became peripheralized upon reaching maturity. However, he continued to receive support from adult female kin members of the Bera matrilineage.

This second instance of severe aggression also seemed to occur as the result of social conflicts between central and peripheral males who had had a long-term background history of conflict between their matrilineages. In general, any social relationship between fully mature adult males and young adult males may involve conflicts to some extent. The attack from Mn to Ba, however, appeared to be accelerated by matrilineal conflicts, in addition to conflicts between fully mature adult males and young adult males. A fully mature sixteen-year-old male, such as Mn, would not have needed any alliance with the alpha male Rn to attack a young seven-year-old male such as Ba. The alliance between Mn and Rn was probably formed by Mn's motivation to expel the young adult male of the highest ranking matrilineage. This was probably facilitated by the conflicts that characteristically occurred between Rn and sons of the Bera matrilineage, such as was seen between Rn and Be in 1976.

CONCLUSIONS

1. Long-term observations of the Katsuyama group of Japanese monkeys indicate that severe antagonism has occurred only infrequently among the adult males. The maintenance of the social organization of the group seemed to be largely dependent upon the inhibition and resolution of severe antagonism among the adult males.

2. One of the factors contributing to the inhibition and resolution of severe antagonism among adult males is their age-matriline composition. In general, the category of high-ranking central males contained a greater proportion of fully mature males of older age than did the category of lower ranking peripheral males. This may imply that fully mature, older adult males possess age-related advantages—in terms of their physical characteristics and experiences—for survival. Furthermore, the dominance of fully mature older males over younger adult males seemed to contribute to the inhibition and resolution of severe antagonism among males and other individuals in the group.

3. The age composition of adult males was exceptional in the period following the change of the alpha male in 1970. In this period, the alpha male was much younger than the lower ranking males. This composition seemed to create social conflicts among adult males and among females, and these conflicts were perhaps one of the causes of the group's fission in 1973.

4. The high-ranking adult males of the group were very diversified in terms of their matriline background. It was found that their individual dominance ranks did not correlate in a simple manner with their matriline dominance ranks. This diversity in the age-matriline composition of adult males may normally create multifarious social relationships among the males and between the males and other individuals in the group with respect to alliances, leader-follower relationships, and mating. The diversified age-matriline composition of adult males had the effect of maintaining the social organization and of making the social organization dynamic in this large-sized, multimale group.

5. The results of an analysis of spatial proximity among the adult males suggested that the maintenance of proper interindividual distance is important for the inhibition and resolution of severe antagonism among the males and among other individuals in the group. Interindividual spacing-out was typically observed during the course of the group's fission. The adult males split into two groups and maintained discreet distances between themselves, thus contributing to the termination of severe fighting among the adult females.

6. The high-ranking adult males tended to form a central subgrouping by remaining in proximity with one another. The central subgrouping was not seen in periods such as those following a change in the alpha male, when social relationships among the adult males were unstable. The central subgrouping was normally composed of high-ranking males, and among these the fully mature adult males of similar age tended to remain close to one another, irrespective of large differences in dominance ranks. This may indicate that long-term male-to-male associations are more important than simple hierarchical arrangements of dominance and subordination in maintaining social relationships among the males themselves, as well as with other individuals in the group.

7. Severe antagonism occurred among adult males, although infrequently. Severe aggression was mostly directed by central males toward peripheral males who approached the central part of the group. On these occasions, alliances were probably formed between central males on the basis of male-to-male associations. It was estimated that fathers might direct severe aggression to sons by forming alliances with central males who were not blood relatives. The severe aggression from fathers to sons seemed to promote the separation of maturing sons from their mothers and from their adult female kin.

8. The severe antagonism among adult males appeared to contribute to maintaining the social organization of the group

chiefly by expelling maturing males from their natal group into the outer world. On the other hand, severe antagonism among adult females, such as was seen in the course of the group's fission, seemed to contribute to the separation of the group into two reproductive units.

REFERENCES

Fujii, H. (1975) A psychological study of the social structure of a free-ranging group of Japanese monkeys in Katsuyama. In Kondo, S., Kawai, M., Ehara, A. (eds.), "Contemporary Primatology." Basel: S. Karger. 427–436.

Hayashi, C. (1954) Multidimensional quantification with the applications to analysis of social phenomena. *Annals of the Institute of Statistical Mathematics* 5:121–143.

Itoigawa, N. (1973) Group organization of a natural troop of Japanese monkeys and mother-infant interactions. In Carpenter, C. R. (ed.), "Behavioral Regulators of Behavior in Primates." Lewisburg: Bucknell University Press. 229–250.

Itoigawa, N. (1975) Variables in male leaving a group of Japanese macaques. In Kondo, S., Kawai, M., Ehara, A., Kawamura, S. (eds.), "Proceedings from the Symposia of the Fifth Congress of the International Primatological Society." Tokyo: Japan Science Press. 233–245.

Kawai, M. (1958) On the rank system in a natural group of Japanese monkeys, 1 and 2. *Primates* 1:111–148.

Kawamura, S. (1958) The matriarchal social order in the Minoo-B group: A study on the rank system of Japanese macaque. *Primates* 1:149–156.

Koyama, T.; Fujii, H., Yonekawa, F. (1981) Comparative studies of gregariousness and social structure among seven feral *Macaca fuscata* groups. In Chiarelli, A. B., Corruccini, R. S. (eds.), "Primate Behavior and Sociobiology." Berlin, Heidelberg: Springer Verlag. 52–63.

Mori, A. (1977) Intra-troop spacing mechanism of the wild Japanese monkeys of the Koshima troop. *Primates* 18:331–357.

CHAPTER SEVEN

The Physiology of Dominance in Stable versus Unstable Social Hierarchies

ROBERT M. SAPOLSKY

For most species of primates, particularly those living in large groups, group members can be ranked within a pecking order or dominance hierarchy. For such species, few aspects of social life would appear to be less equivocal or more readily established than this relationship of dominance and subordination. For example, it is fairly easy to determine that one individual consistently defers to another by yielding a food source or giving up a desirable resting place, thus identifying itself as the subordinate in the relationship.

As this chapter demonstrates, however, the ease with which a relationship of dominance and subordination can be operationally defined and measured is no indication that the underlying processes referred to by these concepts are equally simple and straightforward. The focus of the chapter is on the physiology of social dominance. Animals that are near the top of the dominance hierarchy might be expected to differ in physiology—as well as behavior—from those occupying lower positions. This is often confirmed. At the same time, however, the indications are that no single physiological profile is associated with either dominance or subordination.

A major factor in the variations in the interrelations between endocrine function and dominance is social instability. Social instability can be viewed as the outcome of sustained and pervasive social conflicts between individu-

---

Robert M. Sapolsky—Department of Biological Sciences, Stanford University, Stanford, California als—*interpersonal conflict*—which inevitably reduce predictability and increase the likelihood of intraindividual conflict, uncertainty and stress.

The research presented in this chapter indicates that, during periods of social stability, basal testosterone concentrations are similar in dominant and subordinate males, although dominant males tend to be more resistant to stress-induced suppression. In contrast, during periods of instability, basal testosterone levels are considerably higher in dominant than in subordinate males. The differences between dominant and subordinate males in gonadal activity during stressful situations in stable groups is associated with altered regulation of pituitary-adrenal activity. Specifically, dominant males have lower basal cortisol concentrations than do subordinates, and they exhibit greater sensitivity to negative feedback regulation, more rapid response to stress, and quicker return to basal values. Dominant males during stable conditions also exhibit metabolic, cardiovascular and immune profiles that are potentially less pathogenic than those of subordinates. These adaptive and quiescent features of adrenocortical function that characterize dominant males during periods of social stability are lost during periods of instability.

The vast intertwining of behavior and physiology can be disputed by few biologists, and it influences disciplines ranging from behavioral ecology to neuroimmunology. This intertwining is often its most convincing when at its extremes, and when in the realm of the pathologic facets of behavioral physiology—the florid hallucinations induced by a neurotransmitter agonist, the decline in aggressivity following castration, the pronounced adrenocortical secretion stimulated by profound grief, and the improvement in an affective disorder with the use of antidepressants. What has been much more difficult has been the study of how normal, subtle behaviors influence and are, in turn, influenced by physiology. This has rarely been a more difficult—yet more interesting—task than when examining the physiology of individual differences in social primates.

The question that has been posed most frequently in this regard has been "Does the physiology of individual primates differ according to their social rank?" If the answer is *yes*, additional questions typically follow, such as "Is the distinctive rank and the distinctive physiology causally related?" Or "Is one animal's distinctive physiological profile more adaptive than another's?" In general, the answer has indeed been *yes*. There are distinctive physiological correlates of social rank in primates. This has emerged from the work of many scientists examining varied species in different social settings. The problem, however, has been the lack of consistency in these studies. Thus, it is most difficult to decide what precisely those physiological correlates are. This chapter aims to resolve some of these contradictions. Its main point is that there is no single physiological picture of social dominance. Rather, the critical factors appear to be the way in which dominance is expressed, and the social environment in which it occurs. Specifically, dominance in a stable dominance hierarchy appears to have markedly different physiological concomitants than dominance in a primate society filled with social conflict, instability, and hierarchical reorganization. This chapter will present the data supporting this view, and consider the mechanisms by which physiological correlates of rank are changed so dramatically as primate societies shift between social stability and instability.

PHYSIOLOGICAL CORRELATES OF RANK IN A STABLE PRIMATE SOCIETY

Methods of Study

A stable hierarchy can be characterized by three related features. As calculated by most investigators, rank expresses the number of individuals that one dominates in the category being considered. Thus, within any dyadic pair, the dominant individual—the winner of the majority of the contested bouts—is determined, and the number of individuals that each subject dominates totaled. The more individuals that are dominated, the higher is the rank for that subject (Packer 1979). In a stable hierarchy, the ranks of individuals are relatively unchanging as rank is a relative, rather than an absolute measure.

As a second feature, the outcomes of the dyadic relationships that form the basis of that rank are also unchanging. For example, in a population of, say, five males, one individual of third rank might dominate two other males in food competition. Under stable conditions, not only will his rank not change—that is, he will continue to dominate two males—but which two males he dominates will not change. If such shifts were occurring, it would indicate unstable relationships between individuals that, when tallied, just happen to balance each other out.

A final feature of stability could be the consistency of the direction of dominance in each particular dyadic relationship. For example, the dominant male in a dyad may win 100% of contested bouts with the other member of the dyad, or he may win 51%, with 49% being reversals by the subordinate individual. A high rate of reversals indicates an individual that is barely holding on to dominance within that dyad. A stable hierarchy is thus also characterized by a low rate of reversals within dyads. These three features are obviously related. If the rate of reversals rises above the 50% mark, the direction of dominance in the dyad, by definition, changes. If the direction of dominance in sufficient dyads changes, the overall dominance rankings are likely to change.

Physiological data on individuals with known ranks in stable hierarchies have come from three different types of studies of primates. The most common type has been to form a group of captive primates and maintain the membership unchanged for a long period of time. As will be further discussed, after a substantial percentage of a year—or perhaps even years—the social relations will take on the stable characteristics already described. In a unique study that constitutes the second approach, an entire troop of Japanese macaques (*Macaca fuscata*) that had become agricultural pests were trapped and transported to a primate center in Oregon where, housed in an outdoor corral, they have been under study for decades. Since these individuals have been together as cohesively as any wild population, a high degree of stability in their relations is observed (Eaton 1976).

Finally, my own studies constitute a third and rather different approach. I have studied the behavioral physiology of troops of wild olive baboons (*Papio anubis*) living freely in the Serengeti ecosystem of East Africa. This approach has involved the classical field techniques of habituating known individuals and observing them at intermittent periods over their lives. Coupled with this, I am able to approach these animals close enough to, on occasion, shoot them with anesthetic-loaded darts from a blowgun. This darting is done under conditions in which no animal sees the event, so there is no anticipatory stress. Furthermore, an anesthetic is used which does not distort the hormone values being monitored.

Animals must be darted at the same time of day to control for circadian fluctuations, and when their basal hormonal function has not been disrupted by illness, injury, a recent fight, or mating. Finally, an initial blood sample must be obtained within a few moments of the onset of the anesthetic effects, in order to examine hormone concentrations under these still-basal conditions. A variety of standard clinical tests of endocrine, metabolic, and immune function can then be conducted at a field laboratory before releasing the animals when they have recovered from the anesthetization.

With this approach, it has been possible to gain insights into the behavioral physiology of a primate population living in a completely natural state (Sapolsky 1982). These studies have yielded information about these animals during times of social stability and hierarchical stasis, as well as during periods of hierarchical reorganization and social conflict.

The Testicular Axis

Considerable interest has focused on the relationships among aggressiveness, social dominance, and testosterone concentrations. Study of these varied primate populations—newly formed captive groups, well-established captive groups, and natural troops—has produced some strikingly similar results. To make the most sense of these observations, it is important to discuss briefly the role of testosterone in aggression.

In general, high rates of aggression tend to be positively correlated with elevated testosterone concentrations in the primate. This can be observed correlatively, given the gender difference in testosterone concentrations and rates of aggression in many species; the increase in aggressiveness and testosterone concentrations at puberty; and the parallel seasonal fluctuations of both aggressiveness and testosterone in many species. Causality is implied by typical decline in rates of aggression following castration and its reinstatement with exogenous testosterone administration. However, this relationship between aggression and testosterone is fraught with caveats. For one thing, the hormone is neither necessary nor sufficient for aggression to occur. Castration need not always decrease aggressiveness, and a postcastration decline in aggressiveness is not always reversed by testosterone administration. Furthermore, extremely small quantities of the steroid can reinstate aggression as efficaciously as does the reestablishment of precastration concentrations, implying a nonlinear relationship between hormone concentrations and rates of aggression. Therefore, subtle intra- and interindividual variations in testosterone concentrations within the normal range are not particularly predictive of rates of aggression among primates. Among primates, social factors and learning play tremendous roles. Thus, testosterone seems to have a modulatory rather than activational effect on aggression, and the relatively weak association between high testosterone concentrations and high rates of aggression is essentially a population fact with low predictability when applied to the individual (Dixson 1979).

Given these caveats, is high rank in a stable hierarchy associated with high rates of aggression and high testosterone concentrations? Neither appear to be correlates. In such instances, high-ranking males appear not to be the most aggressive (Dixson 1979; Eberhart et al. 1980; Jolly 1985; Rowell 1972; Sapolsky 1982). This observation is teleologically quite satisfying, as one would think that a perquisite of dominance in a stable hierarchy should be not having to constantly reaffirm one's dominance with fighting (Jolly 1985). In cases in which dominant males are in frequent agonistic interactions, they are rarely the initiators. For example, during a stable period in my study population, it was typically middle-rank juvenile males who initiated the highest rate of escalated aggressive encounters. Among these animals, the initiator lost the fight in 80% of cases (Sapolsky 1983a).

There is also no association between high basal testosterone concentrations and high rank in these stable populations. For example, Gordon and colleagues (1976) studied a stable mixed-sex population of rhesus monkeys (Macaca mulatta) which was maintained undisturbed in an outdoor corral for three years. The researchers began blood sampling six months after group formation and found no relationship between a male's social rank, as determined by agonistic interactions, and his basal testosterone concentrations. In their studies of the population of Japanese macaques in Oregon, Eaton and Resko (1974) also failed to find an association between rank and testosterone levels. Finally, in my own studies of wild olive baboons, there was also no relationship between basal testosterone concentrations and rank, as determined by either reproductive or approach-avoidance criteria (Sapolsky 1982, 1986b).

However, the testicular axis of high-ranking males in stable hierarchies does seem to have at least one distinctive feature. In essentially every mammalian species examined, testosterone secretion is inhibited dramatically by stress, whether of a somatic or psychogenic nature (Rose 1985). Among the wild olive baboons that I have studied, testosterone concentrations were suppressed by a stressor as acute as the process of darting and anesthetization (Sapolsky 1982), or by one as sustained as a drought (Sapolsky 1986a). While high-ranking males did not have the highest concentrations of testosterone under basal conditions, they were relatively resistant to the inhibitory effects of stress. Thus, following darting and anesthetization, testosterone concentrations promptly plummetted in subordinate males and continued to decline over at least the next eight hours. In contrast, high-ranking males actually *elevated* testosterone concentrations during the first post-stress hours, but these concentrations ultimately declined (Sapolsky 1982, 1986b). The mechanisms for this rank-related phenomenon are becoming understood. The inhibition of testosterone concentrations during stress arose from a decline in secretion of the hormone, rather than an enhanced clearance from the blood (Sapolsky 1985). The declining secretion is due to stress-induced release of opiates (acting principally through the μ-opiate receptor) which suppress LH secretion from the pituitary (Sapolsky and Krey 1988), and to stress-induced release of glucocorticoids which inhibit testicular sensitivity to LH (Sapolsky 1985). High-ranking males were resistant to this phenomenon in that their testes were less sensitive to the suppressive actions of glucocorticoids (Sapolsky 1985). In addition, stress-induced release of sympathetic catecholamines increased testosterone secretion only in high-ranking males (Sapolsky 1986b). This could arise either from greater sympathetic outflow or greater target tissue sensitivity in dominant males, with catecholamines either acting directly to stimulate testicular steroidogenesis or to vasodilate testicular parenchyma, thus increasing the quantity of LH reaching the testes (Frankel and Ryan 1981). In any case, while the high-ranking male in a stable baboon hierarchy does not have the highest basal testosterone concentrations, he is uniquely capable of maintaining—and even elevating—those concentrations during stress. The implications of this rank-difference for aggression and muscle metabolism are discussed elsewhere (Sapolsky 1990a).

The Adrenocortical Axis

While a large percentage of behavioral endocrine studies of male primates have focused upon the testicular axis, the relationship between social rank and functioning of the adrenocortical axis has also attracted interest. From the standpoint of understanding the physiology of stress, the mechanisms mediating stress-related disease, and the bases of individual differences in coping with stress, the interest in the adrenocortical axis is quite understandable. The adrenal cortex secretes glucocorticoids in response to stress, and these steroids are central to the adaptations that the body must carry out in order to survive a physical stressor. At the cost of energy storage, the hormones mobilize circulating glucose in order to provide readily available energy substrates for muscular activity. Glucocorticoids also increase cardiovascular tone and inhibit a variety of costly anabolic processes—including digestion, growth, reproduction, and the immune and inflammatory responses—until after the stressor has abated (Munck et al. 1984). All of these glucocorticoid actions are critical for surviving an acute physical stressor. Just as clearly, however, they are all costly catabolic actions. Thus, it is not surprising that the damaging effects of *chronic* stress are substantially mediated by *overexposure* to these same glucocorticoids. Pathophysiologic consequences of stress are, essentially, mere extensions of the glucocorticoid actions just discussed—suppression of energy storage to the point of myopathy; increased cardiovascular tone eventuating in damaging hypertension; and suppression of anabolism to the point of peptic ulceration, impairment of growth and tissue repair, and increased risk of disease, impotency, and amenorrhea. The catastrophic consequences of being unable to secrete glucocorticoids in response to stress demonstrate the importance of this hormone during a stressful emergency. But just as importantly, the multisystem deterioration induced by chronic exposure to the hormone underlines the importance of controlling glucocorticoid secretion. Optimally, concentrations of these potent hormones should be minimized during nonstress periods. Secretion should be elevated rapidly with the onset of stress, and terminated promptly at the end.

Studies of dominant male primates in stable hierarchies have shown their adrenocortical axes to possess precisely these desirable attributes. In general, they have the lowest basal concentrations of cortisol, which is the principal glucocorticoid of the primate. Manogue and colleagues (1975) reported this pattern in a population of captive squirrel monkeys *(Saimiri sciureus)* whose membership and dominance hierarchy had not changed for seven years. Keverne and colleagues (1982) reported the same observation for a group of captive talapoin monkeys *(Miopithecus talapoin)* after their hierarchy had stabilized. Finally, I have observed the same pattern of the lowest basal cortisol concentrations in high-ranking males among wild baboons (Sapolsky 1982, 1989). Subsequent study of these animals showed that these low concentrations were due to a low secretion rate of the hormone, rather than to its enhanced clearance from the circulation as measured by the clearance rate of 3H-cortisol (Sapolsky 1983b). In support of the general picture of an adrenocortical system that is relatively quiescent under basal conditions, Shively and Kaplan (1984) reported that, among cynomolgus macaques *(Macaca fascicularis),* dominant males tended to have smaller adrenal glands.

Intentionally excluded from this discussion is a study purporting to show that dominant crabeating macaques had heavier adrenal glands (Hayama 1966). In this rather poor study, male and female data were pooled for both dominance rankings—a process which is ethologically suspect—as well as for adrenal weights—which is endocrinologically unjustified, given the substantial gender differences in adrenocortical function (Bondy 1985).

Not only did these dominant individuals appear to have the adaptive pattern of low basal cortisol concentrations but, when a stressor occurred, they showed the greatest and/or most rapid elevations of concentrations. This was observed in the responses of the stable squirrel monkey population studied by Manogue and colleagues (1975) to ether exposure, chair restraint, or exposure to a snake. The same outcome was observed among my baboons when the stressor was the process of darting and anesthetization (Sapolsky 1982).

Glucocorticoids are secreted by the adrenals in response to pituitary release of ACTH which, itself, is secreted in response to hypothalamic release of CRF and other minor secretagogues. Subsequent work has shown that the rapid rise in cortisol concentrations during stress among dominant baboons was a central phenomenon. That is to say, cortisol concentrations rose because of more rapid release of CRF and/or related secretagogues, rather than enhanced pituitary or adrenal responsiveness (Sapolsky 1989, 1990b).

Finally, dominant males under these stable conditions appear to be more sensitive to the feedback regulation which helps terminate the stress response. Normally, the adrenocortical axis—as with most endocrine axes—is subject to negative feedback control in which elevated levels of the end product hormone inhibit subsequent activity of the axis. During a stressor, this inhibitory regulation is overridden, and hypersecretion ensues. At the end of the stressor, such sensitivity to negative feedback is reinstated (Keller-Wood and Dallman 1984). Baboons with low basal cortisol concentrations—that is, predominantly high-ranking males—were more sensitive to such feedback regulation than were subordinates. They also showed more rapid suppressions of cortisol concentrations in response to the inhibitory effects of the synthetic glucocorticoid, dexamethasone (Sapolsky 1983b).

Thus, dominant males in a stable hierarchy tend to have low resting cortisol secretion, an ability to rapidly increase secretion during stress, and enhanced sensitivity to the feedback regulation that terminates secretion. As will be seen, this is a rather different picture than that of dominant males in unstable hierarchies.

Metabolism, Vascular Physiology, and Immune Profiles

Additional studies have focused on individual rank-related differences in other aspects of primate physiology, and have uncovered some striking patterns. As part of a large body of studies on the effects of diet, stress, and activity on cardiovascular disease, Kaplan and colleagues (1982) found that dominant male cynomolgus monkeys in a stable hierarchy had slightly less atherosclerotic occlusion of their coronary and aortic vessels than did subordinates. In a look at another aspect of metabolism that yielded a supportive finding, we have found that, among baboons in the wild, dominant males had higher concentrations of high-density lipoprotein cholesterol, the form of cholesterol which is considered to be antiatherogenic (Sapolsky and Mott 1987). Finally, preliminary data demonstrate that such dominant baboons had higher circulating lymphocyte counts than did subordinates. As noted, glucocorticoids are immunosuppressive, and a hallmark of their action is to decrease circulating lymphocyte concentrations by either lysing the cells or sequestering them out of the circulation (Cupps and Fauci 1982). If, in these dominant males, cortisol concentrations were raised to levels seen in subordinates, lymphocyte counts were decreased significantly (Sapolsky unpublished data).

PHYSIOLOGICAL CORRELATES OF RANK
IN THE UNSTABLE PRIMATE SOCIETY

The studies discussed above suggest that in a stable hierarchy, high-rank has a number of consistent attributes. First, testosterone concentrations are not necessarily high but are maintained during stress. Second, an adrenocortical axis, thanks to greater sensitivity to feedback regulation, is quiescent during nonstress periods yet responds rapidly to stress. Third, metabolic, cardiovascular, and immunological profiles are potentially less pathogenic than those of subordinates.

In general, these are all fairly adaptive traits. But are they also observed among dominant individuals in an unstable hierarchy? The punchline of this chapter is the emphatic *no* to this question.

Three types of studies will be considered, all of which involve either the initial organization or the reorganization of dominance hierarchies. The first involves the initial formation of groups of captive primates which are then allowed to remain as a cohesive population for some period thereafter. In a number of these studies, endocrine determinations were made both before and after group formation. In a second type of study of captive primates, instability and reorganization of the hierarchy is provoked by repeated changing of group membership. Finally, in my own studies of wild baboons, a period of hierarchical reorganization arose following the crippling of the alpha male and the complete disintegration of the cooperative coalition that unseated him.

The Testicular Axis

In marked contrast to the situation with stable hierarchies, high rank during times of social instability is associated with both high rates of aggression and high concentrations of basal testosterone. The most natural demonstration of this was in the troop of wild baboons which I studied, and which will be discussed in some detail.

In 1981, I returned to this population after an absence of nine months. Rankings of males had previously been made by reproductive activity and were highly correlated with rankings by approach-avoidance and agonistic criteria. Upon return, it became clear that the male who had been the highest ranking or alpha male nine months before was still in that position, but was now being challenged persistently by a cohort of younger males of ranks two to seven. While the alpha male was still capable of winning the majority of approach-avoidance interactions with these males when facing them individually, the rate of reversals by the younger cohort against him was rising, and members of the cohort were forming cooperative coalitions that were challenging him. Shortly into the season, a coalition of all six subordinates fought the alpha male and crippled him. The alpha male immediately dropped out of all social interactions, remained for a week in the forest in which the troop slept, and lay on his side unable to feed himself. He passed the rest of the three-month observational season with little social contact with the rest of the troop, unable to keep up with the daily foraging progressions and walking on three limbs. A physical examination suggested that his shoulder had been dislocated and the arm broken in at least one place.

In the aftermath of this fight, the coalition promptly disintegrated. While all six of these males dominated the half dozen or so additional males in the troop, the dyadic relationships within this cohort were dramatically unstable for the remainder of the season. As defined earlier in this chapter, this involved frequent shifting of rank among this group, often on a daily basis. In addition, even when the direction of dominance in a dyad would become consistent, there were still significantly elevated rates of reversals of the direction of dominance within the dyad.

The period was also characterized by a number of additional features. The dominance hierarchy became less linear (Landau 1951), indicating more circularities of dominance—A dominates B dominates C dominates A. The rate of approach-avoidance interactions among high-ranking males increased, while the rate between high- and low-ranking males did not, nor did that among low-ranking males. The rate of aggressive interactions among high-ranking males also rose, as did the rate of redirected aggression by high-ranking males onto subordinate innocent bystanders. However, rates of aggression among low-ranking males did not rise.

As a result of these increased rates of aggression, among high-ranking males, they were now initiating the most agonistic encounters in the troop, in contrast to prior and subsequent stable seasons. The number of cooperative coalitions rose, as did their rate of disintegration. Finally, there were increased rates of harassment of sexual consortships as well as of consortships breaking up.

During this unstable period, basal testosterone concentrations within the troop were significantly lower than in stable years. Given the likely stressfulness of this period and the effects of stress upon the mammalian testicular axis, this observation was not surprising. In agreement with data already discussed, high-ranking males were less vulnerable to this suppression. As such, there was a dramatic decline in basal testosterone concentrations among subordinates, and a lesser decline among dominant individuals. The result was that dominant males now initiated aggressive interactions most frequently and had the highest basal testosterone concentrations (Sapolsky 1983a).

A similar conclusion emerges from study of group formation in captive primates. In the wild baboons just discussed, the transition from stable to unstable conditions involved a decline in basal testosterone concentrations in subordinate animals. From that decline, there emerged a positive correlation between rank and testosterone concentration that was not apparent in stable seasons. In some of the cases with captive primates, the transition from individual isolation to social groups involved an increase in testosterone concentrations in dominant males as well as a decline in subordinates. The net result, again, was an association between rank and testosterone concentrations during the period immediately following group formation which was not apparent following long-term stabilization of the group.

This pattern was reported by Rose and colleagues (1975) as they studied rhesus monkeys living in an outdoor corral. Four adult males were initially housed individually, and were then simultaneously released into a corral with thirteen adult females. Behavioral observation and intermittent blood sampling were conducted over the next forty-six days. Prior to group formation, there was no relationship between testosterone concentrations and the eventual rank of males upon group formation. However, within a week of formation, the new alpha male had shown an approximate doubling of testosterone concentrations, and had higher concentrations than did subordinates for nearly all subsequent determinations. In contrast, the lowest ranking male showed such a large decline in testosterone concentrations during the first week following group formation that he had the lowest values of the four. By the second week, he joined the males of ranks two and three in presenting essentially preformation levels for the remainder of the study. Thus, the alpha male showed a sustained rise in testosterone concentrations, while the omega male had a transient decline. An association between high rank and high testosterone concentrations emerged during this seven-week study. No rates of initiating aggressive interactions were given.

A related, earlier study by Rose and colleagues (1971) had given a similar result. In this study, a group of thirty-four male rhesus was formed. No females were present, and preformation testosterone values were not obtained. Behavioral data were collected during months two through four and eight through nine after group formation. Blood was collected during month seven. It was found that the highest ranking quartile of males had significantly higher testosterone concentrations than did subordinates, whose values did not differ among themselves. In addition, highest ranking males also had the highest frequencies of aggressive interactions. It was not clear if this involved initiation of agonistic interactions.

Thus, during this early postformation period, high rank was associated with high testosterone concentrations and aggressiveness. Mendoza and colleagues (1979), studying three trios of captive male squirrel monkeys showed the identical pattern during group formation. As with the Rose study (1975), testosterone concentrations prior to group formation did not predict postformation rank. During the first month after formation of all-male groups, testosterone concentrations rose in the alpha male and declined in the gamma male. Consequently, a positive correlation between rank and testosterone concentrations emerged. The addition of females to the group for a month accentuated the pattern only further. Additional work by this group (Coe et al. 1979) showed dominant males to have elevated testosterone concentrations during the first few months after pair formation.

Finally, Eberhart and Keverne (1979) and Keverne and colleagues (1982), working with captive male talapoin monkeys, showed that preformation testosterone concentrations did not predict postformation rank. Following formation of multisex groups—examined over the next seven to eight weeks—testosterone concentrations rose in dominant individuals so that they had significantly higher concentrations than did subordinates.

Thus, during the immediate period following group formation and organization of the dominance hierarchy, or during the reorganization of the hierarchy in a preexisting group, high rank was associated with high testosterone concentrations and high rates of initiating aggressive interactions. Yet, just as clearly—long after group formation or with the stabilization of the new hierarchy—neither of those traits were associated with dominance. This contrast is strengthened by the fact that it is often the same investigators studying, in at least one case, the same animals who have reported the correlation during the unstable period of hierarchical organization or reorganization, and the lack of correlation at other times.

The Adrenocortical Axis

Just as with the testicular axis, the physiological correlates of dominance during social instability differ radically from those of more stable times. First, during unstable periods, dominant males no longer have the lowest basal concentrations. In fact, in their study of group formation in squirrel monkeys, Mendoza and colleagues (1979) found dominant males to have higher cortisol concentrations than did subordinate males. This was apparent with the formation of all-male groups and was accentuated even further with the addition of females. Looking at pairs of squirrel monkey males within the first two months of group formation,Coe and colleagues (1979) also found the highest cortisol concentrations in the dominant males.

A number of studies, while failing to show dominant males as having the *highest* cortisol concentrations during this time, nevertheless showed that they have concentrations no lower than those of subordinates. This is in contrast to the stable situation, and was observed by myself in the olive baboons during the hierarchical reorganization already discussed (Sapolsky 1983a), as well as by Keverne and colleagues (1982) following group formation of talapoin monkeys, and by Chamove and Bowman (1976) shortly after formation of multisex groups of rhesus monkeys.

Thus, during instabilities, dominant males no longer have the lowest glucocorticoid concentrations. In addition, they no longer show the most dramatic increases in cortisol concentrations during stress. This was observed in newly formed groups of squirrel monkeys being stressed by capture (Coe et al. 1979), as well during the period of social reorganization of the baboons which I study (Sapolsky 1983a). Finally—and somewhat in contrast to the rest of these findings—Shively and Kaplan (1984) have reported that, with sustained social instability due to repeated changing of group membership, dominant cynomolgus males still have somewhat smaller adrenals than do subordinates, just as during stable conditions.

Metabolism, Vascular Physiology, and Immune Profiles

As discussed previously, in stable hierarchies, high-ranking males have less atherosclerosis, more high-density lipoprotein cholesterol, and higher lymphocyte counts. Unfortunately, only the question of atherosclerosis has been considered during social instability. Kaplan and colleagues (1982), induced sustained instability in social groups of cynomologus monkeys by constantly changing group membership, and found that, under those conditions, dominant males had considerably *more* atherosclerotic occlusion than did subordinates.

WHAT FACTORS UNDERLIE THE
DIFFERING PHYSIOLOGICAL CORRELATES OF DOMINANCE
IN STABLE VERSUS UNSTABLE HIERARCHIES?

Study of rodents has produced a consistent picture of the endocrine correlates of competitive social behavior, best summarized as a relationship in which "higher social status and aggressiveness are associated with elevated testosterone levels, whereas submissiveness and defeat are related to increased output of adrenal corticoids" (Coe et al. 1979, 633). The first and most important point to be gleaned from the data discussed in this chapter is that a similar consistency does not apply to the behavioral endocrinology of primates. The variability and complexity of social interactions, the emotional and cognitive capacities of the individuals, and the potentials for distinctive demographic, group, and personal histories combine to make it implausible that social dominance would have a monolithic set of physiologic correlates. Clearly, a single endocrine profile of social dominance in primates does not exist. What, then, is the reason for the distinctive endocrine correlates of dominance that emerge during hierarchical organization or reorganization? Is it a cause, a consequence, or a mere correlate of the social status?

Do Elevated Testosterone Concentrations Cause *the Emergence of Dominance During Instabilities?*

It appears unlikely that high testosterone causes dominance. In every study in which testosterone concentrations were measured prior to group formation, hormone values failed to predict a male's eventual rank (Keverne et al. 1982; Mendoza et al. 1979; Rose et al. 1975). Demonstrating this point even more explicitly, Rose and colleagues (1975) found that, on the day that their alpha male achieved dominance for the first time—day one after group formation—he had the *lowest* testosterone concentrations of any of the males in the group. Thus, high testosterone concentrations cannot be the *cause* of the emerging dominance.

Is the Testicular Profile a Mere Correlate of Dominance?

Is there some factor which gives rise to both dominance and to rank-related endocrine profiles independently? Consider, for example, the wild baboons that I study.

Year after year, high-ranking males were typically prime-aged. The subordinate cohort was filled with older adults, subadults, and juveniles. Moreover, high-ranking males were typically in the best of health, having the highest proportion of meat in their diets and spending the smallest part of each day foraging for food. Do the endocrine correlates of dominance really have anything to do with dominance? Could it be, instead, that, in this population, if you are prime-aged, in good health, and/or well-fed, you are likely to be both dominant and to have certain endocrine features? Could it also be that the dominance and the endocrinology have no direct mechanistic connection to each other?

The data from the unstable 1981 season (Sapolsky 1983a) argues against this correlative interpretation. As described, the coalition of six males crippled the alpha male. Their coalition then disintegrated, and the period of instability ensued. These six were all prime-aged. During this period of instability, they were seemingly as healthy as the top six-ranking males in any stable season. Furthermore, they dominated predation as disproportionately as did other dominant cohorts. Finally, their time expenditure on feeding was no different than for dominant males in other seasons. In both the unstable and the many stable seasons, high-ranking males always had these same physical and nutritional characteristics. Yet, the endocrine correlates of dominance in the unstable season were dramatically different than in the other more stable seasons. Thus, these characteristics were unlikely to be critical.

If the endocrine profiles of dominance are going to be mere correlates of dominance, there must be some factor which can give rise to high rank and, at the time that the high rank is first being attained, also gives rise to the distinctive physiology of dominance during hierarchical instability. Aggression may play such a role. As already discussed, rates of initiation of aggressive interactions by high-ranking males are not particularly high in a stable hierarchy, but are quite frequent during hierarchical organization or reorganization. Thus, during the period of social instability, males who are emerging with high-rank are often the most aggressive. By definition, they are winning a large proportion of those aggressive interactions. Could the successful aggression which paves the way for their high-rank also be elevating their testosterone concentrations? This appears to be possible.

Among primates, both exercise and aggressive interactions increase testosterone concentrations (Davies and Few 1973; Dessypris et al. 1976; Elias 1981; Kuoppasalmi et al. 1976; Perachio 1978; Sutton et al. 1973). In addition to its gradual anabolic effects on muscle growth, testosterone can also potentiate muscle metabolism rather rapidly (Max and Toop 1983). Thus, one might indulge in the teleological speculation that testosterone concentrations are enhanced during times of sustained muscle activity to metabolically support such enhanced activity. This could be due to enhanced secretion—and/or to diminished clearance—of the steroid from the blood, perhaps arising from the decrease in hepatic blood flow that occurs during exercise (Wahren et al. 1971).

While there have been reports that among nonhuman primates, aggressive interactions elevate testosterone concentrations in both winners and losers (Perachio 1978), the bulk of studies have shown the rise to be restricted to winners, with losers tending toward suppressed concentrations (Bernstein et al. 1974; Rose et al. 1972, 1975). In a number of cases, those males who were emerging as subordinate—and were losing fights—were injured during agonistic interactions. The stressfulness of losing the interaction, coupled with the somatic stressor of injury could account for the suppression of testosterone concentrations (Rose 1985). In a subordinate male, the increased testosterone secretion due to increased aggressiveness could be offset by the inhibitory effects of the stressors of defeat and injury upon testosterone secretion. It would be extremely interesting to know whether a male who was emerging as a dominant individual and winning fights in the process would still show a rise in testosterone concentrations even if injured in the course of the successful agonistic interactions. Such data are not explicitly presented in any of these reports.

The rank-related patterns of testosterone concentrations during formation of hierarchies cannot be due merely to the increased rates of aggression, nor even to the increased rates of successful aggression in the emerging dominant cohort or of unsuccessful aggression in the subordinates. As evidence, in one study of captive rhesus monkeys the investigators (Rose et al. 1974) periodically removed and then returned certain members of the social group. In some cases, the removal of a dominant male produced escalated aggression among the remaining members and a shift in rank. In some instances, however, the number-two-ranking male simply assumed the alpha position *without* overt aggressive interactions, and testosterone concentrations still rose in this newly dominant individual.

These studies allow some conclusions about the involvement of aggression in the unique rank-related profiles of testosterone concentrations during troop instability.

1.   The sheer muscular exertion involved in the aggressive interactions that become more prevalent during these times might have enhanced testosterone secretion.

2.   This enhancement is probably offset in subordinate males by the somatic stressor of injury that often accompanies the defeat. Thus, concentrations tend to rise in the emergent dominant cohort, decline in the subordinates.

Superimposed upon these more physiological mechanisms to explain enhanced or suppressed testosterone secretion is the likelihood that the intrinsic psychological state of assuming one's new rank can alter rates of testosterone secretion. This view is supported by the augmentation of secretion even when dominance is attained without overt aggression. The power of this psychogenic component could be tested in a number of interesting ways. For example, in marathon runs, successful participants tend to show increased testosterone concentrations, while losers tend toward declines (Dessypris et al. 1976).

Would a runner who, although losing a race, managed to beat his own best prior time still show a decline in testosterone concentrations? If this individual did not show the suppression expected of a subordinate, it would indicate the importance of internalized standards and self-rankings that may not be apparent to the observer. In an athletic competition in which success is determined by judges' scores, would experimental distortion of scores alter the testosterone response to the competition? In effect, this asks whether the body needs external confirmation as to whether or not it is functioning well.

These general conclusions about the role of aggression in both helping to attain dominance and elevating testosterone concentrations during group formation were reached by Coe and colleagues in their studies of the squirrel monkey, as stated in their concluding sentence. "Therefore, the association between dominance and [high testosterone concentrations] in the squirrel monkey appears to be due primarily to an effect of behavior on the endocrine system; and one would expect a decreased correlation as the social relationships become more relaxed under stable conditions" (Coe et al. 1979, 637).

Is the Testicular Profile a Consequence of the New Rank?

With the attainment of dominance comes a number of advantages in most competitive primate systems, typically revolving around increased ease of access to disputed resources. Among these rewards is increased access to sexually receptive females and increased frequency of copulation. There is good evidence that part of the explanation for the elevated testosterone concentrations in high-ranking males during rank organization or reorganization is newly increased access to females.

In the males of numerous species, copulation can lead to increased circulating concentrations of testosterone and/or LH (Kamel et al. 1975; Purvis and Haynes 1974, for the rat; Harding and Feder 1976, for the guinea pig; Katongole et al. 1971, for the bull; Saginor and Horton, 1968, for the rabbit). A similar pattern holds in primates. Specifically, access to a sexually receptive female (Bernstein et al. 1972) and copulation with her (Herndon et al. 1980) augments testosterone secretion. However, the relationship here is subtle. For example, it was observed that upon placement with receptive females, a similar rise in testosterone did not occur in a male rhesus who had been raised in isolation in childhood (Rose et al. 1972). This points to the importance of early experience in adult sexual competence and drive. The role of advanced cognitive skills in the matter is demonstrated with the report that mere *anticipation* of sexual behavior can stimulate testosterone secretion in the human (LaFerla et al. 1978).

If, upon their attainment of dominance in a re-forming hierarchy, males have a sudden increase in sexual activity, this would add to their pattern of augmented testosterone concentrations. As such, Mendoza and colleagues (1979) observed that, upon formation of all-male groups of squirrel monkeys, testosterone concentrations rose in the alpha males. Then, when placed with females with whom the alpha males were able to mate, the augmentation of testosterone secretion became even more pronounced. This enhancement with access to females seems to reflect the stimulatory effects of actually mating, rather than some intrinsic feature of dominance. As evidence, when males of a social group were allowed individual access to receptive females, subordinate males had as large of a rise in testosterone concentrations as did dominant individuals (Eberhart and Keverne 1979; Keverne et al. 1982).

Is, in this instance, the critical feature about being with females the copulation? This appears to be likely. In studying talapoin monkeys, Eberhart and colleagues (1980) showed that being able to see and smell females had no effect on testosterone concentrations, regardless of the reproductive state of the females. Similarly, in a social setting, subordinate males will be able to see females and smell their pheromones without resultant testosterone secretion. Although close visual and olfactory proximity to the female may be what is critical, this was not allowed in the Eberhart study and, in the wild, is often not possible for subordinate males. In further support of the role for copulation, social access to lactating females did not lead to potentiation of testosterone secretion in rhesus males (Bernstein et al. 1977). The importance of the copulation itself seems to be contradicted by the study by Mendoza and colleagues (1979). In that case, animals were grouped six months after the usual annual mating season. Thus, females were not sexually receptive at that time. However, there was a rapid behavioral induction by the males of female receptivity, as evidenced by the occurrence of and timing of subsequent pregnancies; it was not clear in the study whether the rise in testosterone concentrations in the alpha males that occurred in association with females preceded the first matings.

While all of these data emphasize the importance of mating as a facilitator of testosterone secretion, the frequency of the act may not be important. It is true that, after group formation, there can be a dramatic increase in sexual activity (Bernstein and Mason 1963; Coe and Rosenblum 1978). However, in the baboons that I have studied, the pattern of a positive correlation between rank and testosterone concentrations during social instability occurred despite the fact that the rate of copulations and consortships by the dominant cohort was no higher than during more stable seasons (Sapolsky 1983a).

Finally, the relative power of females to enhance testosterone secretion was assessed in rhesus (Rose et al. 1975). In this study, previously dominant males were placed with females and a sufficiently large number of other males to guarantee the defeat of the newcomer. Testosterone concentrations plummeted in these animals. This was interpreted as showing that the suppressive effects of defeat upon testosterone outweighed the stimulatory effects of the novel presence of females. However, it was unclear in this study, if the new males were ever able to copulate with the females or even if the females were sexually receptive.

Thus, while the distinctive gonadal profile of dominance during hierarchical instability can arise, at least in part, from the successful aggression that contributes to the emergence of dominance, the subsequent increase in sexual activity following the attainment of dominance can also potentiate testosterone secretion.

Why Does Adrenocortical Function in Dominant Males
Differ during Unstable Times?

As discussed, dominant males have the lowest resting cortisol concentrations in a stable hierarchy. They also have high—if not the highest—concentrations during unstable times. Furthermore, their ability to rapidly increase cortisol secretion during stress is lost during unstable times. This pattern can be interpreted as arising from the stressfulness of the unstable period.

Chronic stress elevates cortisol secretion. Furthermore, such sustained stress will blunt the sensitivity of the adrenocortical axis to feedback regulation, making it harder to suppress basal concentrations of the hormone, and more difficult to terminate secretion at the end of stress (Keller-Wood and Dallman 1984). Such blunting is speculated to involve glucocorticoid-induced reductions in the number of glucocorticoid receptors in certain regions of the limbic system in the brain that mediate feedback regulation (Sapolsky and McEwen 1987). In any case, the unique adrenocortical profile of dominant males during stable times is commensurate with an organism that is under only minimal stress.

In the same hierarchies, subordinate males have enlarged adrenals; a sluggish response to stress and relative insensitivity to feedback regulation; and basal cortisol concentrations that are frequently in the range of those seen in response to an overt stressor. For those familiar with primate social behavior, this profile of subordinate males as being under stress is not surprising. When they obtain a highly desirable food item—such as a prey species—they are the least likely to be able to hold on to it. Even when they are feeding upon something as commonplace as a tuber, they frequently lose the item. Their sexual consortships and grooming bouts are the most likely to be disrupted.

Finally, even if they choose not to involve themselves directly in overt agonistic interactions, they are subject to the highest rates of being attacked as innocent bystanders, as in redirected aggression (Sapolsky 1983a). Thus, their social lives are filled with lack of control or predictability over contingencies, hallmarks of a psychologically stressful milieu (Weiss 1970). In agreement with this interpretation, dominant individuals in stable hierarchies among nonprimate species have the lowest basal glucocorticoid concentrations or the smallest adrenals (Archer 1970; Barnett 1955; Bronson and Eleftheriou 1965; Davis and Christian 1957; Louch and Higginbotham 1967; Popova and Naumenko 1972; Southwick and Bland 1959, for rodents; Fox and Andrews 1973, for wolves; Frank, Glickman and Sapolsky unpublished, for the spotted hyena).

During periods of social instability, in contrast, dominant males lose the adrenocortical profiles suggestive of their not being under stress. This fits well with what is known about the psychological components of a stressor. For example, among the baboons that I have studied, during the period of instability, the rate of dominance interactions increased among the high-ranking cohort, and the predictability of outcome declined. Stability and linearity of the rankings declined. The rate of harassment of sexual consortships rose, as did the rate at which the consortships were successfully disrupted. Finally, coalitions formed frequently, but proved to be unstable. In all regards, the switch from dominance in a stable to an unstable hierarchy involved the loss of control and predictability over contingencies in the social environment. Clearly, dominance atop a stable hierarchy is a very different behavioral and psychological position than dominance atop a shifting, unstable one, and the physiology of the latter reflects the increased stressfulness of the position.

CONCLUSIONS

These studies have generated a great deal of data concerning the ways in which the physiological correlates of social rank differ in differing social contexts. It is important to view these, however, with a number of caveats in mind. These data are derived from a number of different species, spanning both the New and Old World primates. One of the main facts that fuels both the productivity and the pleasure of primate field work is the variability of primate social behavior. Thus, the conclusions here about the generic primate should be taken cautiously.

Furthermore, substantial physiological differences are likely to exist among the primate species. For example, there are dramatic differences in adrenocortical function between Old and New World primates (Chrousos et al. 1982; Klosterman et al. 1985). As additional problems in considering these studies en masse, they differ in their ranking criteria. Some studies use reproductive success. Others rely on success in approach-avoidance interactions, or in agonistic interactions. Some consider spontaneous occurrences, while still others provoke them with food competition. Furthermore, the studies differ as to whether the rate

Table 7.1 Physiological Correlates of Dominance in Stable versus Unstable Hierarchies

STABLE	UNSTABLE

THE TESTICULAR AXIS

Dominant males have the highest basal testosterone concentrations

STABLE	UNSTABLE
No: *P. anubis* (Sapolsky 1982, 1986a)	Yes: *P. anubis* (Sapolsky 1983a)
No: *M. mulatta* (Gordon et al. 1976)	Yes: *M. mulatta* (Rose et al. 1971, 1975)
No: *M. fuscata* (Eaton and Resko 1974)	Yes: *M. talapoin* (Eberhart et al. 1979)
	Yes: *S. sciureus* (Coe et al. 1979; Mendoza et al. 1979)

Dominant males maintain testosterone concentrations during stress

STABLE	UNSTABLE
Yes: *P. anubis* (Sapolsky 1982)	No: *P. anubis* (Sapolsky 1983a)

THE ADRENOCORTICAL AXIS

Dominant males have the lowest basal cortisol concentrations

STABLE	UNSTABLE
Yes: *M. talapoin* (Keverne et al. 1982)	No: *M. talapoin* (Keverne et al. 1982)
Yes: *P. anubis* (Sapolsky 1982)	No: *P. anubis* (Sapolsky 1982)
Yes: *S. sciureus* (Manogue et al. 1975)	No: *S. sciureus* (Coe et al. 1979; Mendoza et al. 1979)
	No: *M. mulatta* (Chamove and Bowman 1976)

Dominant males have largest increases in cortisol secretion during stress

STABLE	UNSTABLE
Yes: *S. sciureus* (Manogue et al. 1975)	No: *S. sciureus* (Coe et al. 1979)
Yes: *P. anubis* (Sapolsky 1982)	No: *P. anubis* (Sapolsky 1983a)

Dominant males are more sensitive to glucocorticoid feedback regulation

STABLE	UNSTABLE
Yes: *P. anubis* (Sapolsky 1982)	Not examined

Dominant males have smaller adrenal glands

STABLE	UNSTABLE
Yes: *M. fascicularis* (Shively and Kaplan 1984)	Yes: *M. fascicularis* (Shively and Kaplan 1984)

Table 7.1 *(continued)*

METABOLISM, VASCULAR PHYSIOLOGY, IMMUNE PROFILES

Dominant males have less atherosclerotic occlusion

Yes: *M. fascicularis*	No: *M. fascicularis*
(Kaplan et al. 1982)	(Kaplan et al. 1982)

Dominant males have more high density lipoprotein cholesterol

Yes: *P. anubis*	Not examined
(Sapolsky and Mott 1987)	

Dominant males have higher circulating lymphocyte counts

Yes: *P. anubis*	Not examined
(Sapolsky unpublished)	

of initiation of agonistic interactions is considered versus mere involvement in such interactions. Finally, very few of the studies have indicated whether circadian and stress effects upon hormones have been controlled by obtaining all blood samples at the same time of day, and at the same time interval after investigators first disturbed the animals.

Given these caveats, there is still striking agreement in the results obtained allowing some conclusions (table 7.1; see table 7.2 for descriptions of the particular studies discussed).

1. Upon group formation, stability of dominance relations and a low rate of dyadic reversals appear to require months or even years to emerge.

2. Under stable conditions, dominance is not associated with high rates of aggression nor with high testosterone concentrations. This can probably best be interpreted as reflecting the status quo of such dominance, obviating the need for aggressive reassertion of rank. Adrenocortical function in these dominant males is highly efficient, with low cortisol concentrations under nonstressed conditions, and a rapid ability to activate the system at the time of stress. This is a pattern typically observed in unstressed mammals, and is less predisposing toward pathology than the profiles of subordinate males.

Table 7.2 Criteria Used in Designating Population as Stable or Unstable

STABLE POPULATIONS		
Authors	Species	Condition
Eaton and Resko 1974	*M. fuscata*	Group intact for a number of decades at time of study
Gordon et al. 1976	*M. mulatta*	Group intact for three years at time of study
Kaplan et al. 1982; Shively and Kaplan 1984	*M. fascicularis*	Group intact for twenty-two months at time of study, low rates of aggression and reversals, as stated by authors
Keverne et al. 1982	*M. talapoin*	Group intact for fifteen months at time of study, absence of dominance reversals, as stated by authors
Manogue et al. 1975	*S. sciureus*	Group intact for seven years with no change in hierarchy
Sapolsky 1982, 1983b, 1986b	*P. anubis*	Group intact indefinitely; linear unchanging hierarchy with low rates of dominance reversals

3. During periods of hierarchical organization or reorganization, dominance is associated with both high rates of aggression and with high testosterone concentrations. Subordinance is associated with suppression of testosterone concentrations. This testosterone pattern seems to arise in at least three ways: (a) from the stimulatory effects of the successful aggression that paves the way for dominance coupled with the inhibitory effects of unsuccessful aggression among the emerging subordinate cohort; (b) from the increased frequency of sexual behavior that comes as a reward of dominance; and (c) from some intrinsic psychological features that come with the attainment of high or low rank, separate from aggression or sexual activity.

Table 7.2   *(continued)*

UNSTABLE POPULATIONS

Authors	Species	Condition
Bowman and Chamove 1976	*M. mulatta*	Group intact for one hour at time of study
Coe et al. 1979	*S. sciureus*	Group intact for less than eight weeks at time of study
Eberhart and Keverne 1980	*M. talapoin*	Group intact three to seven weeks at time of study
Kaplan et al. 1982; Shively and Kaplan 1984	*M. fascicularis*	Group membership changed every three months for a year, then every month for ten months
Keverne et al. 1982	*M. talapoin*	Group intact for less than three months at time of study
Mendoza et al. 1979	*S. sciureus*	All-male group studied after one month; male-female group studied after additional month
Rose et al. 1971	*M. mulatta*	All-male group intact for nine months at time of study
Rose et al. 1975	*M. mulatta*	Membership shifted during fifty-day study
Sapolsky 1983a	*P. anubis*	Intact wild group undergoing three-month hierarchical reorganization following crippling of the alpha male

---

4. In addition, during periods of instability, the adaptive and quiescent features of adrenocortical function in dominant males in a stable hierarchy are lost. This is most readily interpreted as representing the increased stressfulness of attempting to attain and maintain dominance in an unstable society.

5. Finally, there is no single physiologic profile of dominance or of subordinance. The physiology seems exquisitely sensitive to the behavior, and the critical features of behavior seem not to be rank

per se, as much as the style by which the rank is expressed and the social milieu in which it occurs. The marvelous complexity and individuality of the primates that we study should make this conclusion anything but surprising.

REFERENCES

Archer, J. (1970) Effects of aggressive behavior on the adrenal cortex in laboratory mice. *Journal of Mammology* 51:327–332.

Barnett, S. A. (1955) Competition among wild rats. *Nature*175:126–127.

Bernstein, I. S.; Mason, W. A. (1963) Group formation by rhesus monkeys. *Animal Behaviour* 11:28–31.

Bernstein, I. S.; Rose, R. M.; Gordon, T. P. (1974) Behavioral and environmental events influencing primate testosterone levels. *Journal of Human Evolution* 3:517–525.

Bernstein, I. S.; Rose, R. M.; Gordon, T. P. (1977) Behavioural and hormonal responses of male rhesus monkeys introduced to females in the breeding and non breeding seasons. *Animal Behaviour* 25:609–614.

Bondy, P. K. (1985) Disorders of the adrenal cortex. In Wilson, J., Foster, D. (eds.), "Textbook of Endocrinology" 7th ed. Philadelphia: Saunders. 816–890.

Bronson, F. H.; Eleftheriou, B. E. (1965) Adrenal response to fighting in mice: Separation of physical and psychological causes. Science 147:627–628.

Chamove, A. S.; Bowman, R. E. (1976) Rank, rhesus social behavior, and stress. *Folia Primatologica* 26:57–66.

Chrousos, G. P.; Renquist, D.; Brandon, D.; Eil, C.; Pugeat, M.; Vigersky, R., Cutler, G. B., Jr.; Loriaux, D. L.; Lipsett, M. B. (1982) Glucocorticoid hormone resistance during primate evolution: Receptor-mediated mechanisms. *Proceedings of the National Academy of Sciences USA* 79:2036–2040.

Coe, C. L.; Rosenblum, L. A. (1978) Annual reproductive strategy of the squirrel monkey *(Saimiri sciureus)*. *Folia Primatologica* 29:19–42.

Coe, C. L.; Mendoza, S. P.; Levine, S. (1979) Social status constrains the stress response in the squirrel monkey. *Physiology and Behavior* 23:633–638.

Cupps, T. R.; Fauci, A. S. (1982) Corticosteroid-mediated immunoregulation in man. *Immunological Review* 65:133–155.

Davies, C. T. M.; Few, J. D. (1973) Effects of exercise on adrenocortical function. *Journal of Applied Physiology* 35:887–891.

Davis, D. E.; Christian, J. J. (1957) Relation of adrenal weight to social rank of mice. *Proceedings of the Society for Experimental Biology and Medicine* 94:728–731.

Dessypris, A.; Kuoppasalmi, K.; Adlercreutz, H. (1976) Plasma cortisol, testosterone, androstenedione and leuteinizing hormone in a non-competitive marathon run. *Journal of Steroid Biochemistry* 7:33–37.

Dixson, A. F. (1979) Androgens and aggressive behavior in primates: A review. *Aggressive Behavior* 6:37–67.

Eaton, G. G. (1976) The social order of Japanese macaques. *Scientific American* 253:96–106.

Eaton, G. G.; Resko, J. A. (1974) Plasma testosterone and male dominance in a Japanese macaque *(Macaca fuscata)* troop compared with repeated measures of testosterone in laboratory males. *Hormones and Behavior* 5:251–259.

Eberhart, J. A.; Keverne, E. B. (1979) Influences of the dominance hierarchy on LH, testosterone and prolactin in male talapoin monkeys. *Journal of Endocrinology* 83:42–43.

Eberhart, J. A.; Keverne, E. B.; Meller, R. E. (1980) Social influences on plasma testosterone levels in male talapoin monkeys. *Hormones and Behavior* 14:247–266.

Elias, M. (1981) Serum cortisol, testosterone, and testosterone-binding globulin responses to competitive fighting in human males. *Aggressive Behavior* 7:215–224.

Fox, M. W.; Andrews, R. V. (1973) Physiological and biochemical correlates of individual differences in behavior of wolf cubs. *Behaviour* 46:129–140.

Frankel, A. I.; Ryan, E. L. (1981) Testicular innervation is necessary for the response of plasma testosterone levels to acute stress. *Biological Reproduction* 24:491–495.

Gordon, T. P.; Rose, R. M.; Bernstein, I. S. (1976) Seasonal rhythm in plasma testosterone levels in the rhesus monkey *(Macaca mulatta)*: A three year study. *Hormones and Behavior* 7:229–243.

Harding, C. F.; Feder, H. H. (1976) Relation between individual differences in sexual behavior and plasma testosterone levels in the guinea pig. *Endocrinology* 98:1198–1205.

Hayama, S. (1966) Correlation between adrenal gland weight and dominance rank in caged crab-eating monkeys. *Primates* 7:21–26.

Herndon, J. G.; Perachio, A. A.; Turner, J. J.; Collins, D. C. (1980) Fluctuations in testosterone levels of male rhesus monkeys during copulatory activity. *Physiology and Behavior* 26:525–528.

Jolly, A (1985) "The evolution of primate behavior," 2nd ed. New York: Macmillian.

Kamel, F.; Mock, E. J.; Wright, W. W.; Frankel, A. I (1975) Alterations in plasma concentrations of testosterone, LH, and prolactin associated with mating in the male rat. *Hormones and Behavior* 6:277–288.

Kaplan, J. R.; Manuck, S. B.; Clarkson, T. B.; Lusso, F. M.; Taub, D. M. (1982) Social status, environment, and atherosclerosis in cynomologus monkeys. *Arteriosclerosis* 2:359–368.

Katongole, C. B.; Naftolin, F.; Short, R. V. (1971) Relationship between blood levels of luteinizing hormone and testosterone in bulls, and the effects of sexual stimulation. *Journal of Endocrinology* 50:457–466.

Keller-Wood, M. E.; Dallman, M. F. (1984) Corticosteroid inhibition of ACTH secretion. *Endocrine Reviews* 5:1–24.

Keverne, E. B.; Meller, R. E.; Eberhart, A. (1982) Dominance and subordination: Concepts or physiological states. In Chiarelli, A. B., Corruccini, R. S. (eds.), "Advanced Views in Primate Biology." New York: Springer-Verlag. 81–94.

Klosterman, L. L.; Murai, J. T. Siiteri, P. K. (1985) Cortisol levels, binding, and properties of corticosteroid binding globulin in the serum of primates. *Endocrinology* 117:424–434.

Kuoppasalmi, K.; Naveri, H.; Rehunen, S.; Harkonen, M.; Adlerkreutz, H. (1976) Effect of strenuous anaerobic running exercise on plasma growth hormone, cortisol, luteinizing hormone, testosterone, androstenedione, estrone, and estradiol. *Journal of Steroid Biochemistry* 7:823–829.

LaFerla, J. J.; Anderson, D. L.; Schlach, D. S. (1978) Psychoendocrine response to sexual arousal in human males. *Psychosomatic Medicine* 40:166–172.

Landau, H. G. (1951) On dominance relations and the structure of animal societies: I. Effect of inherent characteristics. *Bulletin of Mathematical Biophysics* 13:1–19.

Louch, C. D.; Higginbotham, M. (1967) The relation between social rank and plasma corticosterone levels in mice. *General and Comparative Endocrinology* 8:441–444.

Manogue, K. R.; Leshner, A. I.; Candland, D. K. (1975) Dominance status and adrenocortical reactivity to stress in squirrel monkeys *(Saimiri sciureus). Primates* 16:457–463.

Max, S. R.; Toop, J. (1983) Androgens enhance in vivo 2-deoxyglucose uptake by rat striated muscle. *Endocrinology* 113:119–126.

Mendoza, S. P.; Coe, C. L.; Lowe, E. L.; Levine, S. (1979) The physiological response to group formation in adult male squirrel monkeys. *Psychoneuroendocrinology* 3:221–229.

Munck, A.; Guyre, P. M.; Holbrook, N. J. (1984) Physiological functions of glucocorticoids during stress and their relation to pharmacological actions. *Endocrine Reviews* 5:25–44.

Packer, C. (1979) Intertroop transfer and inbreeding avoidance in *Papio anubis. Animal Behaviour* 27:1–36.

Perachio, A. A. (1978) Hypothalamic regulation of behavioural and hormonal aspects of aggression and sexual performance. In Chivers, D., Herbert, J. (eds.), "Recent Advances in Primatology," vol. 1. New York: Academic Press. 549–565.

Popova, N.; Naumenko, E. (1972) Dominance relations and the pituitary-adrenal system in rats. *Animal Behaviour* 20:108–111.

Purvis, K.; Haynes, N. B.. (1974) Short-term effects of copulation, human chorionic gonadotropin injection and non-tactile association with a female on testosterone levels in the male rat. *Journal of Endocrinology* 60:429–439.

Rose, R. M. (1985) Psychoendocrinology. In Wilson, J.; Foster, D. (eds.), "Textbook of Endocrinology," 7th ed. Philadelphia: Saunders. 653–681.

Rose, R. M.; Bernstein, I. S.; Gordon, T. P. (1975) Consequences of social conflict on plasma testosterone levels in rhesus monkeys. *Psychosomatic Medicine* 37:50–61.

Rose, R. M.; Gordon, T. P.; Bernstein, I. S. (1972) Plasma testosterone levels in the male rhesus: Influences of sexual and social stimuli. *Science* 178:643–645.

Rose, R. M.; Holaday, J. W.; Bernstein, I. S. (1971) Plasma testosterone, dominance rank and aggressive behavior in male rhesus monkeys. *Nature* 231:366–368.

Rose, R. M.; Bernstein, I. S.; Gordon, T. P.; Catlin, S. F. (1974) Androgen and aggression: A review and recent findings in primates. In Holloway, R. (ed.), "Primate Aggression, Territoriality and Xenophobia." New York: Academic Press. 275–304.

Rowell, T. E. (1972) "Social Behaviour of Monkeys." Middlesex: Penguin.

Saginor, M.; Horton, R. (1968) Reflex release of gonadotropin and increases in plasma testosterone concentration in male rabbits during copulation. *Endocrinology* 82:627–630.

Sapolsky, R. M. (1982) The endocrine stress-response and social status in the wild baboon. *Hormones and Behavior* 16:279–292.

Sapolsky, R. M. (1983a) Endocrine aspects of social instability in the olive baboon *(Papio anubis)*. *American Journal of Primatology* 5:365–379.

Sapolsky, R. M. (1983b) Individual differences in cortisol secretory patterns in the wild baboon: Role of negative-feedback sensitivity. *Endocrinology* 113:2263–2267.

Sapolsky, R. M. (1985) Stress-induced suppression of testicular function in the wild baboon: Role of glucocorticoids. *Endocrinology* 116:2273–2278.

Sapolsky, R. M. (1986a) Endocrine and behavioral correlates of drought in the wild baboon. *American Journal of Primatology* 11:217–227.

Sapolsky, R. M. (1986b) Stress-induced elevation of testosterone concentrations in high-ranking baboons: Role of catecholamines. *Endocrinology* 118:1630–1635.

Sapolsky, R. M. (1989) Hypercortisolism among socially-subordinate wild baboons originates at the CNS level. *Archives of General Psychiatry* 46:1047–1051.

Sapolsky, R. M. (1990a) Stress in the wild. *Scientific American,* January. 116–123.

Sapolsky, R. M. (1990b) Adrenocortical function, social rank, and personality among wild baboons. *Biological Psychiatry* 28:862–878.

Sapolsky, R. M.; Krey L. C. (1988) Stress-induced suppression of luteinizing hormone concentrations in wild baboons: Role of opiates. *Journal of Clinical Endocrinology and Metabolism* 66:722–726.

Sapolsky, R. M.; McEwen, B. S. (1987) Why dexamethasone resistance? Two possible neuroendocrine mechanisms. In Schatzberg, A. F., Nemeroff, C. B. (eds.), "Hypothalamic-Pituitary-Adrenal Physiology and Pathophysiology." New York: Raven Press. 155–169.

Sapolsky, R. M.; Mott G. E. (1987) Social subordinance in a wild primate is associated with suppressed HDL-cholesterol concentrations. *Endocrinology* 121:1605–1610.

Shively, C.; Kaplan, J. (1984) Effects of social factors on adrenal weight and related physiology of *Macaca fascicularis.* Physiology and Behavior 33:777–782.

Southwick, C. H.; Bland, V. P. (1959) Effect of population density on adrenal glands and reproductive organs of CFW mice. *American Journal of Physiology* 197:111–114.

Sutton, J. R.; Coleman, M. J.; Casey, J.; Lazarus, L. (1973) Androgen responses during physical exercise. *British Medical Journal* 163:520–522.

Wahren, J.; Felig, P.; Ahlborg, G.; Jorfeldt, L. (1971) Glucose metabolism during leg exercise in man. *Journal of Clinical Investigations* 50:2715–2725.

Weiss, J. M. (1970) Somatic effects of predictable and unpredictable shock. *Psychosomatic Medicine* 32:397–408.

CHAPTER EIGHT

Temperament and Mother-Infant Conflict in Macaques: A Transactional Analysis

WILLIAM A. MASON, D. D. LONG, AND SALLY P. MENDOZA

Ordinary human experience suggests that people differ among themselves in the tendency to create social conflict, and in the ways in which they deal with conflictual situations. Although it is less widely appreciated, animals show similar contrasts, particularly when different species are compared. Questions regarding the sources, forms, generality, and consequences of such differences are extremely difficult to answer conclusively in the human case, in spite of their relevance to personality and social relationships. These questions are obviously more amenable to systematic study with animal subjects.

In this chapter, the focus is on conflicts between mothers and infants in two species of macaques that differ in temperament. Rhesus and bonnet mother-infant dyads were compared during the period from the infant's birth through its sixth month of life. The basic unit in the analysis was a social interchange or transaction between parent and offspring. The species did not

William A. Mason—Department of Psychology and California Regional Primate Research Center, University of California, Davis

D. D. Long—California Regional Primate Research Center, University of California, Davis

Sally P. Mendoza—Department of Psychology and California Regional Primate Research Center, University of California, Davis

Preparation of this chapter and the research reported herein was supported by the U.S. National Institutes of Health grants HD06367 and RR00169. We thank D. M. Lyons for comments on this manuscript.

differ in total number of transactions. They differed markedly, however, in the relative contributions of mothers and infants to the initiation of transactions and in the incidence of interpersonal conflict. Rhesus mothers initiated more transactions than did bonnet mothers, whereas bonnet infants initiated more transactions than rhesus infants. Conflict was more prominent in rhesus mother-infant dyads than in bonnet dyads, and this was largely a reflection of the more frequent resistance of rhesus mothers to transactions initiated by their infants. Thus, bonnet infants more often exercised the initiative in their relationship with their mothers, and fewer of these initiatives resulted in interpersonal conflict. It is reasonable to attribute these contrasts in mother-infant relations to broad differences between the species in temperament and, more generally, to consider that temperamental factors contribute to many of the differences in life-history patterns among closely related species of macaques.

The most complete and systematic information on the development of the mother-infant relationship in nonhuman primates has been obtained on the macaques. For general reviews of this topic, see Higley and Suomi 1986; Mason 1965; Pereira and Altmann 1985; Rosenblum 1971b. Although only a few of the sixteen or so species have been studied as often or as closely as the rhesus monkey (*Macaca mulatta*), the content and timing of major developmental epochs appear to be much the same within the entire genus.

During the first few days of postnatal life, the infant's movements are guided by such basic reflexes as clinging, rooting, and sucking. It remains in virtually continuous contact with its mother. These behaviors are complemented by the mother, who cradles, protects, transports, grooms, and in other ways contributes to the infant's safety and well-being.

The mobility of the infant increases rapidly over the first few weeks of life and, by one to three weeks, it is regularly attempting to move away from the mother. She responds to these early attempts by restraining or retrieving the infant, so that its forays are usually quite limited in distance and duration. Later, the infant spends an increasing amount of time away from its mother. These excursions are associated with more time spent interacting with the environment, particularly with other animals. By three months, the infant displays a varied repertoire of play patterns with other infants. While these changes are proceeding in the infant, the mother shows a corresponding reduction in protective behaviors. Over the same period, the rate of maternal rejections of the infant's attempts to maintain contact tends to increase.

Despite basic commonalities in the ontogeny of the mother-infant relationship among macaques, species differ considerably in the details of this developmental process. This was first described by Rosenblum and Kaufman in their comparative studies of two species of macaques, bonnets (*Macaca radiata*) and pigtails (*Macaca nemestrina*).

In the immediate postpartum period, pigtail mothers were more likely than bonnet mothers to withdraw from contact with other members of their social group, and to ward off attempts by others to direct attentions to their infant with threats and aggression. Subsequently, as the infant attempted to break contact and move away from its mother, pigtail mothers were more likely to restrain their infants or to retrieve them. Somewhat later in development, however, pigtail mothers were also more forceful and aggressive in their efforts to prevent their infants from maintaining contact with them (Kaufman and Rosenblum 1969; Rosenblum and Kaufman 1967). Thus, at each stage of development, pigtail mothers played a more active and assertive role in the relationship than did the relatively passive bonnet mothers.

Similar interspecific contrasts are reported by Thierry (1985b), based on comparisons of mother-infant relations in rhesus (*Macaca mulatta*), crabeaters (*Macaca fascicularis*), and Tonkean macaques (*Macaca tonkeana*) during the first ten weeks following birth. Tonkean mothers did not initiate contacts with their infants as frequently as did mothers of the other two species, and they prevented the infants from interacting with other members of the group less often. They are described as permissive, in contrast to the more restrictive and protective rhesus and crabeater mothers.

Interpersonal conflict is inevitable in any ongoing, complex relationship, and it is particularly likely to occur in relationships that change rapidly, as between mother and infant. It is curious, however, that conflict appears to be more prominent in some species of macaques than others. We have no adequate explanation for the sources of these differences. They may reflect such psychosocial attributes as the persistence of mothers and infants in the pursuit of their individual goals; how far each member of the dyad goes in imposing its will on the other, and the degree to which compliance occurs freely, under duress, or not at all. Species may also differ in the extent to which mother and infant are engaged in the relationship, in the occasions that engender conflict, or in the manner in which conflict is expressed or resolved. Other possibilities might be imagined.

QUALITATIVE ASPECTS OF MOTHER-INFANT RELATIONSHIPS: METHODOLOGICAL CONCERNS

The differences between macaque species in the relationship between mother and infant appear to have a large qualitative component. If we are to achieve a better understanding of the factors that contribute to differences among macaque species in mother-infant conflict, it seems necessary to focus more systematically and directly on this aspect of their relationship. Although the quality of an interaction is readily observable, it is poorly conveyed by quantitative analyses of discrete behavior categories derived from a species-specific ethogram (Hinde and Simpson 1975). For example, a record of the number of times that a mother cradles her infant provides no information on the quality of the act or about the larger social interchange of which it is a part. As a category of maternal behavior, the act of cradling may be initiated by the mother without regard to what the infant is seeking, or it may be a response to the perceived intention of the infant. It may also be accomplished gently or forcefully. Likewise, the infant may seek to be cradled. It may also accommodate willingly if the mother initiates the act, or it may resist her. These qualitative aspects of the interchange between mother and infant are the best source of information on the affective tone of the interchange, including the presence and source of interpersonal conflict. Furthermore, the quality of the interchange is probably more significant to the participants and has a more important effect on their relationship than the discrete behavior categories by which parent-offspring interactions are usually described.

Based on such considerations, we reasoned that a new method

was needed to describe the differences in mother-infant conflict among species of macaques. Instead of basing conclusions on a large number of specific behavior categories and indices derived from calculations after the fact, we elected to use the entire interchange or social episode itself as our basic unit of analysis. This unit—the transaction—is based on the assumption that social interactions almost always include an element of expectancy or intent. It can usually be inferred that the individual initiating a transaction is seeking a particular outcome from the recipient that impinges on the latter's interests or ongoing activities. The recipient's response to a social overture may allow the actor to attain what it seeks, or it may prevent this from happening. This formulation provides the basic tripartite structure of the transactional unit—Initiation, negotiation, and outcome.

Here, we present the results of a first attempt to develop a transactional approach to mother-infant conflict in a comparative study of rhesus and bonnet macaques. We assumed that much of the behavior that goes on between mother and infant represents a type of steady state. For example, an infant may sit for long periods encircled in its mother's arms. Although such information is pertinent to describing important aspects of their relationship, it has little to say about the dynamic qualities that are the greatest potential sources of interpersonal conflict. We reasoned that conflict was most likely to occur when one individual (the actor) sought to change its ongoing association with the other (the target or recipient). If this change required a complementary response from the target or elicited a response directed toward the actor, the episode was scored as a transaction. The response of the target was considered either to accept and/or accommodate to the sought-after change in association, or to resist it. In the latter case, the transaction was conflictual.

SUBJECTS AND PROCEDURES

Data on mother-infant conflict were based on observations of twelve rhesus (*Macaca mulatta*: six males; six females), eleven bonnet (*Macaca radiata*: eight males; three females) and four stumptail (*Macaca arctoides*: one male; three females) infants and their mothers. All animals lived in social groups with fifty to sixty other animals in outdoor enclosures (0.2 hectare). Over the first six months of life, mother–infant pairs were observed three times a week for the first twelve weeks, twice a week during weeks thirteen to sixteen, and once a week thereafter through week twenty-six. The observation period for each dyad was ten minutes, and observations were balanced between mornings and afternoons.

The data were collected by two individuals, one serving as observer and the other as recorder. Any aspect of the ongoing relationship between mother and infant that persisted for ten seconds or more was termed an association. A transaction was initiated when one member of the dyad attempted to change an association in ways that either required or elicited some response from the target. Among the most common behavioral elements in a transaction were suckle, cling, embrace, groom, restrain, retrieve, and reject.

Some of these categories were represented by specific behavior patterns. Others encompassed several discrete behaviors. However, except with reference to the entire interchange in which it occurred, the form of the behavior was not the critical datum. Any of the foregoing behaviors (and others) could serve to initiate a transaction if they changed an association, but they might also occur in response to a transaction initiated by another.

For example, if an infant was sitting near its mother and she retrieved it upon the approach of a third animal, the transaction was considered to be initiated by the mother. If the infant responded to this by clinging to the mother, the transaction was considered to be without conflict. On the other hand, in similar circumstances, if the infant moved away from the mother toward an approaching animal and was retrieved by the mother, the infant was considered to have initiated the transaction by a change in its association with the mother. Because the mother resisted the infant's attempt to depart, the transaction was conflictual. It was possible in most instances to decide which animal initiated the transaction, and to classify the target's response as one of accommodation to the change or resistance. The transaction was considered to end with a return to a steady state, which was defined as ten seconds with no change in the association. In the following analysis, we will focus on the frequency of transactions initiated by the mother and by the infant, and on the incidence of transactions that were conflictual. Detailed analysis is limited to the bonnet and rhesus dyads because of the small number of stumptail subjects.

COMMONALITIES IN THE DEVELOPMENT OF
THE MOTHER-INFANT RELATIONSHIP

We consider first the similarities between species. Figure 8.1 shows mean total frequencies of transactions across the entire period of observation. (The data for stumptail dyads are included for comparison only in this figure.) The pattern was similar across the three species: Transactions were relatively infrequent during the first two weeks of life, increased over the next four to six weeks, then declined gradually.

Bonnet and rhesus dyads did not differ in the total number of transactions. This was assessed in two ways. One way was to compare the number of ten-minute observation periods in which transactions occurred during the first twelve weeks, when observations were made three times per week. In both species, transactions were recorded on more than 70% of the periods, and they did not differ significantly. A second assessment compared the mean frequency of transactions per observation period across the entire twenty-six weeks of the study. Although frequencies were slightly higher in rhesus dyads (3.03 versus 2.57 per ten-minute period), this difference was also not statistically significant (all reported statistical outcomes are based on t-tests). In both species, the frequency of conflictual transactions was relatively low. More than 85% of all transactions were without conflict (p <.001 based on frequencies). The percentage of total transactions that were conflictual tended to increase with age, but at no point did they exceed 35%.

Similarities were also evident in the relative contributions of mothers and infants. Figure 8.2 shows the percentage of total transactions for the dyad that were initiated by infants. In both species, infants initiated significantly more transactions than did mothers (p <.001). This difference was evident as early as the first two weeks of

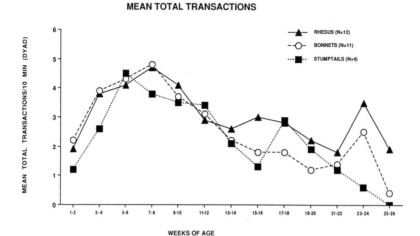

Figure 8.1   Mean total transactions between mothers and infants during 10-min observation periods.

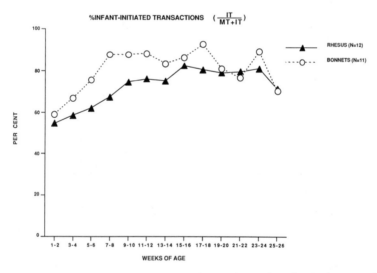

Figure 8.2   Percent of total transactions for mother-infant dyads that were initiated by infants.

life, increased progressively through the end of the second or third month, and remained fairly stable thereafter. Overall, more than 70% of all transactions were initiated by infants. It was also the case that the majority of conflictual transactions in both species (more than 80%) was precipitated by mothers in response to transactions initiated by infants (p <.05 based on frequencies).

CONTRASTS BETWEEN SPECIES

In spite of broad similarities between bonnets and rhesus in the development and patterning of mother-infant transactions, there were notable differences in the quality of mother-infant relationships in the two species. Conflict was clearly a more prominent feature of the relationship in rhesus. Mean frequency of conflictual transactions was significantly higher in rhesus dyads (mean per ten-minute period = 0.42 versus 0.16, p <.01). The difference between species was also significant when the comparison was based on the percentage of total transactions that were conflictual (13.8% versus 6.4%, p <.01).

Further contrasts became evident when we examined the separate contributions of mothers and infants. Comparison of rhesus and bonnet mothers indicated that rhesus mothers initiated more transactions than did bonnet mothers. During the first twelve weeks they initiated transactions on 48.6% of the observation periods, as compared to 36.1% for bonnets (p <.05). Overall, the mean frequency of transactions initiated by mothers was significantly higher for rhesus (mean per ten-minute period = 0.85 versus 0.49, p <.01). Separate comparisons of conflictual and nonconflictual transactions initiated by mothers were also significant (mean per ten-minute period, rhesus versus bonnets: conflictual =.07 versus .02, p <.01; nonconflictual =.78 versus .47, p <.05). However, even though rhesus mothers initiated more conflictual transactions in terms of absolute frequencies, the percentage of total transactions initiated by the mother that were conflictual did not differ reliably between the species. Values were somewhat higher for the rhesus (8.5% versus 4.8%), but the overlap between species was considerable. Thus, the strongest difference between rhesus mothers and bonnet mothers in this analysis is seen in the greater tendency of rhesus mothers to initiate transactions with their infants. The proportion of these transactions that resulted in conflict, however, did not reliably differentiate the species.

Comparisons of infant-initiated transactions between the two species were also instructive. Bonnet infants did not differ reliably from rhesus infants in the number of transactions they initiated (mean per ten-minute period: rhesus = 2.18; bonnets = 2.08). In relation to total transactions for the dyad, however, the percentage of transactions initiated by bonnet infants was reliably higher than the comparable value for rhesus infants (81.3% versus 71.1%, p <.05). As can be seen in figure 8.2, this difference was most consistent during the first four to five months of life. Transactions initiated by rhesus infants were also more likely to result in conflict than were transactions initiated by bonnet infants. This was the case whether comparisons were based on mean frequencies of conflictual transactions (rhesus = 0.35 versus bonnet = 0.14, p <.01), or percentage of total transactions initiated by the infant that were conflictual (rhesus = 16.0% versus bonnets = 6.6%, p <.01. See figure 8.3).

It thus appears that the primary differences between infants of the two species were that proportionately more transactions were initiated by bonnet infants than by rhesus infants, and that, of infant-initiated transactions, proportionately more of those initiated by rhesus infants were conflictual. Rhesus infants were apparently more likely than bon-

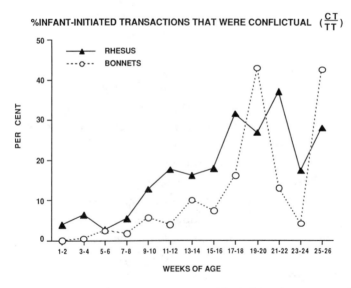

Figure 8.3   Percent of transactions initiated by infants that were conflictual.

net infants to meet some resistance when they attempted to change their association with the mother, whereas the infants themselves were about equally likely to resist their mothers' attempts to change their association with them. This interpretation is supported by a further analysis in which the percentage of total transactions initiated by the infant that produced conflict was compared with the percentage of total transactions initiated by the mother that did so. In the bonnets, these values did not differ reliably between infants and mothers (infants = 6.6% versus mothers = 4.8%); whereas for rhesus, transactions initiated by infants were reliably more likely to produce conflict than those initiated by mothers (infants = 16.0% versus mothers = 8.5%, p <.01).

METHODOLOGICAL COMPARISONS

The transactional approach was planned as an adjunct to traditional methods of studying social relationships rather than a substitute for them. As such, it is expected to provide data that are different from those obtained with traditional methods, but complemental to them. This seems to be the case.

When our results are compared with data based on specific behavior categories, we find several areas of clear agreement, as well as some new information. In agreement is the overall downward trend in the frequency of transactions after the first few weeks of life. This corresponds well with the decreases in approaches, physical contact and proximity that are generally reported for macaques (Berman 1980; Hansen 1966; Hinde 1974; Kaufman and Rosenblum 1969).

The relatively low frequencies of transactions in our observations during the first two weeks are not directly reflected in measures of specific behaviors, but can be readily reconciled with these data. During this initial period, mother and infant are in virtually continuous ventro-ventral contact—a steady state. Hence, few transactions are initiated.

Another point of similarity with data based on specific behaviors is the finding that infants increasingly exercised the initiative in their relations with mothers. In apparent contrast to most of these data, however, our findings suggest that this was the case—although weakly so—even during the first two weeks of life. We suspect that this reflects the explicit concern of the transactional method with the identity of the initiator of an interchange.

For example, nipple contacts are always made by the infant. If this was the first behavior occurring after a steady state, it was scored as initiating a transaction. However, if it occurred in response to being moved into the cradle position by the mother it was considered to be an accommodation to the mother's attempt to change the association. Infants also attempt to leave the mother very early in life, even though the mother may prevent them from doing so. Attempts were explicitly recognized in our method and, whether they were thwarted or not, were considered to initiate a transaction.

Our data indicate that the general decline in frequency of transactions over the first six months of life was accompanied by an increase in the percentage of total transactions that were conflictual. After the age of two months, virtually all conflictual transactions were initiated by the infant, indicating that the infant's attempts to change its association with the mother were increasingly met with some form of resistance. This corresponds roughly with the pattern reported by other investigators. For example, Berman compared mother-infant relations between free-ranging rhesus on Cayo Santiago and captive social groups at Madingley during the first thirty weeks of life. In both settings, time on the mother's ventrum and on the nipple declined with age, whereas the relative and absolute frequencies of maternal rejections of nipple contacts increased (Berman 1980).

No previous studies have made direct comparisons of the development of mother-infant relations in bonnets and rhesus. Based on the research of Rosenblum and Kaufman, we expected that interpersonal conflict would be more prominent in rhesus than in bonnets, and this proved to be the case. The particular pattern of similarities and differences between rhesus and bonnets revealed by the transactional record, however, were not anticipated.

Contrary to what might be expected, the overall frequency of transactions during the first six months of life did not differentiate species. Although the level was slightly higher in rhesus, the difference did not approach statistical significance. Based on these findings, we can say that differences between the two species were not reflected in the number of occasions in which one member of the dyad initiated some action in which the other became involved. The more limited data on stumptails are consistent with this conclusion.

The principal contrasts between rhesus and bonnets emerged when the structure and quality of transactions were examined. A comparison between mothers yielded several unexpected findings. Rhesus mothers were more likely than bonnet mothers to initiate transactions with their infants. This was true for transactions that produced no conflict, as well as for those that did. The proportion of transactions initiated by rhesus mothers that were accompanied by conflict, however, did not differ reliably from the comparable value for bonnet mothers. Thus, rhesus mothers appeared to show more overall initiative toward their infants, but the transactions they initiated were no more likely to be resisted by the infants than those initiated by bonnet mothers.

Transactions initiated by infants also differentiated species in ways that we did not anticipate. Bonnet infants initiated a reliably larger proportion of total transactions than did rhesus infants, in spite of the fact that, in absolute terms, rhesus infants initiated transactions somewhat more frequently than did bonnet infants. Moreover, a significantly smaller proportion of the transactions initiated by bonnet infants was accompanied by conflict, as compared to those initiated by rhesus infants. In further contrast, the relative contributions of parent and offspring to conflictual transactions were closer to parity in the bonnet group. The proportion of conflictual transactions initiated by bonnet infants was only slightly (and nonsignificantly) higher than the proportion of conflictual transactions initiated by their mothers. However, rhesus infants initiated a reliably greater proportion of conflictual transactions than did their mothers. Thus, in relative terms, bonnet infants more often exercised the initiative than did rhesus infants, and fewer of these initiatives were resisted by the mother.

CHARACTERIZING SPECIES DIFFERENCES: TEMPERAMENT

Similar contrasts to those between bonnet and rhesus mother-infant dyads are suggested for other macaque species. It is likely that these differences extend well beyond the parent-offspring relationship. In this section, findings on mother-infant dyads are considered within a broader perspective on the proximal sources of interspecies contrasts in responsiveness.

In ordinary circumstances, the mother-infant dyad does not exist as an isolated unit, but within a network of ongoing relationships. The birth of a baby is a significant event, not only for the mother but for her previous immature offspring and for other members of the group. The reactions of these individuals to the infant—and the mother's responses to them—are potential sources of conflict that impinge on the mother-infant relationship. The few comparative data available provide a clear suggestion that species differ in the mother's response to such intrusions. Pigtail, rhesus, and crabeater mothers are apparently more likely than their bonnet or Tonkean counterparts to prevent others from interacting with their infants. Consequently, bonnet and Tonkean infants have earlier, more frequent, and more varied contacts with animals other than their mothers (Kaufman and Rosenblum 1969; Rosenblum 1971a; Thierry 1985b). It thus appears that the tolerance and permissiveness ascribed to bonnet and Tonkean mothers—in contrast to the diligence, protectiveness and apprehension attributed to the other species—refers not only to the mother-infant relationship, but also to the mother's reactions to other members of the group.

Furthermore, it is clear that interspecies contrasts in the tendency to create interpersonal conflict are not dependent on the special relationship between mother and infant. On the contrary, they appear in virtually every facet of social life. For example, Hawkes (1970) compared matched, newly formed, mixed-sex groups of rhesus, bonnet, crabeater, and stumptail macaques. The bonnet group scored highest of the four species in social contact and grooming, and lowest in the frequency of agonistic behavior. Hawkes also found that rhesus showed the highest level of agonistic behavior of the four species on the first day of group formation and a significant increase in the daily percentage of agonistic behavior after day one. In contrast, scores for agonistic behavior in the bonnet group showed a significant daily decrease. Hawkes' findings for the bonnets are in keeping with the relatively high levels of social tolerance in this species noted by many other investigators (Brandt et al. 1970; Caine and Mitchell 1979, 1980;

Rahaman and Parthasarathy, 1969; Shively et al. 1982; Simonds 1965; Small 1982; Srinath 1980; Sugiyama 1971). Similar interspecies contrasts have been noted in other studies based on direct comparisons. Tonkean macaques show less intense aggression than do crabeaters. The rhesus exceeds both species in this dimension (Thierry 1985a; Zumpe and Michael 1983).

Differences in aggressivity are also evident in responses to humans. Rhesus have the reputation of being pugnacious and difficult to handle, as compared to other species of macaques (Kling and Orbach 1963; MacDonald 1971; Orbach and Kling 1964). In formal tests, rhesus were much more likely than bonnets to direct threats and other agonistic behaviors toward a passive human observer, and less likely to show friendly behavior (Clarke and Mason 1988).

Differences between species are, by no means, restricted to agonistic interactions. For example, contrasts have been reported in social development. Pigtail infants spend less time in social play than do bonnet infants, and they are more likely to play with members of their own family groups (Kaufman 1975; Rosenblum 1971a). Juvenile rhesus females show more interest in infants than do bonnet females (Caine and Mitchell 1980). During their first two years, bonnet females extend their social network by interacting with an increasing number of female peers, whereas young rhesus females show the opposite trend (Caine and Mitchell 1980). Species also vary in aspects of adult social behavior and social relationships that are linked only indirectly, if at all, to agonistic conflict. These include interanimal distance, grooming, copulatory patterns, the salience of kinship and rank in social dynamics, and the frequency and quality of adult males' involvement with young.

Such variations cannot readily be attributed to differences in behavioral repertoires, which are closely similar in all macaques. Moreover, there are clear suggestions that, considered as a whole, the particular attributes of each species form a consistent pattern. As Rosenblum and Kaufman noted with respect to bonnets and pigtails, these species appear to differ in the "whole tone of social behavior" (Rosenblum and Kaufman 1967, 35).

In fact, the profile of differences among macaques is even

broader than this conclusion suggests. Differences have been reported for visual monitoring, manipulation of the inanimate environment, motor activity, responses to food and food deprivation, sleeping postures, and adaptation to novel situations and procedures (Caine et al. 1981; Caine and Mitchell 1980; Clarke et al. 1988a, 1988b; Davis et al. 1968; Glickman and Sroges 1966; Rosenblum et al. 1964, 1969; Schrier 1965; Symmes and Anderson 1967).

It is of particular interest that macaque species also differ in the activity of the pituitary-adrenal and autonomic nervous systems (Clarke 1985; Clarke et al. 1988a, 1988b; Kling and Orbach 1963). Both systems are known to be responsive to environmental events, and to have direct effects on the organism's allocation of resources. They influence normal activity patterns, prepare the individual to act, and affect the type or quality of the action it takes.

In summary, there is no serious question that broad differences exist among macaque species. To be sure, almost all direct comparisons have been based on no more than two or three species. Nevertheless, there are a fair number of instances in which the same species have been compared by different investigators over a range of situations. Although the evidence for the generality of differences between a given set of species is incomplete, it is impressive. The pattern of contrasts is evident in physiology and in behavior. Differences are based on direct quantitative comparisons between species observed under essentially identical conditions. There is good reason to suppose, therefore, that they constitute reliable and significant distinguishing features of each species. Moreover, the apparent consistency of the pattern of differences across age/sex-classes, and across situations suggests that the underlying processes are operating at a fairly basic integrative level.

The term usually applied to such broad dispositional attributes is temperament (Allport 1937; Goldsmith et al. 1987; Rothbart and Derryberry 1981). Although generally reserved for discussions of differences between individuals, temperament is the most plausible conceptual device for linking the different profiles of macaque species to an appropriate functional level. Temperament refers to an individual's characteristic stance toward the world, as well as to general and abiding dispositions toward certain modes of responding that are manifest over a wide range of situations, particularly those that constitute a significant departure from the status quo.

The transaction is a departure from the status quo, and thus represents an appropriate level of analysis for the social manifestations of temperament. As we have shown, the initiation of a social transaction more often meets with resistance from rhesus than from bonnets. Our thesis is that differences between these species in the structure of social transactions and the incidence of interpersonal conflict in the mother–infant dyad reflect differences in temperament. This interpretation is consistent with our finding that rhesus mothers initiated more transactions than did bonnet mothers. It also provides a reasonable explanation for the apparent paradoxical tendency for high levels of maternal diligence or protectiveness early in development to be associated with high levels of rejection or punitive deterrence at later developmental stages. Furthermore, we suggest that temperamental influences are not peculiar to the special relationship between mother and infant, but extend to virtually every aspect of social behavior, including contrasts between the two species in the frequency and severity of interpersonal conflict.

CONCLUSIONS

Interpersonal conflict is bound to emerge in any continuing multifaceted relationship, and it is particularly likely to occur in those that are undergoing change, such as the relationship between mother and child. In this instance, the sources of change reside in both parties. On theoretical grounds it can be argued that the social agenda of mother and infant are not wholly congruent, and that occasions for discord will increase as the infant moves toward greater self-sufficiency and the mother prepares to resume her normal reproductive cycle (Trivers 1974. See also Silk chapter 3, this volume). This pattern is characteristic of most mammals, of course, and the primates in the genus *Macaca* are no exception.

Nevertheless, it is a matter of considerable interest that significant variations exist within the genus in the frequency and intensity of interpersonal conflict between mother and infant. Such differences were first demonstrated by Rosenblum and Kaufman in their comparative studies of bonnets and pigtails (Rosenblum and Kaufman 1967). Subsequent comparisons of rhesus, crabeaters, and Tonkean macaques (Thierry 1985b), and our own findings on rhesus and bonnets, also show reliable interspecies differences in the form and frequency of mother-infant conflict.

We interpret our findings on mother-infant conflict as part of a broader profile of interspecies contrasts indicative of differences in temperament. Temperament refers to a level of psychobiological organization that modulates behavior rather than determining it directly. In closely related species, differences in temperament will not be manifest in the specific motor patterns that make up behavioral repertoires, but in such general parameters as thresholds, latencies, and the tempo, vigor, and direction of an individual's responses to environmental events. These parameters seem to differentiate rhesus and bonnets in most aspects of their social lives, and they are reflected in physiological responsiveness as well as behavior.

More generally, the concept of temperament offers promise of helping to explain the proximal sources of the differences among various species of macaques in life-history patterns. Characteristics of sexual behavior, male involvement with young, intermale tolerance, the incidence of social conflict, and the salience of kinship and rank in social dynamics tend to co-vary within species. So, too, do such sociodemographic attributes as adult sex ratios, rate of emigration, and group size (Caldecott 1984, 1986; Fooden 1982, 1986; Shively et al. 1982). It is reasonable to suppose that the covariations within species and the contrasts between them are heavily influenced by the sort of basic predispositions that the concept of temperament implies.

These predispositions are not invariant in spite of their persistence and generality. In fact, the evidence suggests that major differences in temperament can be brought about by early experiential influences in rhesus monkeys (Mason 1978; Mason and Capitanio 1988; Singh 1969; Wood et al. 1979. Similar effects have been produced with other mammalian species (Denenberg 1970). Experientially induced modifications in temperament may also be one factor contributing to intraspecific variations in social systems, at least in mammals, in which a major accompaniment of change is in the degree of tolerance toward unrelated conspecifics (Lott 1984).

The susceptibility of temperament to the influences of early experience raises some intriguing possibilities concerning the tempo and mode of evolutionary change between closely related species. Ontogenetic plasticity of temperament can be regarded as a preadaptation. Given the strong likelihood of genetically based variations in temperament among individuals within a population, selection for a particular variant could presumably proceed very quickly since no fundamental organismic changes are required at any functional level. The result, however, would be a population that showed significant departures from the parent population in many aspects of its social behavior and ecology, while retaining a high degree of similarity to the parental stock in its genetic make-up, its behavioral repertoire, and its basic physiology. Variations in qualities of temperament could influence the suites of socioecological variables that appear to differentiate macaque species.

In this context, it is significant that a parallel process appears to occur during artificial selection and domestication. Phenotypic change occurs rapidly, and is reflected at the behavioral level in alterations in thresholds and responsiveness to the environment rather than modifications in motor patterns (Price 1984). These are precisely the types of changes most clearly associated with the concept of temperament. It is reasonable to consider that similar changes have been brought about by natural selection on more than one occasion.

REFERENCES

Allport, G. W. (1937) "Personality: A Psychological Interpretation." New York: Henry Holt.

Berman, C. M. (1980) Mother-infant relationships among free-ranging rhesus monkeys on Cayo Santiago: A comparison with captive pairs. *Animal Behaviour* 28:860–873.

Brandt, E. M.; Irons, R.; Mitchell, G. (1970) Paternalistic behavior in four species of macaques. *Brain, Behavior and Evolution* 3:415–420.

Caine, N.; Mitchell, G. (1979) The relationship between maternal rank and companion choice in immature macaques (*Macaca mulatta* and *M. radiata*). *Primates* 20:583–590.

Caine, N. G.; Mitchell, G. (1980) Species differences in the interest shown in infants by juvenile female macaques (*Macaca radiata* and *M. mulatta*). *International Journal of Primatology* 1:323–332.

Caine, N. G.; Caine, C.; Davidson, C.; Maddock, J.; Thompson, V.; Mitchell, G. (1981) Extra-troop orientation in captive macaques. *Biology of Behaviour* 6:119–128.

Caldecott, J. (1984) Coming of age in Macaca. *New Scientist* 101:10–12.

Caldecott, J. O. (1986) Mating patterns, societies and the ecogeography of macaques. *Animal Behaviour* 34:208–220.

Clarke, A. S. (1985) Behavioral, cardiac, and adrenocortical responses to stress in three macaque species (*Macaca mulatta*, *Macaca radiata*, and *Macaca fascicularis*). Doctoral dissertation. University of California at Davis.

Clarke, A. S.; Mason, W. A. (1988) Differences between three macaque species in responsiveness to an observer. *International Journal of Primatology* 9:347–364.

Clarke, A. S.; Mason, W. A.; Moberg, G. P. (1988a) Differential behavioral and adrenocortical responses to stress among three macaque species. *American Journal of Primatology* 14:37–52.

Clarke, A. S.; Mason, W. A.; Moberg, G. P. (1988b) Interspecific contrasts in responses of macaques to transport cage training. *Laboratory Animal Science* 38:305–309.

Davis, R. T.; Leary, R. W.; Smith, M. D. C.; Thompson, R. F. (1968) Species differences in the gross behavior of nonhuman primates. *Behaviour* 31:326–338.

Denenberg, V. H. (1970) Experimental programming of life histories and the creation of individual differences: A review. In Jones, M. R. (ed.), "Miami Symposium on the Prediction of Behavior, 1968: Effects of Early Experience." Coral Gables, Fla.: University of Miami Press. 61–91.

Fooden, J. (1982) Ecogeographic segregation of macaque species. *Primates* 23:574–579.

Fooden, J. (1986) Taxonomy and evolution of the Sinica group of macaques: 5. Overview of natural history. *Fieldiana Zoological Publication* 1367, 29:1–22.

Glickman, S. E.; Sroges, R. W. (1966) Curiosity in zoo animals. *Behaviour* 26:151–188.

Goldsmith, H. H.; Buss, A. H.; Plomin, R.; Rothbart, M. K.; Thomas, A.; Chess, S.; Hinde, R. A.; McCall, R. B. (1987) Roundtable: What is temperament? Four approaches. *Child Development* 50:505–529.

Hansen, E. W. (1966) The development of maternal and infant behavior in the rhesus monkey. *Behaviour* 27:107–149.

Hawkes, P. N. (1970) Group formation in four species of macaques in captivity. Doctoral dissertation. University of California at Davis.

Higley, J. D.; Suomi, S. J .(1986) Parental behaviour in non-human primates. In Sluckin, W., Herbert, M. (eds.), "Parental Behaviour." London: Basil Blackwell. 152–207.

Hinde, R. A. (1974) Mother/infant relations in rhesus monkeys. In White, N. F. (ed.), "Ethology and Psychiatry." Toronto: University of Toronto Press. 29–46.

Hinde, R. A.; Simpson, M. J. A. (1975) Qualities of mother–infant relationships in monkeys. *Ciba Foundation Symposium* 33 (new series). 39–67.

Kaufman, I. C. (1975) Learning what comes naturally: The role of life experience in the establishment of species typical behavior. *Ethos* 3:129–142.

Kaufman, I. C.; Rosenblum, L. A. (1969) The waning of the mother–infant bond in two species of macaque. In Foss, B. (ed.), "Determinants of Infant Behaviour IV." London: Methuen & Co., Ltd. 41–59.

Kling, A.; Orbach, J. (1963) Plasma 17-hydroxycorticosteroid levels in the stump-tailed monkey and two other macaques. *Psychological Reports* 13:863–865.

Lott, D. F. (1984) Intraspecific variation in the social systems of wild vertebrates. *Behaviour* 88:266–325.

MacDonald, G. J. (1971) Reproductive patterns of three species of macaques. *Fertility and Sterility* 22:373–377.

Mason, W. A. (1965) The social development of monkeys and apes. In DeVore, I. (ed.), "Primate Behavior: Field Studies of Monkeys and Apes." New York: Holt. 514–543.

Mason, W. A. (1978) Social experience and primate cognitive development. In Burghardt, G. M., Bekoff, M. (eds.), "The Development of Behavior: Comparative and Evolutionary Aspects." New York: Garland Press. 233–251.

Mason, W. A.; Capitanio, J. P. (1988) Formation and expression of filial attachment in rhesus monkeys raised with living and with inanimate mother substitutes. *Developmental Psychobiology* 21:401–430.

Orbach, J.; Kling, A. (1964) The stumped-tail macaque: A docile Asiatic monkey. *Animal Behaviour* 12:343–347.

Pereira, M. E.; Altmann, J. (1985) Development of social behavior in free-living nonhuman primates. In Watts, E. S. (ed.), "Nonhuman Primate Models for Human Growth and Development." New York: Alan R. Liss, Inc. 217–309.

Price, E. O. (1984) Behavioral aspects of animal domestication. *Quarterly Review of Biology* 59:1–32.

Rahaman, H.; Parthasarathy, M. D. (1969) Studies on the social behaviour of bonnet monkeys. *Primates* 10:149–162.

Rosenblum, L. A. (1971a) Infant attachment in monkeys. In Schaffer, H. K.

(ed.), "The Origins of Human Social Relations." New York: Academic Press. 85–109.

Rosenblum, L. A. (1971b) The ontogeny of mother-infant relations in macaques. In Moltz, H. (ed.), "The Ontogeny of Vertebrate Behavior." New York: Academic Press. 315–367.

Rosenblum, L. A.; Kaufman, I. C. (1967) Laboratory observations of early mother-infant relations in pigtail and bonnet macaques. In Altman, S. A. (ed.), "Social Communication among Primates." Chicago: University of Chicago Press. 33–41.

Rosenblum, L. A.; Clarke, R. W.; Kaufman, I. C. (1964) Diurnal variations in mother-infant separation and sleep in two species of macaque. *Journal of Comparative and Physiological Psychology* 58:330–332.

Rosenblum, L. A.; Kaufman, I. C.; Stynes, A. J. (1969) Interspecific variations in the effects of hunger on diurnally varying behavior elements in macaques. *Brain, Behavior and Evolution* 2:119–121.

Rothbart, M. K.; Derryberry, D. (1981) Development of individual differences in temperament. In Lamb, M.; Brown, A. L. (eds.), "Advances in Developmental Psychology," vol. 1. Hillsdale, N.J.: Lawrence Erlbaum Associates. 37–86.

Schrier, A. M. (1965) Pretraining performance of three species of macaque monkeys. *Psychonomic Science* 3:517–518.

Shively, C.; Clarke, S.; King, N.; Schapiro, S.; Mitchell, G. (1982) Patterns of sexual behavior in male macaques. *American Journal of Primatology* 2:373–384.

Simonds, P. E. (1965) The bonnet macaque in South India. In DeVore, I. (ed.), "Primate Behavior: Field Studies of Monkeys and Apes." New York: Holt, 175–176.

Singh, S. D. (1969) Urban monkeys. *Scientific American* 221:108–115.

Small, M. F. (1982) A comparison of mother and nonmother behaviors during birth season in two species of captive macaques. *Folia Primatologica* 38:99–107.

Srinath, B. R. (1980) Husbandry and breeding of bonnet monkeys (*Macaca radiata*). In Anand-Kumar, T. C. (ed.), "Non-Human Primate Models for Study of Human Reproduction." Basel: Karger. 17–22.

Sugiyama, Y. (1971) Characteristics of the social life of bonnet macaques (*Macaca radiata*). *Primates* 12:247–266.

Symmes, D.; Anderson, K. V. (1967) Comparative observation on *Macaca speciosa* and *Macaca mulatta* as laboratory subjects. *Psychonomic Science* 7:89–90.

Thierry, B. (1985a) Patterns of agonistic interactions in three species of macaque (*Macaca mulatta, M. fascicularis, M. tonkeana*). *Aggressive Behavior* 11:223–233.

Thierry, B. (1985b) Social development in three species of macaque (*Macaca mulatta, M. fascicularis, M. tonkeana*): A preliminary report on the first ten weeks of life. *Behavioral Processes* 11:89–95.

Trivers, R. L. (1974) Parent-offspring conflict. *American Zoologist* 14:249–266.

Wood, B. S.; Mason, W. A.; Kenney, M. D. (1979) Constrasts in visual responsiveness and emotional arousal between rhesus monkeys raised with living and those raised with inanimate substitute mothers. *Journal of Comparative and Physiological Psychology* 93:368–377.

Zumpe, D.; Michael, R. P. (1983) A comparison of the behavior of *Macaca fascicularis* and *Macaca mulatta* in relation to the menstrual cycle. *American Journal of Primatology* 4:55–72.

CHAPTER NINE

Impact of Foraging Demand on Conflict within Mother-Infant Dyads

MICHAEL W. ANDREWS, GAYLE SUNDERLAND,
AND LEONARD A. ROSENBLUM

Intraindividual conflicts frequently give rise to interpersonal conflicts. For example, in nature, animals must successfully meet the often conflicting demands of social activities and foraging requirements. The outcome of this strategic conflict regarding the use of time and energy may generate interpersonal conflicts which could have long-range consequences.

In the case of mothers and their dependent offspring, social conflicts resulting from a mother's foraging demands might have important influences on the mother-infant relationship, as well as significant effects on the infant's psychosocial development. These possibilities can be investigated in the laboratory by systematically manipulating foraging demands. This was done in the research described in this chapter.

The results indicate that nonhuman primates adjust significant aspects of their individual, dyadic, and group behaviors in response to different foraging requirements. Subjects appear to be particularly affected by unpredictably varying foraging demands. Individuals alter meal patterns. Groups may become more hostile. Mothers may be more rejecting and less responsive to infants and, in response to particular demand patterns, infants may be more disturbed and less independent in their development.

Michael W. Andrews—State University of New York Health Science Center, Brooklyn.

Gayle Sunderland—State University of New York Health Science Center, Brooklyn

Leonard A. Rosenblum—State University of New York Health Science Center, Brooklyn

Among the various aspects of the physical environment which may influence individual and group behavior, feeding conditions clearly have the potential to alter both acute and chronic patterns of social interaction (Hall 1963; Loy 1970; Southwick 1967). A major goal in our laboratory has been the illumination of the social and developmental consequences of environmental manipulations which alter the nature of the foraging demand confronting group-living primates. In particular, we have been interested in the impact which such manipulations have on the relationship between mother and infant and on infant development.

We anticipated that increasing the foraging demand placed on a mother would increase the conflict between her need to attend to the feeding environment and her infant's need to have a mother responsive to its changing needs and interests (Bowlby 1973). On the one hand, from an evolutionary perspective, there is a sound theoretical basis for explaining the conflict of interest between mother and infant which might derive from increasing foraging demands on the mother (Trivers 1974). On the other hand, there has been little empirical evidence regarding the influence on the mother-infant relationship and infant development of factors which increase the amount of time and effort which a mother must expend in foraging (but see Johnson 1986). Our studies over the last several years, however, have begun to provide evidence for substantial effects of the foraging environment on mother-infant relationships and infant development.

The potency of the foraging task in affecting the mother-infant interaction is clearly evident in a recently completed study with squirrel monkeys, *Saimiri sciureus* (Sunderland and Rosenblum unpublished). The subjects in this study were divided into two groups. One group consisted of six mother-infant dyads and two adult females and the second group contained five dyads and one additional adult female. The infants ranged in age from one week to four months at the start of the study. Each group lived in one side of a double pen, each side totalling approximately 84 cubic feet. The sides were connected by a floor-level door. This door remained locked except during the test hour when it was open to allow subjects to pass from the living pen to the adjoining foraging test pen. Each pen contained several levels of shelves, a watering spigot, and a one-way observation screen in the front door. For foraging test purposes, the floor of the test pen was covered with approximately six inches of wood-chip bedding, and food—standard New World monkey chow—was buried beneath the bedding's surface.

Subjects were tested four days each week (in counterbalanced order) on one of two conditions. On provisioned test days, subjects received a ration of food two hours prior to the opening of the test-pen door. On hunger test days, subjects received no food prior to the one-hour foraging period. The two groups received the feeding conditions in alternating two-day blocks, so that one group was tested in the provisioned condition while the other was tested under the hunger condition. Tests were conducted in the mornings and, on all test days, subjects received vitamin supplements in their home pens in the late afternoon according to standard Primate Laboratory procedures. In addition, two days of ad lib feeding were provided each week. Normal health and infant growth was maintained in all subjects throughout the study.

Maternal responses to the different conditions of testing were quite clear. Upon entrance to the test pen on hunger days, eager searching for food began almost immediately. The heightened focus on foraging was accomplished at the cost of marked changes in adult social behavior and maternal patterns. The usual gregariousness of females of this species was significantly diminished, as the adult females reduced the time spent in contact-huddles. Maternal rejection of the infants and maximal spatial separation of the dyad increased significantly during the hunger trials, and the duration of dyadic contact dropped precipitously, in spite of increases in infant approaches. Not surprisingly, the temporarily abandoned infants showed marked increases in affective disturbance. None of these changes showed any signs of abatement during the sixteen test trials, spread across eight weeks of the infant's life. Subsequent follow-up observations of these animals were done five months later, when infants were spending little time in dorsal contact with their mothers and were largely feeding independently. As might be expected with these older, more independent infants, little conflict within the dyad was in evidence, and infant disturbance was no greater during the hunger than the provisioned trials.

Thus, these squirrel monkey data reflect the capacity and readiness of primate mothers to adjust or, at times, dramatically alter their maternal patterns in response to acute changes in the demand quality of the foraging environment. We as yet know very little about the factors affecting maternal decisions regarding continued transport or rejection of the infant in the effort to meet demands; the continued capacity of a mother to respond to certain critical needs of her infant during such periods (one presumes that the attack of a predator, for example, would shift the mother from foraging to maternal defense); or the relative cost to the mother's foraging efficiency when maternal strategies are followed. As we shall see, mothers can change both their response to their infants and to the foraging demands in a variety of ways. For example, mothers can more often carry their infants while foraging. They can also increase foraging speed, or they can shift the focus of their attention back and forth between maternal and foraging tasks. The specific maternal response depends on a number of factors which we can manipulate experimentally to some degree.

Although our studies with squirrel monkeys have been quite fruitful, a majority of our foraging work has focused on the bonnet macaque, *Macaca radiata*. Based largely on studies from this laboratory that began more than twenty-five years ago (Rosenblum et al. 1964a, 1964b, 1969), we can make the following generalizations about the mother-infant relationship in the bonnet macaque. Within dyads living in stable groups with constant and easy access to abundant food and water, mothers are highly permissive and accepting—restraint, rejection, and punishment of infants are minimal. Most of the making and breaking of contact between mother and infant is due to the comings and goings of the developing infant. Overall, the mother-infant relationship in the bonnet macaque under stable and undemanding conditions shows little evidence of conflict. Such a positive dyadic relationship offered a promising baseline against which to analyze the impact of foraging tasks which place greater demands upon the mother. We will examine our findings thus far by looking first at the acute effects on bonnet dyads of an imposed foraging task, and then at studies which were aimed at illuminating the chronic effects of living in an environment posing substantial foraging demands.

ACUTE EFFECTS OF A DEMANDING FORAGING TASK

We have recently completed a series of studies aimed at exploring the dynamics of the mother-infant interaction when the mother is acutely confronted with a demanding foraging task. Among our goals in this series of studies was the examination of the effects of manipulating the motivation of either one or both partners for dyadic contact. We are still in the process of analyzing the data from these studies. Although the entire picture is not yet complete, some major findings are already clear.

The subjects in this series of studies were housed as a group in a 2 x 4 x 2.1 meters high pen. The group consisted of four mother-infant dyads with two of the infants male and two female. The infants ranged in age from 3.5 to 4.5 months at the beginning of the series of studies.

Foraging opportunities were restricted to a single hour on week-days. By temporally localizing all foraging to a single hour we hoped to create a task which was quite urgent and thereby enhance any changes in the dynamics of the mother-infant interaction which might result from a demanding foraging task. From Friday afternoon through Monday morning, subjects were given food ad lib. These feeding procedures were quite adequate for maintaining the health of mothers and infants, as revealed by daily health checks and weekly weighings. During the one-hour restricted foraging periods, food was buried in pans within two feeding structures such as the one dia-grammed in figure 9.1. The two feeders were placed end to end creat-ing a structure that was 2.6 meters long with sixteen holes on each side providing the only access to the sixteen enclosed food pans (see figure 9.2). An amount of food—which was 20% in excess of that consumed each day—was equally distributed among the food pans and covered with 12 centimeters of clean wood-chip bedding.

As a first step, we wished to compare mother and infant behav-ior during the foraging hour with their behavior under three other conditions: The hour immediately before the foraging hour (preforag-ing), the hour immediately after the foraging hour (postforaging), and a comparable hour of the day during ad lib feeding. Observations were made Tuesday through Friday with the exception of ad lib hours

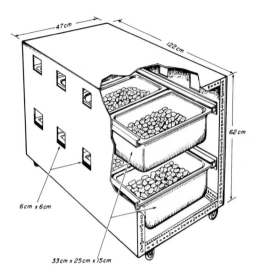

Figure 9.1   Diagram of an 8-pan foraging structure.

which were recorded on weekends. All observations were made between 11:00 A.M. and 1:00 P.M. Eight repetitions for each of the conditions were balanced for observation time and day of the week, with the exception of ad lib feeding. The one-hour feeding began between 10:00 A.M.and 2:00 P.M., depending on the scheduled observation conditions for the day. The observations were conducted over a two-month period.

The imposed task was clearly manageable for the subjects. Across the observation days, nearly all the foraging activity (more than 92%) was completed before the last quarter of the foraging hour. The task did require considerable effort during active foraging, however. Overall, less than 20% of foraging attempts—that is, reaches into the foraging structure—were successful in producing a food item. In this first study, all food items obtained were consumed or placed in cheek pouches by the end of the foraging hour.

Compared to preforaging, postforaging, or ad lib hours, mothers and their infants were separated by a distance equal to or greater than a foot most often during the feeding hour [Mean per cent of observations in which dyads were separated: mean = 72 for the foraging hour; mean = 39 for the preforaging hour; mean = 41 for the postforaging

Figure 9.2 Subject probing through one of the holes located on the side of the foraging structure.

hour; and mean = 57 for the ad lib hour. Condition effect $F(3, 9) = 38.08$, $p < 0.001$]. Nonetheless, there was little direct evidence of overt dyadic conflict during the foraging hour. Mothers did not break contact with their infants more during the foraging period, nor did they reject their infants more at this time. In fact, the longest bouts of mother-infant separation occurred during the foraging hour [Mean separation bout length in seconds: mean = 81.5 for the foraging hour; mean = 30.9 for preforaging hour; mean = 46.5 for the postforaging hour; and mean = 57.5 for the ad lib hour. Condition effect $F(3, 9) = 9.19$, $p < 0.01$].

There was, however, substantial indirect evidence that the foraging conditions did create tension between mother and infant. Mothers broke contact with their infants more during the preforaging condition than during any of the other conditions. In fact, they broke contact nearly twice as often [Condition effect $F(3, 9) = 8.95$, $p < 0.01$]. The fact that mothers began to separate themselves from their infants in anticipation of the foraging hour is in contrast with previous findings by Rosenblum and colleagues (1969), and may be due, in part, to the urgency of the feeding condition. Another piece of indirect evidence pointing to dyadic tension resulting from the feeding conditions is that mother-infant grooming was higher in the postforaging hour than in any other condition. The values were nearly twice as high, on average, as in any of the other three conditions [Condition effect $F(3, 9) = 9.29$, $p < 0.01$]. This grooming behavior generally appears to have a calming effect for both recipient and groomer and, thus, may indicate an effort to resolve tension in the dyadic relationship. De Waal (1986) has been among those who have also suggested that, under certain circumstances, grooming may provide a means of reducing tension between individuals. As a final piece of evidence pointing to tension in the dyadic relationship during the foraging hour, it should be noted that the mothers were virtually never observed to forage in contact with their infants.

The results suggest to us the tentative view that mothers tended to remain out of contact with their infants while actively foraging. It has been suggested that this may be related to the foraging efficiency of the mother (Johnson 1986). During such periods of active foraging, the infants did not overtly protest the separation. In contrast, during preforaging periods in which mothers more frequently broke contact in apparent anticipation of the foraging hour, separation was contested by the infants. The highest scores for infant return to contact occurred during the prefeeding condition. Although the infants remained apart from their mothers during periods of active foraging, the high levels of mother-infant grooming in the postforaging hour— which may have functioned to reestablish the disrupted relationship and reduce tension in the infant—suggests that the foraging condition (and its anticipation) did place a stress upon the dyadic relationship.

Before moving on to consider the next in this series of studies, mention should be made of the impact of the foraging task upon adult interactions. The foraging task did substantially increase conflict in the form of hierarchical behaviors. These behaviors, however, were largely restricted to the period of active foraging with fairly typical levels during the other conditions. Hierarchical behaviors were more than five times as frequent during the foraging condition than during any other condition [$F(3, 9) = 14.55$, $p < 0.001$]. Positive adult interactions also differed among conditions. Adult affiliation—such as huddling contact—was highest in the postforaging condition, more than 25% greater than that of either the preforaging or ad lib conditions, and more than 200% greater than that occurring during the foraging condition [Condition effect $F(3, 9) = 28.01$, $p < 0.001$]. Overall, the results point to a restriction of group tensions to the period of active foraging and rapid resolution of tensions following the foraging period.

In this first study of the series, mothers foraged only when separated from their infants. Furthermore, mothers and infants separated quite readily during the foraging hour. In the second study, our aim was to increase the motivation to maintain dyadic contact in order to determine what consequences this might have for the mother's foraging efficiency and for the interaction between mother and infant. To this end, we examined the influence of three-hour separations of mother and infant upon behavior of the dyad during the foraging hours. Behavior following separations was compared to behavior following sham separations to control for the effects of handling the animals.

Using the same group as previously studied, with infants now between the ages of 8.5 and 9.5 months, four separations and four sham-separations were completed over a two-week period. On each of the four separation days, the mothers and infants were separated at 8:00 A.M. with the infants removed to individual cages in another room. At 11:00 A.M., the feeding structure was provisioned, and the infants were returned to the group. Observations began immediately upon the return of the infants. On sham-separation days, the mothers and infants were separated at 8:00 A.M. and immediately reunited. At 11:00 A.M., the feeding structure was provisioned and observations began.

The separations were clearly effective in increasing the motivation for contact between infant and mother during the foraging hour. Following separations, mothers and infants were out of contact 43% less than they were following sham separations [$F(1, 3) = 30.08$, $p < 0.05$]. The infants also played less following separations than they did following sham separations. Social play was absent following separations, and exercise play was exhibited significantly less following separations than it was following sham separations [$F(1, 3) = 49.00, p < 0.01$].

From the view that the demands of foraging and demands of infant care were primary in competing for the mother's time and energy in this study, it would be anticipated that a substantial increase in contact would have been accompanied by a substantial decrease in foraging activity by the mother. This was not the case. Despite the substantial increase in contact following separations as compared to sham-separations, foraging activity was decreased overall by only 13%. This small decrease in foraging activity was not statistically significant. Furthermore, virtually the identical amount of food was obtained under the two conditions during the four repetitions. The mothers appeared to be sacrificing little foraging success for the sake of their infants.

With respect to the mother-infant relationship, the data point to the view that the mothers were attempting to be more attentive to their infants which had just returned, but were being distracted by the urgent foraging task at hand. The mothers did not respond in a uniform manner, however. Following separations, the dominant female did not begin to forage immediately following the return of her infant. Typically, she sat for some minutes in contact with her infant without approaching the feeder. When she did begin to forage following separations, it was usually while in contact with her infant. For this female, foraging while in contact with her infant was approximately ten-fold higher following actual separations than it was following sham separations, whereas foraging out of contact with her infant was 65.9% lower. Another female, of midrank, was also hesitant to begin foraging following separations, initially remaining in contact with her infant. Over the entire foraging hour, however, her total foraging activity was little different from that following sham-separations. In fact, her rate of foraging while out of contact with her infant was 57.6% higher following separations than it was following sham-separations.

Following separations, the foraging activity of the two low-ranking females differed sharply. The foraging activity of one of these females appeared to be moderately affected by the separations. (It should be noted that this female had the lowest scores for infant contact in sessions following sham-separations.) Compared to sessions following sham-separations, she exhibited a slight increase in foraging activity while in contact with her infant in sessions following separations. Her foraging activity while out of dyadic contact was 29% lower after separations than after sham-separations, with rate of foraging comparable in both conditions. The other low-ranking female, which was generally excluded from the vicinity of the foraging apparatus during the early periods of access to the feeder, showed a marked response to the postseparation foraging condition. She rapidly separated from her infant and began to forage, apparently taking advantage of the opportunity afforded by the hesitancy of the more dominant females to approach the feeding apparatus. Overall, for this female, foraging scores *while out of dyadic contact* increased 156% in sessions following separations as compared to those following sham-separations. However, the increase in foraging was not accomplished by increases in separation. In fact, for this particular female—as was true for the other females—separation from her infant was less frequent in postseparation sessions than it was in sessions following sham-separations. Her increase in foraging activity was accomplished by an increase in foraging *rate*, reflecting a more intense involvement in foraging during periods of separation from her infant. In light of these results, we propose that the effect of a quantitative increase in dyadic contact following separations may have been mitigated by a decrease in the quality of the maternal attention which the infants were receiving, in and/or out of contact with their mothers, as a consequence of the uncompromising demand of the mother's foraging task.

Thus far, we have presumed that the mother's need to forage was at odds with the infant's need to have a responsive and attentive mother. It is possible, however, that the large amount of separation between mothers and infants and the lack of overt conflict during the foraging hour in the two studies just discussed may have occurred, at least in part, because the infant, as well as the mother, was motivated to obtain some of the provisioned food. When an infant has had limited access to food it may choose to leave its mother to obtain food when it becomes available. (Food was available on the floor during at least the first half of the foraging hour, having been dropped by the foraging mothers.) In effect, conflict between mother and infant may have been reduced by the infant's desire to seek food in the environment at the same time that its mother was doing so. If this view of the lack of overt dyadic conflict is correct, it would be hypothesized that conflict might be increased and/or separation decreased by allowing the infants to eat prior to the foraging period for the mothers.

To test the above hypothesis using the same subject group—infant ages now ranging from ten to eleven months—behavior on days in which infants were given access to food prior to the foraging hour (prefeeding days) was compared to behavior on days without prior access to food. Prior access to food for only the infants was accomplished by placing food inside a small enclosure (0.74 x 1.22 x 0.71 meters high) located within the living pen, which could be entered only by the infants. The infants were habituated to the prefeeding procedure prior to beginning the study. It was established during the habituation period that all infants obtained food from the special enclosure. Over a two-week period, there were four days each with and without prefeeding. On prefeeding days, food was placed in the special infant enclosure at 9:00 A.M. At 11:00 A.M., the food was removed from the enclosure, and the adult feeding structure was provisioned. Observations began at 11:00 A.M., or as soon as the feeding structure had been provisioned. On days without prefeeding, the feeding structure was similarly provisioned at 11:00 A.M. and observations began immediately.

On prefeeding days, an average of 112 grams of food was removed from the food source within the infants' enclosure. Although some food was dropped or stolen, the infants consumed the majority of this food. Consequently, comparing infant eating during the hour following provisioning of the foraging structure on days with and without prefeeding, it was found that eating by infants was scored 19% less often on prefeeding days. Thus, the prefeeding treatment did not eliminate infant feeding during the foraging hour, but it did significantly decrease it [$F(1, 3) = 61.74, p < 0.01$].

Contrary to the prediction of the hypothesis, we found that prefeeding did not change the extent to which mother or infant broke or returned to contact. Furthermore, the amount of mother-infant separation during the foraging hour following prefeeding was actually somewhat greater than when prefeeding did not occur. The greater degree of separation following prefeeding may have been due to increased arousal of the mothers which appeared to accompany the infants' feeding behavior during the two hours of prefeeding. Foraging by the mothers during the foraging hour, in fact, was significantly higher following prefeeding of the infants than it was following no prefeeding [$F(1, 3) = 33.00$, $p < 0.05$]. The results of this study support the view that neither mother-infant separation nor the lack of overt conflict during separation in the previous studies was the result of the infants being attracted to feeding opportunities in the environment. Mother-infant separation appeared to be the result of mothers being actively engaged in foraging which, the results of this and other studies suggest, was incompatible with infant contact. We are compelled to conclude at this time that mothers and infants were separated during the foraging hour because the mother was busy.

EARLY STUDIES OF CHRONIC EFFECTS

These studies compared three groups, each consisting of five mother-infant dyads (Rosenblum and Paully 1984; Rosenblum and Sunderland 1982). The groups were designated LFD (low-foraging-demand), HFD (high-foraging-demand), and VFD (variable-foraging-demand). In both the LFD group and the HFD group there were three male infants and two female infants. There were four male infants and one female infant in the VFD group. Infants in the three groups were matched for age which, with the exception of one infant born half way through the study, ranged from four to seventeen weeks at the start of the study. Group comparisons with respect to infant behavior were based only on the four oldest infants in each group.

The LFD, HFD, and VFD groups each lived in a double pen (2.1 meters high x 11.2 square meters of floor area) which was divided by an opaque wall with one or two pass-through doors (45 x 45 centimeters) that remained open at all times. Each pen contained perches and ad lib access to water spigots. The lights remained on from 7:00 A.M. to 7:00 P.M.. One-way windows, through which all observations were made, were located in the front wall of each pen.

Treatment of the groups differed only in the way in which the daily ration of primate biscuits was provided. For the LFD and HFD groups the food biscuits were randomly distributed among approximately one thousand food holes (5 centimeters in diameter x 15 centimeters deep) contained within a series of foraging panels located along the pen walls. Each day, the HFD group received ninety to one hundred biscuits, which was approximately 10% in excess of their total daily consumption. Animals in this group, therefore, were required to continue searching until nearly all (90%) of the hidden food items were found. The value of each food item to the HFD animals is evident from the fact that they ate all food items which they found.

In contrast, six hundred biscuits were distributed among the food loci for the LFD group. Animals in this group could readily obtain their daily ration by finding only a small fraction of the hidden food. The LFD animals, in fact, ate only about half of what they found. The greater selectivity (or food wastage) of the LFD group than of the HFD group is consistent with other data regarding the effects of search cost on selectivity (Collier 1983).

For the VFD group there were six feeders which were located on the floor, three on each side of the opaque partition. Each feeder was approximately 32 centimeters in diameter and 37 centimeters high. To obtain food from the feeder, an animal was required to reach through a 5-centimeter-diameter hole in the top to rotate an inner chamber which also contained a 5-centimeter-diameter hole. When the inner and outer holes were lined up, the animal could reach into the inner chamber to obtain any food which was there.

For the VFD group, two weeks of low-foraging-demand were alternated with two weeks of high-foraging-demand during the twelve weeks of observations. During the weeks of low-foraging-demand, six hundred biscuits were distributed among the feeders each morning. The biscuits were placed on the upper surface of the inner chamber. Ample food was therefore easily obtained during the low-demand phases for the VFD group by simply reaching into the drum. During the weeks of high-foraging-demand, only ninety to one hundred biscuits were distributed among the feeders each day, and in this phase the biscuits were placed within the rotary inner chamber. No subject in this group or either of the other two groups was food deprived, and repeated weighings indicated that mothers maintained their ad lib feeding weight while infant growth was normal.

Comparing first the LFD and HFD groups, the results clearly indicated that the mothers in the HFD group were confronted with a more demanding task than were the LFD mothers. The foraging scores of the HFD subjects were, on average, about five times those of the LFD individuals. Although the group difference was highest in the morning following provisioning, the foraging scores of the HFD group were higher throughout the day. In addition, whereas the LFD group ceased foraging when the lights went out in the evening, the HFD group often continued to forage for approximately one additional hour. The HFD mothers were clearly spending more time and exerting more effort in foraging than were the LFD mothers.

Despite the clear evidence of increased foraging demand on the HFD mothers as compared to the LFD mothers, there was little evidence of sustained overt conflict between mothers and infants in either group after the first several weeks of stable task conditions. Overall, infant rejection was infrequent and comparable in both groups, although it was somewhat higher in the HFD group. Similarly, although the frequency with which mothers broke contact with their infants was also generally higher in the HFD group—particularly at the start of observations—levels were comparable in both groups following the initial weeks, and differences did not reach statistical significance.

The observations made during the treatment conditions also suggested that there were few apparent negative consequences for either the mother-infant relationship or infant development in the HFD group when compared to the LFD group. The amount of contact between mother and infant gradually decreased in the LFD and HFD groups across the study, and did not differ between groups. In fact, the development of independence in the HFD infants appeared to be advanced relative to that of the LFD infants. HFD infants had significantly longer bouts off their mothers than did LFD infants, particularly early in the study when maternal responsivity was somewhat more negative in the HFD group than it was later on. Furthermore, the HFD infants were more likely to be found on the opposite side of the pen dividing wall from their mothers than were LFD infants.

Subsequent findings, however, indicated that the HFD infants were less able to cope effectively with the temporary loss of their mothers. The findings support the view that the infants in the HFD group were less secure in their attachment to their mothers than they had initially appeared to be. Consequently, there was conflict in the mother-infant relationship at some level. In this follow-up study, infants from the two groups were separated from their mothers (Plimpton and Rosenblum 1983, 1987). Despite the fact that separations occurred during low demand for both groups, infants from the HFD group exhibited more intense and more sustained depression and emotional disturbance following separation, and they recovered more slowly than did LFD infants. This suggested that the HFD infants were less securely attached to their mothers, and less able to cope with autonomous functioning than were the LFD infants. The increased foraging demand on HFD mothers relative to LFD mothers, therefore, did have adverse consequences for infant development in the HFD group, albeit consequences which became manifest only under conditions of sufficient challenge.

Before comparing the VFD group to the LFD and HFD groups, mention should be made of differences between the LFD and HFD groups in adult interactions. Relative to the LFD group, relationships among the adults in the HFD group appeared to be more tense. After the first four weeks of observation, adult conflict in the form of hierarchical behaviors was more than twice as frequent in the HFD group as in the LFD group. In addition, affiliative behavior was more frequent in the LFD group than in the HFD group. Adult contact was about 80% higher in the LFD group than in the HFD group. Grooming was more than twice as high in the LFD group. Thus, heightened group tension in the HFD group may have increased the infant's need for the attention of its mother. As a result, the conflict between the needs of the mother and the needs of the infant may have been increased in the HFD group relative to the LFD group, by changes in the requirements of both mother and infant. The group difference in tension may have contributed to the group difference in infant development subsequently revealed by the challenge imposed by separating mother and infant.

Turning now to the VFD group, it was found that, based simply upon foraging scores during observations, the foraging demand upon the VFD mothers was intermediate between that upon the LFD mothers and that upon the HFD mothers. There were some differences between the low- and high-demand conditions for the VFD group, such as the increases in hierarchical behavior and mother-leave-contact between low and high demand. The differences, however, were neither very large nor very consistent. Behavioral measures during low- and high-demand conditions for the VFD group were therefore combined for comparison with the LFD and HFD groups.

Despite the finding that overall foraging scores were somewhat lower for the VFD group than for the HFD group, the greatest overt dyadic conflict was evident in the VFD group. The mothers in the VFD group broke contact with their infants significantly more than did mothers in either the LFD or HFD group. Furthermore, there was a strong tendency for infants to return to contact with mothers more in the VFD group than either of the other two groups, reflecting the infants' heightened efforts to achieve and maintain contact.

Consistent with the finding that conflict within the mother-infant relationship was most apparent in the VFD group, the relationship of mother and infant and the development of the infant appeared to be most affected in the VFD group. Of the three groups, the VFD infants were the least likely to be on the opposite side of the pen partition from their mothers. The VFD infants also played less than did infants of the other two groups. Finally, the VFD infants exhibited the highest levels of disturbed behavior when off their mothers. Considered together, these results point to a retardation in the behavioral and affective development of VFD infants compared to LFD or HFD infants.

The more dramatic consequences of the imposed foraging conditions for the VFD group than for the HFD group may appear somewhat at odds with the finding of lower mean foraging activity for the VFD group than for the HFD group. We propose, however, that the foraging environment was more psychologically demanding for the VFD mothers than for the HFD mothers. The data indicate that the VFD females were slow to adjust to changes in the demand of the foraging situation when the changes were, in fact, unsignalled. It is likely that such unpredicted changes in the environment would make it more difficult for a forager to adopt a consistent strategy for dealing with the foraging environment. Therefore, we propose that much effort had to be directed toward coping with the unpredictable conditions from week to week. We suggest that the need of the mother to cope with the unpredictable environment—and the consequent variation in her foraging and maternal patterns—increased conflict between her survival needs and her infant's need for the attention of its mother.

It should also be noted that the variable-foraging-demand condition resulted in substantial tension among the adults in the VFD group. The VFD mothers showed significantly more hierarchical behaviors than did either the LFD or HFD mothers. The VFD mothers also groomed each other less often than did mothers of the other two groups. As has already been suggested regarding tension within the HFD group, tension among the adult members of the VFD group may have increased demands on the mother, thereby diminishing her capacity to respond to her infant's changing requirements, while at the same time increasing her infant's need for her attention.

The finding that infant development was retarded in the VFD group despite the observation of the highest levels of mother-infant contact of any group, supports the view that any one or a combination of three mechanisms may have contributed to the observed effects.

First, as we have noted, the markedly elevated tension in the VFD group relative to the other groups may have increased the amount of maternal attention needed by the infants in the VFD group well beyond that which they received.

Second, the high task demands on the VFD mothers, both from the social and foraging aspects of the environment, in addition to reducing the speed and consistency of maternal responsivity when infants were apart from them, may also have adversely influenced the quality of contact time between mother and infant (Andrews and

Rosenblum 1988). That is, although the infant was often in contact with the mother, the mother may have been generally distracted by her tasks and therefore unattentive to the moment-to-moment needs of her infant.

Third, the apparent affective disturbance, with low levels of play, exhibited by VFD infants during their brief bouts off their mothers suggests that the infants were likely to have been in a state of high emotional arousal. High emotional arousal may be presumed to be inimical to the complex learning necessary for the development of autonomous functioning (as stated in the Yerkes-Dodson Law, Yerkes and Dodson 1908).

RECENT STUDY OF CHRONIC EFFECTS

More recently we have sought to replicate and extend these early findings (Andrews and Rosenblum 1988). For this purpose, we again compared the effects on development and social relationships of low- and variable-demand conditions. The high-foraging-demand condition was not utilized in this study because the variable-demand condition had been shown to be more effective in influencing the dyads in the early studies. In the current study, the LFD and VFD groups each consisted of six mother-infant dyads, studied in two cohorts of three mother-infant dyads. Each cohort was housed in a pen that was much smaller than the pens used in the earlier studies (5.2 square meters versus 23.5 square meters) and was not partitioned. The physical environment of this study was, therefore, substantially less complex than that of the earlier studies. With only three dyads per group, the social environment in this study was also considerably less complex than that of the earlier studies which involved groups approximately double in size.

To simplify the physical task of daily provisioning and to accommodate somewhat different living conditions, each cohort was provided with a single feeding structure resembling that used in the study of acute effects (figure 9.1). The feeding structure of this study contained eight food pans which could only be accessed by eight small holes on each side of the structure. The food pans were provisioned daily, except on weekends when adequate food to last through the weekend was supplied after all observations for the week had been completed. For the LFD group, at least twice the necessary daily food ration was equally distributed among the food pans each day during the week. The food was not buried for the LFD group. For the VFD group, two-week periods with provisioning identical to that for the LFD group were alternated with two-week high-demand periods during which the food placed in the pans was covered by approximately 12 centimeters of clean wood-chip bedding. In addition, during the high-demand periods, the amount of food was reduced relative to the low-demand condition such that only 10 to 20% of the total amount provided remained in the pans the following day. As is typical of high-demand conditions, all food removed from the pans each day was consumed during the high-demand condition.

The data reported here reflect sixteen weeks of data collection. For a total of fourteen weeks—beginning and ending with two weeks of low-foraging-demand for cohorts in each treatment group—each animal was observed in its living pen on four days each week for a period of three hundred seconds (six hundred seconds per dyad per day). Observations began approximately 1.5 hours following morning provisioning of the feeding structure.

At the conclusion of the fourteen weeks, cohorts in both groups were placed on ad lib food and, in each of two subsequent weeks (separated by a three-week interval), were given four one-hour sessions in a novel environment. The test environment was a novel pen that was more than three times the size of the living pen and contained climbing cables and other unfamiliar objects. During each hour in the novel room, each individual was the focal animal for two three-hundred-second periods (twelve hundred seconds per dyad). All observations in both the home pen and novel room were made behind one-way-glass observation screens.

We will consider first the results from the initial twelve weeks of observation under the experimental foraging conditions in the home pen. This represents a total of six weeks of low-demand foraging and six weeks of high-demand foraging for the VFD group. Even more than in the previous study, there was little evidence of differences between the alternating foraging conditions for the VFD group. The conditions were therefore combined for comparison to the constant low-demand condition of the LFD group.

Despite the absence of any evidence of group differences in foraging during observations—which, it should be remembered, began 1.5 hours after the day's food ration was made available—and despite low and comparable levels of infant rejection in both groups, there was some evidence of greater conflict in the mother-infant relationships in the VFD group than in the LFD group. Although mothers in neither group were responsible for more than 20% of breaks in contact, the VFD mothers, on average, broke contact with their infants nearly 70% more often than did LFD mothers. This group difference was significant [$F(1, 10) = 6.30, p < 0.05$]. VFD mothers also returned to contact with their infants more than twice as often as did LFD mothers, but the group difference was not statistically significant. It should also be noted that VFD mothers groomed their infants nearly twice as much as did LFD mothers [$F(1, 10) = 6.99, p < 0.05$]. As noted previously, this increase in grooming may also have reflected a greater tension within the dyads in the VFD group. Whereas reduced social grooming prior to and during foraging periods—which, under some conditions, may comprise much of the day—may point to increased tensions, elevated grooming following periods of active foraging may reflect attempts to resolve tensions created by the foraging task.

There was also some evidence that the mother-infant relationship and the development of the infant had been influenced by the foraging conditions. VFD infants were observed a foot or more from their mothers 18% less often than were LFD infants, but this difference was not statistically significant. VFD infants also engaged in shorter bouts off mother than did LFD infants (thirty-six seconds versus sixty-seven seconds), and this difference was significant [$F(1, 10) = 6.03, p < 0.05$]. These data suggest that during early development VFD infants were somewhat less independent of their mothers than were LFD infants. In fact, the higher frequency with which VFD mothers broke contact with their infants compared to LFD mothers may have been due, in part, to the somewhat greater proportion of time the VFD pairs were in contact during observations.

In the previous study, we suggested that chronic group tensions may have contributed to the adverse developmental consequences for infants in the VFD group. In this study, however, there were no apparent group differences in either affiliative or hierarchical behaviors among the adult members of the groups. The low level of tension among VFD adults may help to explain the relative absence of overt signs of developmental retardation obtained during the home-pen observations in this study.

Following two final weeks of low demand for subjects in both groups, the animals were placed on ad lib feeding during the period of novel-room testing. On four consecutive days of each of the two test weeks, animals were transported from their home pen to the novel room for the daily one-hour tests. Observations began as soon as the three dyads of a cohort were released into the room.

Four measures in the novel environment lead to the conclusion that VFD infants were less secure in their attachment to their mothers than were the LFD infants and, consequently, less independent in engaging the environment. First, LFD infants broke contact with their mothers more than three times as often as did VFD infants. This difference was significant [$F(1, 10) = 9.63$, $p < 0.05$]. Second, LFD infants exhibited significantly more play and exploration of inanimate objects than did VFD infants [$F(1, 10) = 5.08$, $p < 0.05$] and [$F(1, 10) = 14.66$, $p < 0.01$] respectively. Third, LFD infants were more than twice as likely to be separated from their mothers by a foot or more than were VFD infants. This was due primarily to the extreme value for one VFD dyad in which the mother frequently initiated separation. However, this difference was not statistically significant. Fourth, LFD infants showed a significant increase from the first to the second half hour of the test sessions in the proportion of time they were a foot or more from their mothers [$F(1, 5) = 28.02$, $p < 0.01$]. There was no significant within-trial increase in separations for the VFD group. In contrast to the LFD infants, infants in the VFD group were apparently unable to use their mother as a secure base (Ainsworth and Wittig 1969) from which to begin to explore the environment during the one hour of each session. For the LFD infants, exploration of the novel environment was achieved, in part, by reducing the bout length off the mother to the extent that there was no group difference on this measure in the novel environment. The bout length off the mother dropped significantly between the home pen and novel environment for the LFD group only [$F(1, 5) = 8.28$, $p < 0.05$]. These results are consistent with the view that VFD infants were less secure in attachment to their mothers than were the LFD infants (Ainsworth and Wittig 1969).

CONCLUSIONS

We may make the following generalizations regarding the findings from our laboratory thus far. Conditions which increase the time and effort which a mother must expend in foraging during a given time period clearly have the potential to increase the amount of separation between mother and infant during that period. The elevated separation in a demanding situation is not necessarily achieved at the cost of overt conflict between mother and infant. Dyads, in fact, often appear quite capable of adjusting their interactions to the prevailing conditions. Even in the absence of overt conflict, however, there are often indirect indications that a demanding foraging environment does produce tension in the mother-infant relationship. When the demanding foraging conditions are a chronic aspect of the environment, they typically result in a less secure attachment between an infant and its mother. We emphasize that the resulting insecurity in the dyadic relationship, and the consequences which this has for the development of the infant, may remain latent until the coping capacities of the dyad are challenged.

REFERENCES

Ainsworth, M. D. S.; Wittig, B. A. (1969) Attachment and exploratory behavior of one-year-olds in a strange situation. In Foss, B. M. (ed.), "Determinants of Infant Behaviour IV." London: Methuen & Co. Ltd. 111–136.

Andrews, M. W.; Rosenblum, L. A. (1988) The relationship between foraging and affiliative social referencing. In Fa, J. E., Southwick, C. H. (eds.), "The Ecology and Behaviour of Food-Enhanced Primate Groups." New York: Alan R. Liss, Inc.

Bowlby, J. (1973) "Attachment and Loss," vol. 2. "Separation Anxiety and Anger." New York: Basic Books, Inc.

Collier, G. H. (1983) Life in a closed economy: The ecology of learning and motivation. In Zeiler, M. D., Harzem, P. (eds.), "Advances in Analysis of Behavior," vol. 3. "Biological Factors in Learning." Chichester: John Wiley & Sons. 223–274.

de Waal, F. B. M. (1986) The integration of dominance and social bonding in primates. *Quarterly Review of Biology* 61:459–479.

Hall, K. R. L. (1963) Variations in the ecology of the chacma baboon, *Papio ursinus. Symposium of the Zoological Society of London* 10:1–28.

Johnson, R. L. (1986) Mother-infant contact and maternal maintenance activities among free-ranging rhesus monkeys. *Primates* 27:191–203.

Loy, J. (1970) Behavioral responses of free-ranging rhesus monkeys to food shortage. *American Journal of Physical Anthropology* 33:263–272.

Plimpton, E.; Rosenblum, L. (1983) The ecological context of infant maltreatment in primates. In Reite, M., Caine, N. G. (eds.), "Child Abuse: The Nonhuman Primate Data." New York: Alan R. Liss, Inc. 103–117.

Plimpton, E. H.; Rosenblum, L. A. (1987) Maternal loss in nonhuman primates: Implications for human development. In Bloom-Feshbach, J, Bloom-Feshbach, S. (eds.), "The Psychology of Separation and Loss." San Francisco: Jossey-Bass. 63–86.

Rosenblum, L. A.; Paully, G. S. (1984) The effects of varying environment demands on maternal and infant behavior. *Child Development* 55:305–314.

Rosenblum, L. A.; Sunderland, G. (1982) Feeding ecology and mother-infant relations. In Hoff, L. W., Gandelman, R., Schiffman, H. R. (eds.), "Parenting: Its Causes and Consequences." Hillsdale, N.J.: Lawrence Erlbaum Associates. 75–110.

Rosenblum, L. A.; Clark, R. W.; Kaufman, I. C. (1964a) Diurnal variations in mother-infant separation and sleep in two species of macaque. *Journal of Comparative and Physiological Psychology* 58:330–332.

Rosenblum, L. A.; Kaufman, I. C.; Stynes, A. J. (1964b) Individual distance in two species of macaque. *Animal Behaviour* 12:338–342.

Rosenblum, L. A.; Kaufman, I. C.; Stynes, A. J. (1969) Interspecific variations in the effects of hunger on diurnally varying behavior elements in macaques. *Brain, Behavior and Evolution* 2:119–131.

Southwick, C. H. (1967) An experimental study of intragroup agonistic behavior in rhesus monkeys (*Macaca mulatta*). *Behaviour* 28:182–209.

Trivers, R. L. (1974) Parent-offspring conflict. *American Zoologist* 14:249–264.

Yerkes, R. M.; Dodson, J. D. (1908) The relation of strength of stimulus to rapidity of habit-formation. *Journal of Comparative Neurology and Psychology* 18:459–482.

CHAPTER TEN

Coordination and Conflict in Callicebus *Social Groups*

CHARLES R. MENZEL

Social conflicts are an integral part of the fabric of everyday life. In their daily ranging, social primates are faced with decision-making conflicts over when and where to carry out important activities and when to follow other members of their group. Unless these conflicts are resolved in a manner that enables maintenance of spatial cohesion among members, the group would quickly fragment. Thus, for members of an established social group, decision-making conflicts can readily become compounded with interpersonal conflicts regarding the timing, content or location of activity. These could interfere with the completion of the daily rounds, as well as destabilize relationships with established companions. The development of social customs or routines is one way to avoid or resolve such conflicts.

Charles R. Menzel—Department of Mammalogy and Division of Zoological Research, National Zoological Park, Smithsonian Institution, Washington, D.C.

Special thanks are due to William A. Mason who served as the chair of my dissertation committee and provided conceptual assistance in all aspects of my research on *Callicebus*; to Peter Rodman and Henry McHenry for their generous support of my graduate efforts; and to William A. Mason, Sally P. Mendoza, Emil Menzel, and Jane Gagne for their suggestions on the manuscript. The research was supported by a National Science Foundation grant and a University of California Regents' graduate fellowship to the author and by NIH grant #RR00169 to the California Regional Primate Research Center. Writing was supported by NSF grant #INT–8603379 to the University of California and by a Department of Health and Human Services NRSA postdoctoral fellowship to the author.

This chapter reports on two family groups of the monogamous and territorial titi monkey (Callicebus moloch) *that were studied while both groups lived in a one-hectare field cage. The cage provided many options for travel, feeding, and other activities, yet was ecologically innocuous compared to the monkeys' native habitat. The tendencies to establish and modify social routines in this environment were examined. A variety of species-typical activities were repeated in similar form from one day to another. Travel, loud vocalizations, retiring to sleeping sites, and other collective activities were organized, in part, through the habitual use of specific structures. Individuals, including juveniles, appeared to anticipate certain spatial endpoints during travel and to recognize locations where choices might be made. In some instances, groups continued to use customary locations even after this became unfavorable due to an ecological change. Thus, in titis, the use of space has a historical aspect.*

At the group level, customary patterns could make everyday social events more predictable and serve as a background for collective travel decisions that were negotiated on a moment-to-moment basis. On many occasions, individual group members would deviate in some respect from these routines, in which case conflicts might arise. Conflicts usually took the form of either the other members of the group having to decide between following the deviant individual or continuing as before, or the deviant monkey having to choose whether to continue on its individual course or remain with the group. The decision was usually on the side of keeping the group together. Social attraction either caused the deviant to abandon its independent course or the group to follow its lead.

In their daily ranging, social primates are faced with decision-making conflicts over when and where to carry out activities and whom to follow. Unless these conflicts are resolved in a manner that enables maintenance of spatial cohesion among the various members, the group would quickly fragment. The development of social customs or routines is one possible way to avoid or resolve such conflicts. Conventional patterns could simplify everyday social events and serve as a background for collective travel decisions that are negotiated on a moment-to-moment basis.

For members of an established social group, decision-making

conflicts readily become interpersonal conflicts regarding the timing, content, or location of activity. These can arise in principle, at any place or moment in the daily rounds. From the perspective of the individual—and its presumed ultimate self-interests of survival and reproduction— prolonged interpersonal conflicts could interfere with the completion of the daily rounds, as well as destabilize relationships with established social partners. In this respect, an animal's ability to resolve social conflicts and to control its partners' movements and activities in a cooperative or nondisruptive manner might have long-term biological benefits. Incompatibility of action among social partners is not simply restricted to aggressive competition for a limited resource. For example, conflicts are likely to arise whenever social partners attempt to choose a common path or destination from among multiple or complicated spatial options. The potential for conflict is obvious when two hungry animals detect a single piece of food. However, the same potential also exists in situations such as the following: (1) two social partners remember and orient toward different (and widely separated) food sources; (2) one member of a travelling group refuses to enter a certain part of the home range; (3) two group members prefer different types of sleeping sites; or (4) a group simply encounters a choice point in its pathway. Thus, interpersonal conflicts can arise even when social partners share similar needs and interests (Deutsch 1973), as a consequence of chance differences in perception or attention and each animal's tendency to organize its own behavior with respect to perceived options in the environment.

One common individual level strategy for organizing activity in the face of multiple, complicated or changing ecological options is to form a customary or routine daily pattern.

> Habit simplifies the movements required to achieve a given result, makes them more accurate and diminishes fatigue. (James 1981, 117)

> Habit is...the enormous fly-wheel of society, its most precious conservative agent. (James 1981, 125)

Individuals of many mammalian species form repetitive patterns in their use of space and resources. Animals often make conservative choices by returning to familiar options, and any animal uses only a small number of the total available opportunities in the environment. This is reflected in a broad range of activities, such as the selection of familiar food items (Rozin 1976) and the use of established travel paths, shelters, and other fixed points in the home range (Hediger 1950). In principle, each individual in a population might establish a distinct and separate pattern. Nevertheless, this same propensity for restricting activity can provide an important basis for social coordination and the establishment of shared patterns.

Animals' tendencies to establish routine daily activities can often be discerned at a collective or group, level. For example, in howling monkeys, coordinated group travel is frequently carried out in single file along specific arboreal pathways. As a group approaches the border of its home range, progress is slowed, animals show signs of hesitation, and the group becomes reoriented toward established pathways and more familiar goals (Carpenter 1934; Milton 1980). A primate group's characteristic ways of using food, space, and other resources can differ from the ways of other groups of the same species that live in similar environments. Such group differences can also persist in spite of environmental change and turnover in group membership (Galef 1976; Itani and Nishimura 1973).

The existence of repetitive daily patterns at the group level suggests that individual participants are fairly effective at resolving (or avoiding) interpersonal conflicts over when and where to go and what to do. It may be hypothesized that they tend to handle these moment-to-moment decisions by adopting conventional solutions—that is to say, by relying on established social routines and customs of the group.

From this standpoint, the central issues are, first, how one can identify the shared patterns or routines that the members of a given group display in their use of the environment; and second, what the relative contributions of social and ecological factors are to these patterns. The latter question implies that an observer has some basis for assessing environmental options that are ecologically equivalent from the standpoint of the individual, but which are habitually responded to differentially by the group. More specifically: Do groups maintain any patterns that are not required by the ecological circumstances? Do these patterns persist in spite of adverse changes in the ecology? At what point do established routines begin to break down? Are some members more responsible than others for establishing and maintaining routines? Do familiar places and familiar routes serve to coordinate or cue individuals as to probable group choices?

In this chapter the general issues of social decision making and tradition are investigated in the titi monkey, *Callicebus moloch,* a species that, by all descriptions, is highly conservative in its use of space—especially when compared to other diurnal primates of similar size (Terborgh 1983). Because of this trait, titi monkeys tend to form a limited number of intimate social relationships. In the wild, titi monkeys live in small family groups comprised of an adult heterosexual pair and one or two offspring. Field data suggest that titi groups are highly cohesive in their movements through space, and that they limit their travel options by ruling out many of the available alternatives in a complex environment and returning to certain locations day after day (Mason 1966, 1968; Robinson 1977; Wright 1985).

A clear expression of this tendency is the group's restricted pattern of travel. The animals typically cross back and forth each day within a small, well-delimited area. Moreover, confrontations between neighboring groups tend to occur daily at predictable locations along the boundary of the home range. Family groups travel directly to such locations soon after arising. Within the home range, only a fraction of the available space is used intensively, and the animals return to certain individual trees repeatedly to feed and rest (Mason 1966, 1968; Robinson 1977, 1979b).

Analogous behaviors shown by captive titis suggest that they are strongly disposed individually to develop and adhere to habitual patterns or routines. They show a stronger tendency than squirrel monkeys to maintain a familiar travel path rather than switch to a shorter but unfamiliar alternative (Fragaszy 1979, 1980). Captive adult heterosexual pairs also show evidence of the strong cohesiveness reported for free-ranging animals. They tend to remain much closer to each other than do comparable pairs of squirrel monkeys, and to show greater social coordination in their responses to food, to novel objects, and to unfamiliar conspecifics (Andrews 1984; Anzenberger et al. 1986; Fragaszy and Mason 1978, 1983; Mason 1971, 1974, 1975; Visalberghi and Mason 1983).

These propensities toward adherence to established routines and close coordination between group members suggest that captive titi monkeys would be particularly suitable for a fine-grain examination of the interplay between intrapersonal and interpersonal conflict as these may arise in the course of a normal day. An appropriate setting in which this might be carried out would be one that was sufficiently spacious to provide a group with multiple options at various choice points. Ideally, the setting would include more than one social group, not only for comparative purposes, but more importantly because the establishment and maintenance of spatial boundaries may be an instance of social convention that operates between, as well as within groups. Until recently, however, territorial behavior has not been observed in captive titi monkeys. A previous report (Menzel 1986a) described the first instance in which titi monkeys—or any nonhuman primates—have been observed to establish and maintain territories in captivity. The purpose of the present report is to provide additional descriptive information on the relations between neighboring family groups and on the way in which family group members travelled together, resolved interpersonal conflicts, and established social routines within the established boundaries of their separate home ranges.

FIELD CAGE AND GENERAL METHODS

Two established family groups of dusky titi monkeys were studied in a one-hectare field cage over a four-year period. The cage was designed to provide as complete and naturalistic a setting as possible in captivity, and to provide the animals with a large number of options for organizing and carrying out their activities. This setting and time frame provided a number of advantages for describing social routines. The animals were permanently housed in the enclosure and were familiar with its opportunities, and their responses to a comparatively stable environment could be observed day after day. They were not pushed to establish any organized group routines, because they were provisioned with food and shelter and other ecological demands were minimal. The size and variety of options in the enclosure allowed many different patterns to develop and, at the same time, the structural simplicity of the environment (compared to the natural habitat) aided in detecting unusual or unexpected patterns that were imposed by the animals. Most of the observations were conducted under routine, everyday circumstances.

In other cases, I used simple (and relatively nondisruptive) experimental procedures to examine certain phenomena in more detail. The primary descriptive questions addressed were: What specific group activities are carried out? Which of these activities are routine and predictable? Under what conditions are they carried out? Who participates? How stable are group activities across time, across locations and across changes in group membership? In this report, I will describe primarily the social aspects of travel, loud calling, and retiring.

Subjects

Subjects at the outset of the study were a total of eight *Callicebus moloch* (two adult males, two adult females, one young adult male, one young adult female, one juvenile male and one juvenile female). The two adult males were wild caught and laboratory habituated. The other subjects were born in the laboratory. The exact membership and age-sex composition of each family group varied over time due to births, maturation of the young, deaths, and several instances of spontaneous intergroup transfer and pair formation, but the typical composition of each group over the four-year period was an adult heterosexual pair plus one or more immatures or young adults. The social groups had unrestricted access to the field enclosure and lived there throughout the year. They were well habituated to human observers.

Environment

The field enclosure provided each social group with a large number of spatial choices, since its surface area (1 hectare) was at least two hundred times larger than the area typically occupied by a social group at a given moment. The enclosure contained an elevated artificial runway system, and this was a preferred substrate for travel (Menzel 1986c). The runway measured 61.0 x 67.1 meters, and these linear dimensions were much longer than the typical spatial dispersion of the group at a given time. In other words, even within the spatial limits of the preferred runway system, the group faced many choices about where to go. The runway included at least twelve branching points, where travel choices could be made between two or more opposed directions. Of course, the group also had the option of reversing its travel direction at any point along the runway.

The field enclosure also contained many potential places for resting and sleeping, including ten rectangular wooden perching grids and ten subenclosures, the latter measuring 1.8 x 3.7 x 2.4 meters high, and enclosing perches and a heated hutch box).

The enclosure also provided a wide variety of feeding options. About 200 domestic fruit trees, vines, and other nonfruiting trees were widely distributed within the enclosure, providing daily choices concerning where and when to feed. The natural food supply changed gradually across seasons, which led to additional choices for groups, such as when to stop visiting a familiar tree whose food supply was diminishing, and when to investigate a new tree whose fruit was ripening. Many of the fruit-producing trees grew near the artificial runway system and provided incentives to depart from this preferred travel substrate.

Each group was fed ample portions of a nutritionally balanced laboratory diet at approximately 2:00 P.M. Food was placed in one or more subenclosures within each group's home range. The field cage was watered by a sprinkler system, and the same system was used for evaporative cooling when temperatures exceeded approximately 32.2°C (90°F). Further details about the field enclosure can be found in works by Fragaszy (1978) and Menzel (1986b).

MAINTENANCE OF TERRITORIES

Normative data were obtained to characterize the use of space by individuals and family groups under everyday conditions. The locational data presented here were collected on fifty-four separate days over a twelve-week period of May through August 1982. At most, three samples were collected per day, one in each of the following time blocks: 6 to 11 A.M., 11 A.M. to 3 P.M., and 3 to 6 P.M.. The total number of scans collected at these times was fifty-four, twenty-one, and seventeen, respectively.

The location scores collected for the adult males are shown in figure 10.1. The males occupied clearly separate portions of the field cage, with very little overlap in their home ranges. The location of each male was also representative of the male's remaining family group members. For example, family members were found in contact with one another on 11% of intervals and within 3.1 meters of one another on 67% of available intervals. At no time during this sample were members of one family group found within 3.1 meters of a member of another family group.

The social and spatial separation of groups revealed by these data was the typical pattern shown during the four years that I observed the animals, despite the fact that there were changes in group membership. The relations between groups appeared to involve a large measure of ambivalence, in which the tendency to approach members of the neighboring group was usually overridden by the tendency to avoid or withdraw from them. The proximity of a neighbor tended to elicit marked visual interest and, frequently, approach. This was often accompanied by arching, piloerection, and vocalization, but rarely resulted in contact. Such displays were particularly prominent at certain boundary regions, and appeared to be intensified by the proximity of the pairmate.

Experimental tests, conducted after territorial boundaries were well established, indicated that adult titis were reluctant to encroach upon the perceived home range of a neighboring group, even when the neighbors were not physically present (Menzel 1986a). These observations suggest that a significant element in the confrontations at territorial boundaries that occur almost daily in free-ranging titis (Mason 1966) is intergroup approach-avoidance conflict. Each group is drawn toward the other, becomes notably agitated as proximity increases, and tends to stop short of physical contact. Thus, at a proxi-

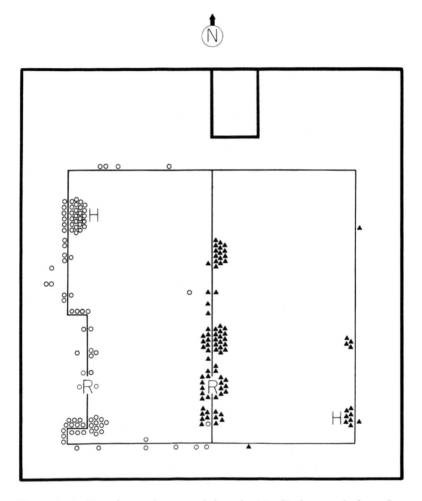

Figure 10.1 Use of space by two adult male titis. Circles = male from Group 1, Triangles = male from Group 2, Heavy Line = perimeter of field cage, Thin Line = runway system, R = release cages, H = hutch boxes used in experiment. Trees and other ecological details omitted for clarity.

mate level, it seems that ambivalence toward neighboring groups, rather than defense of a specific geographic region or resource, is the more important factor in confrontations and the establishment and maintenance of territorial boundaries.

PATTERNS OF COLLECTIVE RESTING AND TRAVEL

The use of space by the titis was intimately related to social coordination and the daily routine. Group members followed one another and travelled together throughout their common home range. Each group used certain specific locations as resting places and returned regularly to specific trees, where the group members fed in close social proximity. In the evening, family group members retired to a shared sleeping site and spent the night in contact. Certain locations were used as sleeping sites much more frequently than others. This will be discussed further in this section.

Travel frequently occurred in discrete bouts or progressions. For example, a group would rest in one location for twenty minutes, then move directly to another location 30 meters away and rest for twenty more minutes. Travel was organized around runways and other artificial structures (Menzel 1986c). Subenclosures and perching grids were the most common endpoints for travel bouts. Travel could end abruptly when the group reached a subenclosure or a perching grid. At this time the animals would frequently gather near one another and sit quietly, as if the activity were concluded without ambiguity.

One could identify other well-defined endpoints to travel. These endpoints included gathering in physical contact, producing loud-calls, and converging upon a familiar food source, such as a fruit-bearing tree. A particular subenclosure, perching grid, or tree could serve as a definite endpoint for travel for many days or weeks in a row, and travel progressions frequently led directly to these structures. Considering the fact that various juveniles and adults would take the lead, range ahead of the others, and stop at the usual endpoint, it seems probable that the animals (as well as the observer) perceived and anticipated many of the patterns.

The onset of travel often occurred smoothly and quickly, without apparent ambiguity or conflict among the individual participants. Feedback among group members in these cases seemed to be almost instantaneous. For example, when one animal stood up, another individual would look over, a third individual would start walking quickly, and within seconds a travel progression—in the direction in which the first individual had oriented originally—would be underway. Progressions frequently appeared to be spontaneous rather than a response to an external stimulus. Quite often, I could see no advantage—in terms of shade, availability of food, proximity to the neighboring group, or the physical structure of the location—obtained by moving to the next location.

Individuals frequently initiated group travel by moving a short distance away from the others and waiting. Any animal who moved ahead on its own was apt to be followed, particularly if it moved along a familiar pathway. Other group members apparently perceived or anticipated the actor's intended route. Recruitment of followers was also supported by other features within a broader context. For example, the sight of the neighboring group moving within its own home range often stimulated the other group to travel in parallel. Animals who led a progression also seemed to take into account such factors in the broad social context and to anticipate what their partners would do next. Nevertheless, many initiations failed to attract an immediate following. If no one followed within several minutes, the unsuccessful leader would often return to sit with the group. Juveniles initiated travel progressions, but young juveniles did not attempt to do so as often as did the adults, nor were they as successful in attracting a following. (See also Robinson 1977.)

An initiator's success frequently seemed to depend upon the perceived consequences of the travel direction. For example, an animal might fail to follow a partner who moved toward the neighboring group, but would then follow the partner quickly when it returned and selected another direction within the home range. Potential followers exerted a degree of passive control over initiators. A single animal sometimes prevented the rest of the group from travelling by simply remaining in his or her original location. When two or more animals waited for another monkey to follow, they might coordinate their own activity around a routine location, as if this avoided a direct choice between continuing along their own course and approaching the partner. In most instances, however, each individual cooperated by following in the proposed direction. Once a collective movement started, the animals typically continued to another distant endpoint. It was also common for one of the followers to leap over the leader's back and take the leading position.

To summarize, titi group members generally followed one another readily. Decision-making conflicts were resolved by a process of negotiation prior to the onset of travel. The animals' perceived choices seemed to include whether or not to travel and in which direction to go. The cohesive quality of the titis' social relationships—reflected in the tendencies to follow or to wait if not followed—and the usual restriction of options to familiar paths leading toward familiar endpoints often gave group movements an aspect of consensus.

The ending of a travel progression appeared to be more loosely organized than the progression itself. A group could linger at several possible locations before it settled in place to feed or rest. Interpersonal conflicts could arise as the group neared a potential endpoint. Signs of conflict or incompatibility between two individuals were often subtle. For example, two animals would oppose one another by simply standing several meters apart in the vicinity of a potential endpoint. Failure by one individual to follow the group all of the way to the endpoint would often cause the others to hesitate and reverse their direction. Even if one animal initiated a new activity at the stopping place—such as feeding or lying on a perch—the group, as a whole, could still abandon the area and embark on another travel progression. The final resolution of an ambiguous or conflictual situation sometimes required several minutes or longer.

Animals were visually attentive to one another during these times, and they were probably alerted to the potential for additional travel by signs of nonconformity or passive resistance, such as the sight of a group member lingering near the edge of the physical structure upon which the group rested. Conflicts over direction and activity were often resolved without extensive vocalizations or emotional displays. The animals were usually tolerant and showed full participation once the travel progression was reinitiated. The last animal to abandon the original destination might even move forward and take the lead. The animals frequently exchanged the lead as they moved along the runway. In other words, the titis often showed minimal signs of intra-group conflict during travel and during the transition from one activity to another. Well-coordinated travel also occurred with great regularity and repetition along established pathways. Many of the conflicts that were observed involved passive resistance or noncooperation, rather than active attempts to move in an incompatible direction.

As long as group members moved toward familiar destinations within their own home range, conflicts surrounding the choice of travel direction seldom escalated to include aggressive manual contact. Competing leaders could show signs of agitation—including repetitive, high-pitched vocalizations—if their conflict over travel direction was prolonged. Vocalizations were especially common if nightfall was approaching and if the group had not yet selected a sleeping site.

One possible exception to the generalizations already given can be mentioned. A juvenile female and a young adult male who had waited for the adult male to follow them toward the sleeping site approached the adult male (from 5 to 10 meters away) and touched him lightly on the head and back using their hands.

Some of the more dramatic instances of interpersonal conflict centered around movement toward the neighboring group. For example, on several occasions, adult males were observed to chase down, grab, and slap their female pairmate as she moved toward the neighboring group or into its home range. A more typical response was for the adult male to follow his pairmate toward the neighboring group, move into contact with her, give low intensity vocalizations or touch her lightly, and then initiate movement back toward the interior of his own home range.

Even after they arrived at a familiar endpoint, group members still faced choices about what to do. They would often converge upon a common shared activity, such as feeding or resting together. This convergence did not seem to result solely from each animal's independent response to an external ecological factor. In other words, individuals seemed to choose an activity, in part, by watching their social partners' activities and tentative choices. An animal would abandon its own initial activity to join another individual. Of course, animals also found a variety of uses for the same general location, and they would embark upon different kinds of activities. For example, one individual would forage on grass, two other individuals would lie on the runway, and one animal would feed in a nearby tree. The animals might gradually spread out 15 meters or more. In principle, such differences could challenge the group's intimate coordination. Nevertheless, the animals could reassemble quickly. If there was an outside disturbance—or if one individual foraged on its own for a long time, initiated an activity in a new location, or oriented toward something at a distance and moved toward it a few steps—this would attract the other animals and draw them into the same place and activity.

Moments of visual orientation toward partners occurred periodically in the course of sustaining an activity and while simply remaining in the same general vicinity. Animals would also wait for partners who were engaged in a different activity. For example, an animal would wait for five minutes or longer on a portion of the runway that was not typically used for resting until its partners finished feeding in a nearby tree. Conversely, an animal who fed alone in a tree would carry a food item to sit nearby the other group members. The animals' abilities to find different uses for a given structure, to converge upon a common activity within a somewhat ambiguous setting, and to wait for social partners who were engaged in different activities seemed to be important in integrating group activities with places and in developing a coherent pattern of home range use.

PERSEVERANCE IN GROUP ACTIVITY

Over the course of several years, I occasionally found group members sitting with their tails twined, in locations that were fully exposed to cold rain or strong wind. The animals seemed to be reluctant to break contact with their partners, in spite of the probable physical discomfort. If I then disturbed the group by tapping its perch, the animals would, in fact, move and reaggregate in a more sheltered location.

Group members would remain together in a specific place for twenty to thirty minutes or longer. The animals often appeared to be absorbed and somewhat unreactive to objects and events in the external setting. Their attention seemed to be directed toward one another and their immediate, small-scale surroundings.

Another example of perseverance at the group level was for animals to pass up potential distractions during travel and to move together quickly and directly from one predictable endpoint to another. The animals were not always easily disrupted by the presence of a convenient food incentive, by a T-shaped intersection in the runway, by the sight of the neighboring group, and so on as they travelled along the runway. The group would often move past an alternative option without pausing, as if it were not even perceived.

A group's tendency to pass up possible incentives during travel was interesting because the same incentives could later become objects of visual attention and frequent approach. Animals showed signs that they perceived some of the forsaken incentives. For example, a travelling group would pause to look at a nearby ripe fruit tree but then continue hurriedly in its original direction. Individuals would lag behind the group to stare at a fruit tree, then run to catch up with the others. Animals who lagged behind in this way showed signs of mild agitation, including piloerection, mild vocalizations, and sneezing, as they glanced back and forth between the fruit and the departing group. An animal who approached the fruit tree could show pronounced hesitation and, if it obtained a piece of fruit, run toward the other animals with the item in its mouth before feeding.

Such individual conflicts of choice during travel were interesting because they could be repeated and eventually resolved by bringing about a change in the group's routine. For example, when an animal hesitated near a particular tree during several travel progressions and then, during the next progression, deviated from the group's line of travel and initiated movement toward the tree, the others would orient quickly in the same direction and prepare to follow. The tree might then become a routine travel objective. In this manner, a group's attention and activity could shift to a variety of places and objects that had been ignored for weeks, including trees of various species, perches, and subenclosures. Once a place became an endpoint, it might continue to be preferred, even if the original incentive for visiting that location was no longer present.

COORDINATION OF DUETS

Spatiotemporal routines probably enhanced social coordination among group members and reduced the chance that prolonged interpersonal conflicts would occur during the daily rounds. Group members would, of course, often deviate from a routine path or activity and, at this point, interpersonal conflicts might arise. Nevertheless, most of these conflicts involved predictable choices and outcomes and, in favor of maintaining group cohesion, were resolved relatively quickly, perhaps within a few minutes or less.

In certain daily situations, however, monkeys faced more substantial difficulties in anticipating their partners' actions, creating the possibility of prolonged conflicts among group members. Ambiguities about where to go and what to do were especially common when one group was close to the neighboring group. In principle, the proximity of the neighbors increased the chance that a monkey's social relationships and daily routines would be substantially disrupted, if not permanently altered. For an adult, this included a risk that its pairmate would interact with, or reorganize its daily routine around, a sexual competitor.

During intergroup confrontations, the monkeys were probably uncertain about how to initiate specific activities and whether their partners would participate. Hesitation, piloerection, spatial vacillations, and other signs of intraindividual conflict were relatively common near territorial boundaries. Adults often approached and followed their pairmates, and when pairmates drew near to one another they often showed arching, gnashing, and other aggressive displays. Adults also directed aggressive postural displays toward members of the neighboring group, particularly toward monkeys of the same sex. The coordinated behaviors shown by adult pairs during intergroup confrontations and their displays against the neighbors often culminated in duet calling.

Loud or duet calling is of special interest to the question of social conflict in titi monkeys in at least two respects. First, from the perspective of intraindividual conflict, loud calls were often preceded by signs of individual hesitation and agitation. The loud-calls per se constituted an extreme form of behavioral agitation (Mason 1966). When monkeys called, their backs shook from the exertion, and they showed piloerection during most of the call. Arching and tail lashing also occurred sporadically. The duration of several minutes or longer and volume of the call suggested that considerable effort was involved.

Second, from the perspective of interindividual, within-group conflict, and coordination, loud-calling constituted an extreme form of spatial and vocal coordination between pairmates, as described by Mason (1966, 1968) and Robinson (1977, 1979a, 1979b, 1981). The calls were usually performed simultaneously by pairmates as a duet and, during the call, they stood in physical contact or in close proximity to one another and faced the neighboring group. If one animal shifted its posture during the call, the other typically shifted to maintain physical contact. The pairmates usually began their loud-calls within a few seconds, or fractions of a second, of one another. The structure of the call also reflected a high level of coordination. Each monkey alternated between two relatively distinct phases of the call. In one time interval, a series of slower, high-pitched vocalizations was given, followed in the next time interval by a series of rapid, low-pitched vocalizations. These two components of the call were alternated between the animals, and each component was given by a different animal at a given moment.

Adult pairmates often coordinated the onset of their loud-calls by returning to specific regions of the field cage and to exact locations from which they had given calls on previous days. For example, during the first weeks of observation one of the adult pairs was observed to give fourteen duet calls. All fourteen of the calls were given from the boundary region of the group's home range. Most of the calls (93%) were restricted in space to an imaginary circle of radius 2.5 meters, which constituted a small fraction of the boundary region, and 79% of the calls were given from a single specific location. Similarly, during the same time period the neighboring pair of monkeys gave each of its twenty-two loud-calls from the boundary region of its home range; 86% of the calls were restricted to an imaginary circle of radius 2.5 meters, and 41% of the calls were given from a single specific location.

Movement toward a familiar calling location by one partner during an intergroup interaction seemed to cue the other partner about what would follow. Adults often responded to a partner's initial movement by following the partner or by jumping over the partner's back and taking the lead to the routine location. As the monkeys approached the location they might begin to walk quickly, to bound forward, and to show arching, part-calls or moans, and piloerection. The partners' mutual arrival at a routine calling location was often followed promptly—usually within one or two seconds—by loud-calling. Adults apparently recognized or anticipated their partner's intended endpoints from a distance, for the two monkeys might converge upon a familiar calling location along lines of travel differing by 90 or 180 degrees, from starting positions 25 meters apart.

To summarize, the monkeys possibly perceived a relationship between their partners' locomotion and displays, on the one hand, and familiar places or landmarks in the environment, on the other, and used this information to coordinate their vocalizations. Routine calling areas often served as focal points for pair behavior during the context of intergroup interactions. The areas which the groups used reliably for calling varied across months. This suggests that the areas were established, in part, by social conventions of the pairmates.

Similar spatial performances were shown by juveniles, even though they did not usually participate in the loud-call per se. The juveniles seemed to pick up the spatial and situational aspects of the routine, for they often ran ahead of the adults and jumped to a familiar calling location when the adults began to move toward the location. The juveniles could also converge upon the location independently, for example, from a travel angle which was different from that of the adults. The juveniles usually stood in contact or in close proximity to the adults during the loud-call.

The monkeys seemed to take into account a variety of factors in the broader setting before initiating their loud-calls which were almost always restricted to situations in which the monkeys had interacted with the neighboring group in some manner. Certain events arising during an intergroup encounter seemed to increase the chances that the partners would follow one another or converge upon a familiar location and give loud-calls within the next few minutes. Some of these specific events included intergroup chases, close approaches between groups initiated by either party, and withdrawal by the pairmates from the boundary area toward the center of their home range.

Vacillation, hesitation, behavioral agitation, and interpersonal within-group conflicts seldom led to duet calling unless they were accompanied by some form of interaction with the neighbors.

The process of initiating a routine on a specific occasion could involve incorrect predictions about a partner's future actions, or predictions which had to be frequently revised. Incompatible choices could lead to hesitation, repeated attempts to initiate an expected activity, and conflicts between group members. When a group was in close proximity to the neighboring group, conflicts for adults often took the form of having to decide whether to approach the neighbors, follow the pairmate, or withdraw into the center of the home range. In general, difficulties in social coordination were particularly common when a monkey was confronted simultaneously with the following events: an external challenge (such as the close proximity of the neighboring group), multiple choices about where to go and what to do, and a desired partner who did not readily participate in an anticipated activity.

The animals seemed to face difficulties in coordinating their vocalizations. For adult titis, the decision of when to duet was probably more difficult than deciding where to duet. During confrontations, monkeys seemed to be uncertain about whether their partner would participate in loud-calls. An adult's behavior was only partially predictable, and there were often clear differences between the pairmates in their preferred travel directions. Thus, one adult might show signs of interest in the neighboring group and vacillate between approaching the neighboring group and approaching its pairmate. The other pairmate, however, might attempt to withdraw into the home range and vacillate between withdrawing and following its partner toward the neighbors. These spatial vacillations probably increased each partner's uncertainty about how or whether to begin its loud-call.

Several factors contributed to the difficulty for pairmates in coordinating their duets. Each pairmate's readiness for calling seemed to fluctuate over short time spans. There were also short-term fluctuations in important external variables, particularly in the behavior of the neighbors. Pairmates often approached one another and exchanged postural displays, gnashing, and part-calls without these interactions culminating in a duet, and one partner might simply move off at any moment. There appeared to be consistent individual differences between pairmates in their readiness to call. If the neighbors approached closely, one pairmate might suddenly withdraw rather than remaining with its partner. The individual monkeys showed different levels of interest in the neighbors, and it was common for a monkey to withdraw toward the core of the home range while a partner was approaching the neighbors. Attempts to initiate travel in opposed directions by the pairmates often led to hesitation and to revisions in direction. It should be remembered that duet calling was not always given during a confrontation, and that it constituted only one of several behavioral options or alternatives for coordinating activity with a partner.

To summarize, loud-calls were not given at any fixed time within the confrontation, and the monkeys' coordination of calls seemed to require some form of communication and mutual anticipation. Routine patterns of movement directed toward familiar locations seemed to enhance the monkeys' ability to extract information about the moment-to-moment inclinations of their partners.

It was relatively unusual for a single monkey to give loud-calls, or at least to sustain calling for very long. The events preceding the loud-call possibly constituted a form of mutual recruitment into the call. In many cases, the male and female began moving together in the direction of a familiar calling area at about the same time. If one monkey (usually the female) did not readily initiate its loud-call, then the other partner might approach the first animal, gnash into the partner's face and chest, give part-calls, touch its partner's face and back, jump over the partner's back from one side to the other, stand in physical contact, tail lash while facing the neighbors, and respond to any vocalizations emitted by its partner by intensifying its own vocalizations.

As with most other coordination problems arising between pairmates, partial conflicts over when and where to call seldom involved any patently aggressive physical contact. Overtly punitive contact—such as slapping—was uncommon, and it did not seem to be specifically elicited by a monkey's failure to participate in loud-calls.

Once a pair of monkeys initiated a duet, they might call continuously for several minutes or longer. If one partner (usually the female) stopped calling, then the other partner might touch its partner's face or back and continue calling alone for ten seconds or longer. This activity often induced the other partner to resume its calling. Nevertheless, these and other partial conflicts over whether to call were not always resolved in favor of duetting. A monkey might respond to its partner's overtures simply by moving closer to the partner, orienting visually toward the neighbors and gnashing. The monkeys might then sit quietly in contact, or resume their vacillations with respect to the neighboring group, in which case the possibility for calling at a later time remained open.

The adults' use of routine calling sites varied sharply according to the social context. Outside of confrontations, the pairmates seldom converged quickly and directly upon a routine calling site. Some of the sites—including small plastic pipes and the door frames of subenclosures—were seldom used for other activities such as resting. A monkey's postural orientation toward such sites was thus uncommon and, when it did occur, probably provided a partner with a distinctive and conspicuous cue for what would follow.

These and other observations suggest that the titis were familiar with the events preceding duet calling, and that they could anticipate the likely outcomes of molar, large-scale units of action, especially of ongoing travel. In other words, the titis' manner of initiating duet calls suggested that they learned the regularities—that is, the most common outcomes—of certain daily social events at a fairly broad spatiotemporal scale and used this information to coordinate their activity with partners. In general, the titis were keenly attentive to deviations from typical activities and spatial routines by their partners. Specific deviations from a routine seemed, in fact, to be a familiar part of the daily rounds in titis, as if the animals recognized a pattern and its major variations.

GROUP PATTERNS OF RETIRING

The manner of sleeping posed a decision-making task for the titis. Monkeys had to decide when and where to retire. They also had to implement their choices within the constraints of the group setting (Norton 1986). In principle, decisions about retiring could have been made independently by the various group members. Individuals had the daily option of leaving their groups and of retiring alone, since the field cage included a large number of perches and hutch boxes. The monkeys were not prevented by their physical environment from retiring at substantially different times or in different locations. Nevertheless, group members almost always retired at about the same time and slept together with their tails twined. Thus, an important set of decisions for individuals concerned how to coordinate their spatial choices with other group members.

One task for the group was deciding when to get ready to retire. In some cases the decision appeared to be simple. One animal moved away from the group after a period of resting, the others immediately followed, and the group travelled directly to a familiar sleeping site. The animals seemed to discriminate the onset of retiring from the previous activity. In other cases, however, no single or conspicuous event initiated retiring. In these situations interpersonal conflicts could arise.

For example, one monkey might make several attempts to initiate travel before it attracted any partners. The group might come to a stop after 5 or 10 meters. Individuals might move at different times, rather than simultaneously, or meander in different directions. Individuals also differed in their apparent readiness to make a direct or near approach to a familiar sleeping site. Differences in travel direction and temporary spatial separations could lead to conflicts of choice. Travel conflicts often took the form of animals having to decide whether to continue along their independent route or follow the other party, and decide how long to wait before making a choice. Choices generally went in the direction of maintaining proximity, and, as I observed under a wide range of other daily situations, the monkeys revised their spatial choices on the basis of the other group members' location, direction, and rate of movement.

Social attractions were strong. Group members frequently approached and followed one another prior to retiring. Two monkeys who sat with their tails twined might abandon their location to approach a third group member who remained at a distance. Intraindividual approach-approach conflicts often arose from a direct choice among two or more available social partners. A typical choice might be between sitting with one partner in a likely retiring site versus following another partner toward another familiar endpoint where feeding might occur. Such choices could lead to signs of agitation, including high-pitched vocalizations, and these behaviors might persist until one of the other monkeys revised its activity and joined the group.

Arbitrariness and Durability of the Routine

Routine use of particular sleeping sites occurred despite the fact that other comparable locations were available. Alternative locations were sometimes visited prior to retiring, but the titis exercised a considerable degree of restraint and goal orientation in the face of such alternatives. Routines were also maintained in the face of mild environmental barriers. In one notable case, a group's preference for retiring outdoors was maintained from summer into fall. This was beyond the time when temperatures dropped below the levels expected to be comfortable for titis.

Group Differences

Family groups living in a common environment and exposed to the same temperatures and sunset times differed reliably in several aspects of their retiring. Thus, groups differed in their times of retiring, in the regions of the field cage they used, and in the type of structure (such as perch versus hutch box) which they selected most often for retiring. In all these respects, there was no strict correspondence between retiring patterns and the structure of the physical environment, suggesting that the animals imposed their own patterns upon the field cage.

Thus, on seventeen of the first twenty-five nights of observation, all of the members of one group retired by making their final entry to the sleeping site before any members of the other group retired, whereas the reverse pattern occurred on only one night. Similarly, the groups differed in the type of location that they typically selected as a sleeping site. The early-retiring group slept outdoors on a particular perch on twenty-three of twenty-five nights, whereas the late-retiring group slept outdoors on a particular perching grid on only four nights and slept indoors in a total of two hutch boxes on twenty-one of twenty-five nights. Social factors were an important determinant of where the animals went at a given moment, and the choice of specific retiring location was apparently based, in part, on convention.

Deciding Where to Retire

The sleeping sites used by the animals could not be described easily along a single physical dimension which distinguished the sites from all other locations in the field cage. Some broad regularities could be specified in the monkeys' choice of location. Thus, the animals always retired to sites that were located in or near subenclosures, and the sites chosen always bore a degree of resemblance to a shelter. Monkeys did not retire to trees, to the ground, nor to the open runway. Nevertheless, of the sleeping sites chosen, subenclosure perches and perching grids were schematic and incomplete shelters, and they only partially enclosed the animals and insulated them from the surrounding environment. The animals' choice of retiring sites was not precisely determined by the physical characteristics of the available locations. The tendency of titi monkeys to establish a specific location for retiring did not seem to depend on their having narrowly defined criteria for what constituted a potential sleeping site. The titis might well come to retire in a particular location even if there was initially a degree of ambiguity about its suitability.

Family group members almost always retired together in a single location (>95% of evenings). Retiring with the group was a higher priority than returning to a customary sleeping site. Even the members of the group that retired consistently to a particular location (92% of evenings) still travelled and entered the sleeping site together. Group members did not simply move off on their own to the usual endpoint.

Deciding When to Retire

Particular retiring sites were used consistently, despite the fact that group activities preceding retiring were often complex and variable and did not always lead in a direct route to the site. In many cases, the customary sleeping site was approached several times before the animals retired there.

Monkeys showed postural orientation toward customary sleeping sites in a manner suggesting that they took into account the response of their partners. For example, an animal would sit apart from the group in the direction of its customary sleeping site for several minutes. It would move ahead toward the sleeping site each time that a partner took a few steps in the same direction.

Narrative Descriptions of Retiring

The process of retiring to a sleeping site was repeated in similar form from one day to the next, and was organized around a specific set of locations for each group of monkeys. The most well organized pattern was shown by one group, which often moved in an abrupt, continuous progression from its feeding subenclosure—where the group had spent much of its afternoon—to another subenclosure located about 60 meters away. A perch inside this subenclosure served as the group's retiring site on twenty-three to twenty-five evenings. It was often possible to identify a particular individual who initiated the travel progression, even though this was not always the same individual from one night to the next. The initiator walked from 1 to 4 meters away from the feeding cage, in the direction of the main runway, and remained there until another animal followed. The other animals, who were typically resting at this time, often followed within thirty seconds. Once movement began, the rank order of travel varied as one animal after another took the leading position by jumping over the backs of the other monkeys. Members travelled together to the sleeping cage, and the group made few extended pauses along the runway. They often entered the cage immediately after arriving in its vicinity, gathered in contact on the top perch of the cage and twined their tails. Interpersonal conflict appeared to be minimal during the retiring sequence.

In contrast to this group's direct and relatively unelaborated manner of retiring, the neighboring group's activities during the same time period were more variable in content and more loosely coordinated. Nevertheless, the neighboring group also showed regularities in its activities. The neighboring group retired, for example, on eighteen of twenty-five nights (72%) in a single specific location. Another consistent aspect of its retiring sequence was that the group would hesitate, vacillate, come to a stop, and frequently reverse its travel direction at certain specific areas prior to the customary sleeping site. Interpersonal conflicts over which way to go, tentative movements, and signs of agitation—especially high-pitched, repetitive vocalizations—were relatively frequent. A third consistent aspect of this group's evening routine was that the monkeys appeared to negotiate a travel direction at these locations and to coordinate their travel closely when they moved past the customary stopping points.

As a result of the group's oscillations and frequent reversals in travel direction, the total travel distance greatly exceeded the minimum straight line distance between the group's starting point and the sleeping site. For example, the group might leave a resting place and travel about 30 meters toward its most commonly used sleeping site, then mill about for several minutes, return to the vicinity of the starting point, mill about again, and repeat the pattern five or more times over the course of an hour before settling at a potential sleeping site. The group's choice of travel direction appeared to be determined more by internal social processes and preferences than by any member's strong attraction to a particular endpoint.

Reversals in direction were especially common at certain locations along the runway. The animals behaved as if they anticipated a delay or a change in travel direction at these locations. For example, the group might travel quickly along the runway but stop suddenly when it reached a customary place. The monkeys might draw into close proximity, then suddenly embark on a travel progression in the reverse direction.

One of the group's reliable travel delays and changes in direction occurred in the vicinity of a mild environmental obstacle, a large growth of pampas grass that overhung a portion of the runway. This delay at the pampas grass constituted a central organizing feature in the group's evening routine. Even though, at other times of day, the animals traversed the pampas grass with little hesitation and seldom carried out the social interactions to be described next, interpersonal conflicts arose here routinely during the time of retiring.

In the evening, the group typically approached and withdrew from the pampas grass one to five times before crossing. The monkeys might sit near the grass for up to twenty minutes before they actually initiated a move to cross. The pattern seemed to result largely from the animals' uncertainty about whether their partners would follow, and from their reluctance to move too far ahead of the group independently. The monkeys were probably not simply cautious of the grass per se, because animals occasionally sat in the grass while waiting for the other group members.

The delay at the pampas grass was accompanied by signs of agitation in group members. For example, animals who sat near the pampas grass would give frequent, high-pitched vocalizations while they looked back toward the other group members, especially if the latter monkeys failed to follow. Similarly, animals who attempted to move away from the pampas grass would often fail to attract an immediate following. This might lead to high-pitched vocalizations, and the conflictual situation was not always resolved quickly. The monkeys' vocalizations and hesitation possibly contributed to group cohesion. Monkeys typically crossed the pampas grass together, with less than two minutes intervening between the first and last animal. Monkeys were very attentive to one another as they made the crossing. One animal would stand motionless in midstride if another stopped following. After crossing the pampas grass together, the animals often picked up speed in the direction of the sleeping site. One or two individuals would run ahead of the group to a perch that was located near the usual sleeping site.

The exact location of the sleeping site appeared to be previously established and familiar to the members. Nevertheless, the group members did not simply enter the site or gather together in physical contact immediately after arriving in its vicinity. The group would approach to within a few meters of the eventual sleeping site but then stop, mill about, and possibly travel up to 55 meters away. The animals seemed to anticipate the possibility of a travel reversal occurring here. The location was a routine choice point.

On most evenings the group rested on a perching grid before entering the sleeping site. This resting period often lasted twenty minutes or longer. During this time the monkeys vacillated between the probable sleeping site and the other group members. They often approached to within about 1 meter of the site but sat without entering it. This pattern of approach and withdrawal might be repeated ten times or more during an evening. The monkeys' initial approaches often failed to include specific postural and visual orientation toward the sleeping site, which suggested that the animals were not actually preparing or expecting to enter it at that time. Individuals who approached the sleeping site were clearly responsive to the movements of their partners. For example, animals would inhibit their final approach to the site if one of the remaining group members moved a few steps in an opposed direction. Animals might cling, head-down, to the vertical wire of a subenclosure or remain poised in other unlikely postures while waiting for another monkey to follow. Tentative interactions of this sort often culminated in two or three group members suddenly moving toward the sleeping site and entering it together. Thus, two monkeys might enter the site within several seconds of one another. The group members settled in contact, twined their tails, and typically remained in place until well after sunset, when observations were terminated.

To summarize the description, this group typically retired in a particular location, but prior to retiring, it tended to travel extensively. The group stopped, vacillated, and rested at certain predictable locations, and the monkeys seemed to be familiar with these aspects of the routine. Thus, the animals watched one another when the group approached particular locations, as if they anticipated further changes in direction and were reluctant to move too far ahead independently or to be left behind. The group traveled in loosely coordinated and reversible stages toward a familiar destination. The use of the structures often bore a clear relation to the social setting, and the structures visited by several monkeys simultaneously often provided few obvious nonsocial benefits or incentives. The degree of the monkeys' involvement with social choices relating to travel was unexpected and did not seem to be required by the ecological setting. The activities selected seemed to be determined in part by the monkeys' social preferences and by group conventions.

DISCUSSION

A major emphasis of this chapter is on the importance of conventional, self-imposed spatiotemporal routines as a means of enhancing social coordination and reducing or constraining potentially disruptive conflicts in titi monkey groups. The generally conservative stance toward the environment by titi monkeys and their tendency to form small cohesive groups are probably conducive to the types of conflicts which titi monkeys experience most frequently and to the ways in which their conflicts are resolved.

In my observation of captive groups, conservatism was expressed in the tendency to stay on familiar pathways, to move toward familiar endpoints, and to adhere to the same daily routine. As the data on retirement times show, this was only partly a response to environmental variables. Individual group members on many occasions would deviate in some respect from these routines, in which case conflicts might arise. Conflicts usually took the form of other group members deciding whether to follow the deviant individual or to continue as before; or the deviant monkey choosing between continuing on its individual course or remaining with the group. The decision was usually on the side of keeping the group together. Social attraction either caused the deviant to abandon its independent course or the group to follow its lead.

Decisions on the side of fragmenting the group could lead to signs of agitation. Such behaviors could attract the attention of the other group members and possibly played a role in reestablishing social proximity. Signs of intraindividual conflict included spatial vacillations and hesitation, repetitive high-pitched vocalizations, and other responses such as sneezing and piloerection.

Most interpersonal conflicts arising in the course of group travel were resolved quickly. The monkeys' tendencies to follow other group members, to wait if not followed, and to restrict options to familiar routes gave travel decisions an aspect of consensus. Groups often selected a particular routine travel direction quickly, as if the other potential options were ignored. It seemed difficult to account for this and other aspects of coordination without taking into account the historical background of group decision-making.

Social conflicts within family groups were generally mild. When adults approached one another they occasionally exchanged agonistic displays before moving into physical contact. In the close proximity of the neighbors, these displays could escalate to duet calling. The presence of food seldom led to conflict among group members. They usually fed amicably in close proximity, at a limited number of the total available food sources. Conflicts arose more often by feeding apart than by feeding together, although these conflicts, of course, took different forms. Conflictual interactions in the presence of food might consist of one monkey grabbing or cuffing another who approached the food source, followed by both monkeys continuing to feed next to one another. If a pairmate was the target, the two monkeys might gnash simultaneously and arch together before they resumed cofeeding (Fragaszy and Mason 1983). If an adult slapped a juvenile or a young adult, the latter might give brief squeals but would remain nearby. More protracted conflicts among group members over where to go could arise in the presence of the neighbors. Touching, hovering, and grasping—as well as more patently aggressive forms of physical contact—could be directed toward a pairmate who moved toward the neighboring group (Cubicciotti and Mason 1978). The most severe forms of interpersonal conflict were correlated with changes in existing social relationships and were between animals of the same sex, such as between an adult male and his young-adult male offspring. These conflicts involved sporadic physical aggression.

Conflicts between groups were constrained in several important ways. The neighboring group was an apparent source of approach-avoidance conflict, and intergroup relations were characterized by a large degree of ambivalence. The presence of the neighboring group evoked keen visual interest and approach, but approaches stopped short of actual contact and were accompanied by marked hesitation, vacillation, and frequent signs of agitation. Confrontations usually occurred at predictable boundary locations. Monkeys showed strong attractions to their pairmates and tended to withdraw from the neighbors toward their own home range.

From the standpoint of coordination, conflict, and decision making, it is interesting that entire family groups, rather than simply individual monkeys, established habitual routes and activities. Game theory predictions suggest that two animals faced with an interpersonal conflict will tend to use any clearly perceived asymmetry as a convention to resolve the conflict if the cost of escalation is high compared with the benefit of winning (Maynard Smith 1982), even if the perceived asymmetry is arbitrary and uncorrelated with modest differences between the animals in power or competitive ability (Hammerstein 1981). The perceived environment is often differentiated according to its history of use by the animals, and this differentiation might serve as a convention for resolving or avoiding conflicts. Similarly, Ullmann-Margalit (1977) suggests that conventions can arise to solve recurrent coordination problems, such as problems in which the participants have some shared interests, are required to converge in space and time to achieve these interests, and in which all participants stand to gain by the solution of the problem. She suggests, in addition, that two characteristics of conventions are that they constitute salient reference points to the participants for organizing their activity and that the conventions are maintained, in part, by a system of mutual expectations.

Strong individual habits of movement possibly become socialized and provide a basis for interpersonal coordination in titis. The monkeys maintained their day-to-day spatial relationships through cooperative travel behaviors. The monkeys not only followed their partners, but they noticed their partners' travel initiation movements and extended the direction of these movements. Conflicts over choice of direction were usually nondisruptive. Monkeys exerted a degree of passive control over partners. The resolution of conflicts depended, in part, on how long opposed parties would wait before one followed the other. Monkeys were fairly effective in attracting and orienting partners, who showed a strong predilection for following their own established patterns.

The tendency for individual titi monkeys to adhere to established habits of movement provided a basis for group coordination and the avoidance of intragroup conflicts in at least two possible ways.

First, monkeys could lead their partners along the partners' preferred routes of travel quite easily. Initiation movements often seemed to be timed to coincide with a partner's momentary orientation and readiness to travel, with the choices made on previous occasions, and with the momentary context. Second, there is the possibility that group members converged upon a common set of spatial preferences and strong habits of movement. In this case, an individual could simply follow its own inclinations and have a reasonable chance of attracting the other group members.

The timing of activity in relation to other group members and the resolution of interanimal conflict over travel direction were important and recurrent daily problems for the titis. Although all social attractions among group members were probably not of equal strength (Mendoza and Mason 1986) and individual monkeys probably varied in their effectiveness as leaders, travel involved a measure of cooperation. Monkeys were sensitive to feedback from others during travel. Animals did not always follow a fixed leader. Any individual might move ahead or take the initiative at a choice point, and the others were likely to follow. The patterns suggested that group members often reached a consensus about where to go (Norton 1986). It seems likely that all anticipated where the group as a whole was likely to go in the near future. The observations suggested that, once an initiator moved beyond a choice point and all animals had started to follow, the outcome was no longer ambiguous. Coordination was probably achieved most easily when travel directions were routine and predictable, and the consensus was probably reinforced through repetition. (See also Robinson 1979b, 1981.)

Maintenance of routines under captive conditions is of interest in relation to imposition of form upon the environment (Holloway 1969). Family groups in the study under discussion here carried out routine patterns of retiring that were not entirely predictable from environmental factors. Groups were not forced to set up the routines to meet stringent ecological demands, nor were individuals forced to retire together as a unit.

In the wild, titi monkeys tend to form repetitive patterns in use of space, travel routes, intergroup confrontations, selection of foods and feeding locations, and the use of resting trees (Kinzey 1981; Mason

1966, 1968; Robinson 1977; Wright 1985). This study demonstrates that well-developed routines can be maintained in environments that are structurally simplified and ecologically innocuous in comparison with the natural habitat. Groups in the field cage occupied well-defined, stable home ranges and had preferred locations within the home range for resting. They gave loud-calls at particular locations within the home range. Confrontations between groups occurred along particular regions of the boundary and much of each group's natural food consumption occurred in a limited number of preferred trees. Groups also retired in characteristic ways. The daily rounds included diverse activities, but some tendency to establish customary patterns could be detected for many of these. The evidence to date suggests that familiar locations, actions, and situations give emphasis and support to intention movements by group members and provide a basis for interpersonal coordination.

Titi monkeys possibly prefer handling choices which are shared, or which have been initiated by a partner, rather than making independent decisions. They might find decision making—at least in the small group context—as straightforward as decision making while alone. Group choices were often made sequentially, through a series of smaller spatial adjustments which had the nature of tentative proposals. It often seemed as if no single animal made the entire decision. Titis tested individually on travel tasks show marked hesitation before selecting a route (Andrews 1984; Fragaszy 1980; Menzel 1986c). This hesitation in the face of a choice seems compatible with the titis' spatial techniques of decision making in a social context and with their strong predilections for remaining in close proximity to group members.

The tendencies to regulate or control other group members in a nondisruptive manner by engaging their attention, leading and following them, anticipating and extending their intention movements, restricting conflict to passive opposition, and so on, helped the individual to maintain a place in the group and in its daily rounds. Under natural conditions, such tendencies presumably contribute to the success of individual titi monkeys in maintaining a close relationship with a partner of the opposite sex and in delimiting and maintaining access to a home range.

The question of how a particular species, group, or individual achieves its characteristic life style at a proximate level is central (Mason 1967). Here the issue is how and why *Callicebus* groups maintained a species-typical life-style in captivity, in the absence of important ecological factors such as predator pressure, interspecific competition, limited food supply, and an extensive canopy. Part of the answer might lie in the proclivities of individuals for forming habitual patterns of movement, forming small cohesive groups, and avoiding prolonged conflict with group members. The titis employed some direct and specific ways of integrating social activities with locations and with partners. Group activities were repeated in similar form from one day to the next and were restricted to a small fraction of the available space. Within these self-imposed restrictions, groups combined and traded off a large number of component activities. Adherence to familiar social routines and attention to the momentary deviations of partners was an important strategy for avoiding disruptive conflicts and for enhancing social coordination in *Callicebus moloch*.

REFERENCES

Andrews, M. W. (1984) Comparative use of space by two species of New World monkeys with special reference to foraging behavior. Doctoral dissertation. University of California at Davis.

Anzenberger, G.; Mendoza, S. P.; Mason, W. A. (1986) Comparative studies of social behavior in *Callicebus* and *Saimiri*: Behavioral and physiological responses of established pairs to unfamiliar pairs. *American Journal of Primatology* 11:37–51.

Carpenter, C. R. (1934) A field study of the behavior and social relations of howling monkeys. *Comparative Psychology Monographs* 10:1–168.

Cubicciotti, D. D., III; Mason, W. A. (1978) Comparative studies of social behavior in *Callicebus* and *Saimiri*: Heterosexual jealousy behavior. *Behavioral Ecological and Sociobiology* 3:311–322.

Deutsch, M. (1973) The Resolution of Conflict. New Haven, Conn.: Yale University Press.

Fragaszy, D. M. (1978) Comparative studies of visuo-spatial performance in squirrel monkeys (*Saimiri*) and titi monkeys (*Callicebus*). Doctoral dissertation. University of California at Davis.

Fragaszy, D. M. (1979) Titi and squirrel monkeys in a novel environment. In

Erwin, J., Maple, T. L., Mitchell, G. (eds.), "Captivity and Behavior."
New York: Van Nostrand Reinhold Co. 172–216.

Fragaszy, D. M. (1980) Comparative studies of squirrel monkeys (*Saimiri*) and
titi monkeys (*Callicebus*) in travel tasks. *Zeitschrift für Tierpsychologie*
54:1-36.

Fragaszy, D. M.; Mason, W. A. (1978) Response to novelty in *Saimiri* and *Callicebus*: Influence of social context. *Primates* 19:311–331.

Fragaszy, D. M.; Mason, W. A. (1983) Comparisons of feeding behavior in captive squirrel and titi monkeys (*Saimiri sciureus* and *Callicebus moloch*).
Journal of Comparative Psychology 97:310–326.

Galef, B. G., Jr. (1976) Social transmission of acquired behavior: A discussion of
tradition and social learning in vertebrates. In Rosenblatt, J. S., Hinde,
R. A., Shaw, E., Beer, C. (eds.), "Advances in the Study of Behavior,"
vol. 6. New York: Academic Press. 77–99.

Hammerstein, P. (1981) The role of asymmetries in animal contests. *Animal
Behaviour* 29:193–205.

Holloway, R. L., Jr. (1969) Culture: A human domain. *Current Anthropology*
10:395–412.

Hediger, H. (1950) "Wild Animals in Captivity." London: Butterworths Scientific Publications Limited.

Itani, J.; Nishimura, A. (1973) The study of infrahuman culture in Japan: A
review. In Menzel, E. W., Jr. (ed.), Symposium of the 4th Congress of
the International Primatological Society, vol. 1: Precultural Primate
Behavior. Basel: Karger. 26–50.

James, W. [1890] (1981) "The Principles of Psychology." Cambridge, Mass.:
Harvard University Press.

Kinzey, W. G. (1981) The titi monkey, genus *Callicebus*. In Coimbra-Filho, A. F.,
Mittermeier, R. A. (eds.), "Ecology and Behavior of Neotropical Primates," vol. 1. Rio de Janeiro: *Academia Brasileira de Ciencias*. 241–276.

Mason, W. A. (1966) Social organization of the South American monkey, *Callicebus moloch*: A preliminary report. *Tulane Studies in Zoology*
13:23–28.

Mason, W. A. (1967) Traditions synthesized. *Contemporary Psychology* 12:589–
590.

Mason, W. A. (1968) Use of space by *Callicebus* groups. In Jay, P. C. (ed.), "Primates: Studies in Adaptation and Variability." New York: Holt, Rinehart and Winston. 200–216.

Mason, W. A. (1971) Field and laboratory studies of social organization in
Saimiri and *Callicebus*. In Rosenblum, L. A. (ed.), Primate Behavior:
Developments in Field and Laboratory Research," vol. 2. New York:
Academic Press. 107–137.

Mason, W. A. (1974) Comparative studies of social behavior in *Callicebus* and *Saimiri*: Behavior of male-female pairs. *Folia Primatologica* 22:1–8.

Mason, W. A. (1975) Comparative studies of social behavior in *Callicebus* and *Saimiri*: Strength and specificity of attraction between male-female cagemates. *Folia Primatologica* 23:113–123.

Maynard Smith, J. (1982) "Evolution and the Theory of Games." Cambridge: Cambridge University Press.

Mendoza, S. P., Mason, W. A. (1986) Parental division of labour and differentiation of attachments in a monogamous primate (*Callicebus moloch*). *Animal Behaviour* 34:1336–1347.

Menzel, C. R. (1986a) An experimental study of territory maintenance in captive titi monkeys (*Callicebus moloch*). In Else, J. G., Lee, P. C. (eds.), "Primate Ecology and Conservation." Cambridge: Cambridge University Press. 133–143.

Menzel, C. R. (1986b) Intergroup relations, home range use and activity patterns in *Callicebus*. Doctoral dissertation. University of California at Davis.

Menzel, C. R. (1986c) Structural aspects of arboreality in titi monkeys (*Callicebus moloch*). *American Journal of Physical Anthropology* 70:167–176.

Milton, K. (1980) "The Foraging Strategy of Howler Monkeys." New York: Columbia University Press.

Norton, G. W. (1986) Leadership: Decision processes of group movement in yellow baboons. In Else, J. G., Lee, P. C. (eds.), "Primate Ecology and Conservation." Cambridge: Cambridge University Press. 145–156.

Robinson, J. G. (1977) Vocal regulation of spacing in the titi monkey *Callicebus moloch*. Doctoral dissertation. University of North Carolina, Chapel Hill.

Robinson, J. G. (1979a) An analysis of the organization of vocal communication in the titi monkey *Callicebus moloch*. *Zeitschrift für Tierpsychologie* 49:381–405.

Robinson, J. G. (1979b) Vocal regulation of use of space by groups of titi monkeys *Callicebus moloch*. *Behavioral Ecological and Sociobiology* 5:1–15.

Robinson, J. G. (1981) Vocal regulation of inter- and intragroup spacing during boundary encounters in the titi monkey, *Callicebus moloch*. *Primates* 22:161–172.

Rozin, P. (1976) The selection of foods by rats, humans and other animals. In Rosenblatt, J. S., Hinde, R. A., Shaw, E, Beer, C. (eds.), "Advances in the Study of Behavior," vol. 6. New York: Academic Press. 21–76.

Terborgh, J. (1983) "Five New World Primates: A Study in Comparative Ecology." Princeton, N.J.: Princeton University Press.

Ullmann-Margalit, E. (1977) "The Emergence of Norms." Oxford: Oxford University Press.

Visalberghi, E., Mason, W. A. (1983) Determinants of problem-solving success in *Saimiri* and *Callicebus*. *Primates* 24:385–396.

Wright, P. C. (1985) The costs and benefits of nocturnality for *Aotus trivirgatus* (the night monkey). Doctoral dissertation. City University of New York.

CHAPTER ELEVEN

Social Conflict in Two Monogamous New World Primates: Pairs and Rivals

GUSTL ANZENBERGER

Among the many problems that confront monogamous primates is what to do about intruders which threaten the integrity of a monogamous group. When a third party is encountered by an established heterosexual pair, social conflicts are inevitable. Two obvious types of conflict can be distinguished. One type is a conflict of choice between two potential sexual partners—the current mate and the intruder. The second type is a competition/aggression conflict between two rivals—the current mate and the intruder—for the same sexual partner.

The present chapter provides a comparative assessment of these types of social conflict in two monogamous species, Callicebus moloch *and* Callithrix jacchus. *Confrontation experiments with captive animals living in family groups brought together unfamiliar heterosexual dyads in the presence or absence of each member's pairmate. This procedure allowed the motiva-*

Gustl Anzenberger—Ethology and Wildlife Research, Institute of Zoology, University of Zurich, Switzerland

The study on common marmosets was conducted at the University of Zurich, and that on titi monkeys was done at the California Regional Primate Research Center, Davis, Calif. (supported by the National Institutes of Health grant RR00169). I thank all my colleagues in Zurich and Davis for their support, assistance, and comments. My special thanks to S.P. Mendoza and W. A. Mason, first for providing a friendly and scientifically stimulating atmosphere during my stay at Davis, and second for advice and assistance in the preparation of this manuscript.

tional structure of both aspects of social conflict to be evaluated. Results were similar for species and sexes with regard to conflicts of choice. Individuals preferred their mates over social alternatives. However, there were obvious differences between species and sexes when pairmates were absent during confrontations.

Under these circumstances male Callicebus *and male* Callithrix *directed affiliative and sexual behaviors toward unfamiliar females. Female* Callicebus *responded positively to these overtures, whereas female* Callithrix *reacted aggressively toward unfamiliar males. Competition conflicts resulted in displays of overt aggression toward rivals by both sexes of* Callithrix. *Although this behavior was less pronounced in either sex of* Callicebus, *males reacted with conspicuous mate-monopolizing behavior. Thus, despite similar evolutionary constraints imposed by strong similarities in their social systems, the solution of comparable social conflicts is quite different in these two monogamous species.*

Picture yourself as a field primatologist in the South American rain forest watching a heterosexual pair of monkeys going about their normal activities. The pairmates are probably sitting side-by-side. Eventually they groom each other. After a while, one gets up and moves away, and the other follows within a few meters. Then, both start feeding in the same tree. They move on, resume feeding, and rest again. All activities seem to be performed in a coordinated fashion and often with pairmates in close spatial proximity. Suddenly, one or both individuals show signs of agitation, as expressed by back arching and piloerection, and they stare at a certain spot in the foliage. There, a third adult individual of the same species appears.

Let us leave our naturalistic scene for a moment and consider the ethological implications of this triadic situation. It is likely that the pairmates will react differently toward the intruder, and the particular response each member of the pair displays will depend on the gender of the intruder. We may expect the intruder to be a social rival for one member of the pair, whereas it may be an incentive or social alternative for the other. For both pairmates, the intruder represents a potential threat to the social integrity of the pair. Thus, the presence of the third individual creates social conflict for both of the pairmates. However, two different types of conflict may be distinguished. One pair-mate is in a conflict of choice between alternatives that are comparable. The other is in a competition conflict with a rival.

Let us return to our naturalistic scene and assume that the intruding individual was a male. This constellation would create male-male competition and female choice, the two major aspects of sexual selection as outlined by Darwin (1871). Our understanding of these evolutionary forces was furthered by Trivers's (1972) theory of "parental investment." According to Trivers, the general rule in animals is that the unequal share in parental investment between genders leads to choosiness in females and to competition among males. (See also Parker 1979 for a comprehensive treatment of "sexual selection and sexual conflict.")

The concept of parental investment provides not only an explanation for the observable asymmetry in reproductive behavior between genders, but predicts that this asymmetry should be less accentuated in monogamous species. Male parental care in these species is often highly developed and thus, parental investment is more evenly shared by both parents. (For literature on mammals, see Kleiman 1977a, 1980, 1981.) Although it is unlikely that females of monogamous species that are not supported by a male would fail completely in rearing offspring, it is likely that, when both parents share in care-giving activities, each enhances its reproductive success considerably. Such evolutionary constraints result in mutual dependency of male and female regarding their reproductive success. Thus, evolutionary pressures are, in a way, balanced for males and females of monogamously living species, and the gap between genders is narrowed as compared to polygamous forms. Nonetheless, there remain two basic risks, one for each gender.

Even when a male helps his mate raise young, copulations with other females may result in a large increase in his reproductive success, while copulations by his mate with other males will tend to have the reverse effect (Trivers 1985). The first possibility poses a risk to females of being deserted by the male helper. (For a theoretical treatment of predictions and probabilities, see Maynard Smith 1977.) The second possibility poses a risk to males of being cuckolded, and being exploited by the female in caring for another male's offspring. In other words, natural selection favors a double standard for males. They pursue for themselves a course that they try to prevent their mates from pursuing. Thus, on theoretical grounds, we may make specific predictions about the behavioral reactions of monogamous males and females in a triadic situation.

1. Males should take vigorous action as soon as a male intruder appears so as to prevent interactions—especially sexual ones—between their mates and the rival.

2. Males should try to copulate with a female intruder.

3. Females should attempt to prevent interactions between their mates and unfamiliar females. The response should be less vigorous than the males' response to an opposite-sex intruder since a female does not suffer a loss of reproductive success when her mate copulates with another female unless his interactions with the strange female curtail his parental duties or lead to desertion.

4. Females, should resist unfamiliar males and stay with the mates with whom they have already successfully raised offspring. This assumption is underscored by studies of monogamous birds which have shown that, with increasing numbers of joint breeding periods, the pair's breeding success increases and dissolution of pairs is most likely to occur after unsuccessful breeding attempts (Coulson and Thomas 1983; Rowley 1983).

These predictions are not entirely independent. For example, as a consequence of male predilection to philandering (item 2 in the preceding list), both genders of monogamous species should develop jealousy-like behaviors which would be expressed when their mates interact with rivals (items 1 and 3). Such behavioral strategies will prevent males from being cuckolded and females from being deserted. Some of the numbered predictions deal with the same function, and one may eliminate the need for another. In other words, there is no need for a male to intervene between his mate and a rival (1) if females are faithful to their mates (4). Of course, it may be advantageous to have a double security system in which both males and females actively thwart attempts of extrapair males to copulate with mated females.

In order to test these hypothesized contributions of males and females to the monogamous social system, we can simulate our initial naturalistic scene under laboratory conditions. This has been done in a series of confrontation experiments during which the social and sexual integrity of a mated pair was potentially threatened by a third individual. This paradigm allows one to provoke reactions from all three parties involved in order to illuminate the social conflicts which emerge and examine how they are subsequently resolved. The subjects of these studies were two monogamous New World primate species: common marmosets (*Callithrix jacchus jacchus*) and dusky titi monkeys (*Callicebus moloch*) (Anzenberger 1983, 1985, 1988). The similarity of design and methods of the experiments allow a direct comparison of the proximate mechanisms contributing to the pair bond in these species.

NATURALISTIC AND EXPERIMENTAL DATA
ON *CALLITHRIX* AND *CALLICEBUS*

Common marmosets (*Callithrix jacchus jacchus*) and dusky titi monkeys (*Callicebus moloch*) are both New World primates and members of the families Callitrichidae and Cebidae, respectively. Although quite different in size—with titi monkeys roughly three times heavier than common marmosets—they have many characteristics in common. Before going into details, it should be stressed that there is an unequal distribution of the type of studies conducted on the two species. Our knowledge of titi monkeys stems from both naturalistic and laboratory studies, whereas the bulk of information about common marmosets is derived almost exclusively from studies of captive monkeys. During recent years, some investigations of social behavior and social dynamics of wild-living common marmosets have been published (Hubrecht 1984a, 1985). However, satisfactory information is still lacking on wild-living common marmosets.

The initial field study of *Callicebus moloch* (Mason 1966, 1968) established that these monkeys have a monogamous and territorial life-style. Extensive laboratory studies by Mason and coworkers have elucidated behavioral characteristics contributing to this social grouping pattern (as summarized in Fragaszy and Mason 1983). Additional field studies on *Callicebus moloch* have been conducted by Wright (1984) and, with respect to vocal territorial behavior, by Robinson (1979, 1981). A comparative review of biology and ethology of the genus *Callicebus* has been provided by Kinzey (1981).

The first systematic studies of captive marmosets were conducted by Epple (1967, 1968). Subsequently, marmosets—especially common marmosets—were subjects of various and detailed studies (Epple 1970, 1974, 1975, 1978, 1981; Kleiman 1977b; Rothe 1975; Rothe et al. 1978). All studies of common marmosets are consistent with the view that the social structure of this species is also monogamous and organized within family units. To date, this seems to be in line with naturalistic findings on this species. Descriptions of group composition indicate that marmoset societies are age-graded and, on this basis, it can be inferred that only one female or one pair per group is reproductive (Hubrecht 1984a; Maier et al. 1982. See also Abbott in this volume).

Both species are highly arboreal and territorial (*Callithrix*: Hubrecht 1985; Lacher et al. 1981; Maier et al. 1982. *Callicebus*: Mason 1966, 1968; Menzel 1986 and chapter 10 this volume; Robinson 1979, 1981). There is little sexual dimorphism in either species and long-lasting pair bonds are formed between one male and one female. In both species, males are heavily involved in rearing the young.

Preambulatory infants are carried most often by males (figure 11.1a and figure 11.2a), and infants are taken by females only for short suckling bouts. This division of labor is well-documented in both species (*Callithrix*: Box 1977; Epple 1975; Ingram 1977. *Callicebus*: Mason 1966; Mendoza and Mason 1986; Wright 1984). Young individuals remain for some time within the parental social unit and, as a consequence, a family-type of group structure emerges with the monogamous pair as the core. In titi monkeys, a typical group consists of parents and one or two offspring (Mason 1966). In common marmosets, an extended family may number up to thirteen individuals (Hubrecht 1984a). Under captive conditions, up to as many as twenty

Figure 11.1a Social characteristics of common marmosets (*Callithrix jacchus*). Father carrying 10 week old infant twins. (Photograph by M. Zollinger).

Figure 11.1b Social characteristics of common marmosets (*Callithrix jacchus*). Extended family consisting of parents and their 8 offspring aged between 9 and 42 months (Photograph by M. Zollinger).

individuals can be maintained in a single family unit (Rothe 1978). The difference in group size between species (figure 11.1b and figure 11.2b) may be partly attributed to the fact that *Callithrix* females normally give birth to twins twice a year, whereas *Callicebus* females give birth to singletons once a year. It is noteworthy that, within family groups, only the parents reproduce, although in either species, individuals reach full maturity around the age of eighteen months. Therefore, groups may contain more than one pair of adults. Sexual suppression of the progeny by parents is likely (*Callithrix*: Abbott 1984, chapter 12 this volume. For a review on reproductive biology and physiology of captive common marmosets, see Hearn 1983).

Experimental studies investigating the pair bond and pair behavior in these species have employed a variety of approaches. These studies, together with the investigations to be presented, provide an extensive and fairly consistent picture of social dynamics during experimental encounters between unfamiliar individuals. In studies of common marmosets, Epple (1970) allowed her subjects to interact freely as a trio. That is, pairmates remained together and, therefore, their individual reactions toward the intruder were not clearly separable from reactions as a pair. The results of this study clearly indicated that intrasexual aggression is well-developed in both sexes, a charac-

Figure 11.2a Social characteristics of titi monkeys (*Callicebus moloch*). Father carrying 9 week old infant.

Figure 11.2b Nuclear family consisting of parents and 2 offspring aged 1 week and 15 months; father is carrying infant.

teristic that has been considered as the foremost contributor to maintaining the pair bond. Evans (1983) tested reactions of common marmosets to strangers of the opposite sex in the presence or absence of the individual's pairmate. In these studies, subjects and strangers were always separated by a screen and, therefore, could not interact physically. Comparable studies of titi monkeys have been conducted. Mason (1971) allowed his test individuals to interact freely as heterosexual trios. In another study, the strength and specificity of attraction between pairmates was examined (Mason 1975). Cubicciotti and Mason (1975) tested behavioral and physiological reactions of titi monkeys to a novel environment when alone, with their pairmates, or with an unfamiliar substitute of the opposite sex. In a subsequent study, Cubicciotti and Mason (1978) examined the changes in reactions of titi monkeys to like-sex strangers that occur as a function of the distance of the strangers from their own pairmates.

Therefore, studies have been conducted on both species in which heterosexual trios have been allowed to interact freely (Epple 1970; Mason 1971). There have also been studies in which pairmates were tested, singly or together, using unfamiliar conspecifics as social stimuli (Cubicciotti and Mason 1975; Evans 1983). In these latter studies, the focus was on the mated individuals as affected by the presence of strangers, and unfamiliar individuals could not interact physically.

CONFRONTATION STUDIES: RATIONALE

My own studies took a different approach. The focus was on confrontations between two unfamiliar individuals of the opposite sex in a situation in which they could interact freely, and under one of two conditions: either they were alone, or the pairmate of one of them was present, but restrained by a screen that prevented it from direct intervention in interaction between the confrontation pair. Two questions were of central concern. First, how does a mated individual behave when alone with an unfamiliar partner of the opposite sex (a confrontation partner)? Second, does this individual's behavior change when both the confrontation partner and familiar partner (pairmate) are simultaneously present?

The conditions examined in the confrontation experiments with each species are presented schematically in table 11.1. Condition 1 provides each individual subject with access to an unfamiliar partner of the opposite sex. Presumably the major source of conflict in this situation resides within the individuals and centers around incompatible tendencies to approach/affiliate versus avoid/attack. This form of conflict is also present in conditions 2 and 3. In addition, however, these situations offer the subjects (condition 2), as well as the confrontation partners (condition 3) the opportunity to indicate a choice between two available alternatives—namely, the pairmate (behind the separation screen) or an unfamiliar monkey of the same gender as the pairmate. These conditions also include for subjects (condition 3), as well as confrontation partners (condition 2) the presence of a stranger of the opposite sex (behind the separation screen), an arrangement that presumably elicits competitive conflict between the like-sex animals. Under these conditions, therefore, the pertinent data include not only the interactions between confrontation pairs (as in condition 1), but also their interactions with the individual behind the separation screen.

Table 11.1 Conditions of Confrontation Experiments

Condition	CJJ	CMO
Dyadic tests		
0 Subject + pairmate	-	+
1 Subject + con partner	+	+
Triadic tests		
2 Subject + con partner (+ pairmate)	+	+
3 Subject + con partner (+ con partner's mate)	+	+

CJJ = *Callithrix jacchus*.
CMO = *Callicebus moloch*.
+ / - = species was tested / was not tested in this condition.
Con partner = confrontation partner.
( ) = present behind separation screen.

CONFRONTATION STUDIES: MATERIAL AND METHODS

In the two studies described here, methods, apparatus, and design were not identical in all respects. However, settings and methods were sufficiently similar to allow a comparative treatment of data. Any methodological differences that might contribute to species contrasts will be clearly indicated in the following discussion.

Subjects

Three adult heterosexual pairs of common marmosets (parent-pairs of three different families) and seven pairs of titi monkeys served as subjects. The animals stemmed from mixed populations of wild- and captive-born individuals and were kept at the Institute of Psychology in Zurich, Switzerland (*Callithrix*) or at the California Regional Primate Research Center at Davis in the United States (*Callicebus*). All laboratory-born individuals had been raised as members of their species-specific family unit, and pairs had been housed together for at least thirty months in the marmosets (mean = 52.0 months) or at least seven months in the titis (mean = 25.7 months) prior to testing. All female marmosets and four female titis were pregnant during the testing periods. (For details of husbandry for marmosets see Anzenberger 1985; and for titis, see Anzenberger 1988.)

Behaviors Recorded

As indicated earlier, one way to detect conflictual situations, is to look at differences in behaviors shown by subjects toward familiar versus unfamiliar individuals. Therefore, it is appropriate to use interaction patterns of established pairs as baseline data to compare with those of unfamiliar confrontation pairs. Mated pairs within family groups of both species characteristically show high degrees of proxemic and affiliative behaviors, low degrees of agonistic behavior, and exclusivity with regard to sexual behaviors. Thus, behavioral data collection focused on these broad categories. Specific behavioral categories are listed in table 11.2. A detailed description of these behaviors is pre-

Table 11.2   Behaviors Records

	CJJ	CMO
Proxemic behaviors		
Social distance	+	+
Proximity (within 30 cm distance)	-	+
Affiliative behaviors		
Huddling	+	+
Tail twining		+ ssp
Sexual behaviors		
Anogenital control	+	+
Mounting	+	+
Thrusting	+	+
Agonistic behaviors		
Chest rubbing		+ ssp
Tail lashing		+ ssp
Arch bristle (CJJ)/Back arching (CMO)	+	+
Threat vocalization	+ ssp	
Intended attack (through Plexiglas)		+
Duet calling		+ ssp

CJJ = *Callithrix jacchus.*
CMO = *Callicebus moloch*
+ / - = behavior recorded / not recorded in this species.
ssp = species-specific behavior.

sented for marmosets in Stevenson and Poole (1976), and for titis in Kinzey (1981) and Moynihan (1966).

Apparatus

Tests were carried out with both species in specially equipped indoor cages. Test cages for marmosets (1.6 X 1.8 X 2.0 meters high) were immediately adjacent to or part of rooms in which marmoset families were kept throughout the year. Testing required only that individuals be released into these areas. Test cages were designed so that subjects could neither see nor be seen by their pairmates; could see them, but not be seen by them; or could interact with them through a wire mesh screen. (For details see Anzenberger 1985.) Titi monkeys were transferred from their home cages to a remote site for testing and were returned to the home cage after testing. They were quite familiar with this mode of handling and readily entered the transport cage that was used to accomplish the transfers. The test cage for titis (2.7 X 0.6 X 1.8 meters high) could be divided into three equal compartments (0.9 X 0.6 X 1.8 meters high), with animals separated from each other by removable Plexiglas screens. In triadic tests, the monkey in the center compartment was always between its pairmate and the confrontation partner. (For details see Anzenberger 1988.)

Design and Procedures

Each subject was tested under the conditions indicated in table 11.1. The two members of a confrontation pair were always of opposite sex. In both species, order of presentation of conditions was approximately counterbalanced across confrontation pairs. Design and procedures differed between the species as follows: (1) For titis condition 0 was part of the experimental series, whereas for marmoset behaviors between pairmates were derived from other studies in our laboratory (Erb 1983) and from the literature (Epple 1975). (2) The triadic tests (conditions 2 and 3) with marmosets were conducted under two different exposure regimens. Pairmates were in view of the subject, but were themselves unaware of the confrontation (behind one-way screens), or pairmates were behind wire mesh and able to observe the encounters. In triadic tests with titi monkeys, pairmates were behind Plexiglas screens and thus able to observe all encounters between confrontation pairs and to be observed by them. (3) Due to kinship relations in titis and/or different numbers of mature members in marmoset families, individuals were tested with different numbers of confrontation partners. In *Callicebus*, each of the seven males and seven females was tested alone with its pairmate (condition 0) in five replicates (for a total of thirty-five trials). In the remaining conditions (1 to 3), each individual was tested once with four to six different confrontation partners in one replicate each (total of thirty-six trials per condition or a total of thirty-six different pair combinations).

In *Callithrix*, the situation was different. Families regularly contained more than one pair of mature members. Therefore, not only mated individuals (parents) but, unmated individuals (mature subdominant offspring) were available as confrontation partners. Accordingly, each of the three mated males and three mated females was tested in conditions 1 to 3 with three to seven confrontation partners in two replicates each (for a total of forty-two trials per condition, or a total of twenty-one different pair combinations). Due to the differences in the social status of confrontation partners, there were three different pair types for *Callithrix*: mated males encountered mated females (four pair combinations); mated males encountered unmated females (nine pair combinations); and mated females encountered unmated males (eight pair combinations). (4) In each species, a single confrontation lasted fifteen minutes. In marmosets, spatial and behavioral data were recorded throughout this period. In titis, confrontations were divided into three five-minute segments. During the first segment, individuals remained visually isolated from each other, and data were not collected (the habituation phase). During the second segment, individuals had visual contact (through Plexiglas) and spatial data were recorded. During the third segment, two individuals could interact directly (with the separation screen removed) and behavioral data were recorded. In triadic tests, the pairmate of one of the subjects always remained behind the Plexiglas screen.

Data Recording and Analysis

In *Callithrix*, each confrontation was recorded simultaneously on audio and video tape marked with time signals. Spoken protocols were used for recording behavioral data, while video tapes were used for recording spatial data of confrontation partners. Behavioral data were based on focal animal sampling (Altmann 1974). The male served as focal subject and all interactions, whether he was initiator or recipient, were recorded. Note that this technique included all female initiated behaviors as well. Data summaries for social interactions were based on absolute frequencies and durations.

In *Callicebus*, only spatial data were recorded in the second five-minute segment (before the Plexiglas screen was raised). The position of all animals was noted at fifteen-second intervals (instantaneous sampling). In the second five-minute segment (after the screen was raised, allowing the confrontation pair to interact freely), behavioral data were recorded using one-zero sampling within fifteen-second intervals. Thus, in both segments, the maximum score for each individual was twenty.

For both species, data were analyzed by analyses of variance. Because individuals were tested repeatedly with several confrontation partners and with different numbers of confrontation partners, analyses used mean scores for each individual in each condition. In titis, two separate analyses were performed—one to determine differences between familiar pairs and unfamiliar pairs (condition 0 versus condition 1); the other to determine the effect of pairmates' presence on confrontation pairs (conditions 1 to 3). Post hoc comparisons, where appropriate, used the Newman-Keuls test. In marmosets, data were analyzed by the Friedman test (conditions 1, 2, and 3). Post-hoc comparisons between conditions used the Dunn-Rankin test, and those between different pair-types used the Dunn Test I (Lienert 1973; Siegel 1976).

CONFRONTATION STUDIES: SIMILARITIES AND CONTRASTS BETWEEN *CALLITHRIX* AND *CALLICEBUS*

Conflict of Choice

For a first appraisal of different tendencies aroused in the individual, we may look at the measure of social distance. In marmosets, the test situation selected for this purpose was one in which the confrontation pair could see, but not be seen by, the pairmate and family members of one member of the confrontation pair. Thus, the family of one member of the confrontation pair was visually present, but these individuals were not themselves aware of the confrontation and, therefore, could not affect the confrontation pair beyond their mere visual presence. Confrontations were recorded on video tape from above and analyzed from still-frame pictures in five-second intervals, yielding 180 data points per individual per fifteen-minute test. From these data, distances of one individual from its family (that is, to the dividing screen) as well as distances between the confrontation partners were calculated. Distances from the family were assigned to one of four possible classes, based on segments (each 30 centimeters wide) parallel to the dividing screen. Distance class 1 was closest to the screen and distance class 4 farthest from the screen.

Spatial distribution of mated *Callithrix* males and females during confrontations in which their families were visually present are shown in figure 11.3. Females stayed nearer to the screen separating them from their mates and families than did males. On 74% of the total observations, females occupied areas in the two closest distance classes (1 and 2), whereas males were located in these areas for only 51% of the observations.

Distances between confrontation partners were divided into six different distance classes (representing increments of 30 centimeters between individuals). Since distance between two freely moving individuals is the sum of mutual tendencies to approach and avoid one another, it is difficult to determine for each individual the extent to which it is attracted to or repelled by the other on the basis of these data alone. However, distances between confrontation partners can illuminate motivational tendencies, especially when comparing conditions in which the same individuals were confronted with partners of different social status. Figure 11.4 presents distances between mated males and either mated or unmated females in the absence of any pairmates or families. When the female was mated, the mean distance between pairs was greater than when the same male was confronted with an unmated female. Because, as we will see, males were equally interested in both types of females, this is an indication that mated

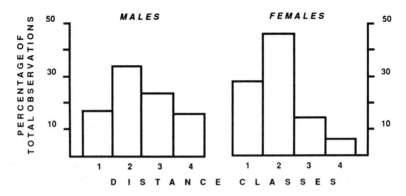

Figure 11.3 Distances of mated *Callithrix* males and females from their own families when present behind one-way screens. Distribution of locations (percentage of total observations) at four different distance classes (30 cm each), with 1 closest to and 4 farthest from dividing screen.

Figure 11.4 Distances between mated *Callithrix* males and mated or unmated females in absence of mates and families. Distribution of locations (percentage of total observations) at six different distance classes (30 cm each), with 1 closest to and 6 farthest from each other.

females keep heterosexual strangers at a greater distance than do mated males.

The relative contribution of male and female confrontees to social distance could be more easily assessed in *Callicebus* than in *Callithrix*, owing to differences in the design of the test cage and the experimental procedures. During the segment in which social distance was measured, the focal animal was always placed in the middle compartment, separated from the outer compartments by Plexiglas screens. A subject could express its motivational tendencies by its spatial position in relation to the screens separating it from the outer compartments (one or both of which were occupied). The center compartment was divided into three areas of equal width (30 centimeters). Locations of the individual in the center compartment were recorded twenty times during the five-minute trial. These data were used to generate composite proximity scores. An animal remaining as close as possible to a given screen throughout the test, attained the maximum score of sixty (twenty positions X location 3). Each subject was tested under three conditions: with the pairmate only, with the confrontation partner only, or with both.

Proximity scores for these conditions are presented as percent of the maximum score in figure 11.5. Male and female titi monkeys differed in their attraction to the social incentives when pairmates (PAI) or confrontation partners (CON) were the only incentive provided. Females stayed closer to their pairmates than did males (PAI in figure 11.5), and males stayed closer to their opposite-sexed confrontation partners than did females (CON in figure 11.5; sex by condition interaction, p <.001). These data indicate that females showed a stronger spatial attraction toward mates than did males, and a weaker attraction to strangers of the opposite sex. When both social stimuli were presented simultaneously, these sex differences were further accentuated. Females stayed even closer to their pairmates when an unfamiliar male was also present (PAI versus P + M in figure 11.5), whereas males stayed farther from their pairmates when an unfamiliar female was present (PAI versus P + F in figure 11.5). This resulted in a significant sex by condition interaction (p <.001) in the analysis of variance. Females decreased proximity to the unfamiliar male in the presence of their pairmates, as compared to the condition when the male stranger was the single incentive (CON versus C + M in figure 11.5). Males also decreased their proximity to the unfamiliar female in the presence of their mates, as compared to the single incentive condition (CON versus C + F in figure 11.5). This effect was significant in both sexes (p <.001). Taken together, these results indicate that females always prefer their mates over strangers. Males, however, seemed to be in a conflict of

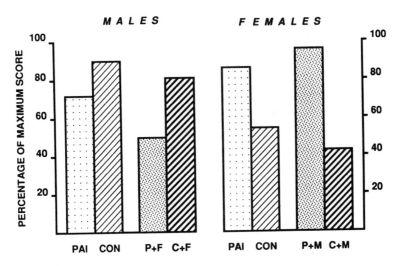

Figure 11.5 Proximity scores (mean percentage of maximum score) of male and female *Callicebus* towards pairmates or confrontation partners. Conditions: PAI = pairmate single incentive; CON = confrontation partner single incentive; P+F = proximity to female pairmate in presence of female confrontation partner; P+M = proximity to male pairmate in presence of male confrontation partner; C+F = proximity to female confrontation partner in presence of female pairmate; C+M = proximity to male confrontation partner in presence of male pairmate.

choice. Presence of both the confrontation partner and pairmate resulted in alternate approaches by the male to the females on either side; the outcome of these back-and-forth movements was that proximity scores to both females were reduced relative to scores when either female was presented alone.

Although useful for a first appraisal, proxemic variables are relatively crude measures of attraction. A high attraction score, for example could result from an attempt to affiliate with the social incentive or from an aggressively motivated approach. Let us turn, therefore, to the quality and pattern of affiliative and sexual behaviors in both species. When unfamiliar heterosexual individuals were allowed to interact directly, males of both species showed immediate interest in their female partners and readily approached them, a behavior never shown by females.

Detailed data will be considered first for *Callithrix*. The immediate interest of marmoset males in unfamiliar females seemed, at least partly, to be sexually motivated as indicated by frequent performance of anogenital controls. Anogenital controls are a sexual behavior of low intensity and are normally followed by nuzzling behavior, mounting, and thrusting. These latter three behaviors, however, were not performed, primarily because females reacted aggressively to the initial sexual overtures of the males. In general, threat vocalizations and cuffing effectively thwarted the males' attempts. Occasionally, the females attacked males vigorously. These female-initiated behaviors were consistent across experimental conditions. That is, females were equally agonistic in their reaction to the unfamiliar male, whether they were alone with him, their mates and families were visible through one-way screens, or they were present behind wire mesh. Such consistency indicates that these behavioral reactions can be attributed to characteristics of mated females as well as the factor that females were not in a strong conflict of choice. They chose not to interact amicably with the unfamiliar male. Under these circumstances, it is obviously difficult for males to proceed.

It is also difficult for an observer to detect whether or not males would have performed extra-pair copulations if the opportunity had been available. Would males have interacted sexually if the unfamiliar females had been more compliant? This question can be addressed by examining their behavior with unmated females (the mature daughters of the families tested).

In encounters with unmated females, males performed the whole array of sexual behaviors and these were not resisted in any way by the females. Nevertheless, these behaviors were reduced as soon as the males' mates and family members were visually present through the one-way screen, and they ceased completely when mate and family were behind the wire mesh. This can be taken as a manifestation of conflict of choice in males because reduction of sexual behaviors is attributable, at least in part, to the fact that males were occupied by interactions with their mates and families. In addition, subdominant female partners seemed to be strongly affected by the male's family and tended to sit motionless during confrontations under these conditions. As a consequence, males literally oscillated between their frozen confrontation partners and their own family members. This was evident even when the family was only visually present, as separated by one-way screens, and therefore unaware of the events in the adjacent cage.

Huddling, an affiliative behavior highly characteristic of mated marmoset pairs (Epple 1975; Erb 1983), was never observed when mated males were confronted with mated females, but was performed by mated males and unmated females. When the male's family was present, this behavior also diminished in the one-way screen condition, or ceased completely in the wire screen condition. This result provides further evidence for a conflict of choice in males.

In summary, when marmoset males and females met unfamiliar individuals of the opposite sex, affiliative and sexual behaviors did not occur if both members of the confrontation pair were mated. There were clear indications that mated females prevented these behaviors from being expressed and did so regardless of the social conditions under which the males were encountered. We may conclude, therefore, that there were no indications of conflict of choice in mated females, whereas mated males showed clear signs of behavioral ambiguity, indicating a high degree of conflict of choice.

Callicebus males also showed a great deal of interest in unfamiliar females. They readily approached unfamiliar females and maintained proximity with them (following or sitting within 30 centimeters). Unfamiliar titi monkey pairs were in proximity in 81% of total observations, and this was significantly more (p <.05) than proximity between pairmates (54%). This finding could be viewed as an indication that both partners were willing to replace their absent mates and to interact with unfamiliar individuals in a quite familiar manner. However, if we further examine the proximity data and determine instances of huddling—animals sitting side-by-side in passive contact—a quite different picture emerges. In mated pairs, when animals were in proximity, they were almost always huddling (54% proximity, 52% huddling), whereas in confrontation pairs, less than a quarter of the proximity data resulted from huddling (81% proximity, 17% huddling). In confrontation pairs, females generally withdrew when the male approached, and proximity was, most frequently, the result of the male following the female. This difference between mated pairs and confrontation pairs was accentuated by the observation that tail-twining—an affiliative behavior highly characteristic of mated pairs—was never performed by confrontation pairs. Furthermore, there was no attempt to initiate this behavior by either sex during the rare occasions when huddling was performed.

Let us turn now to sexual behaviors of unfamiliar titi monkey pairs during encounters when pairmates were absent. Three different behaviors were recorded: anogenital controls, mounting, and thrusting.

Anogenital controls were initiated by all seven males and were observed in thirty of the thirty-six combinations of confrontation pairs. Mounting was performed by twenty-two pairs. Thrusting was performed by only three pairs of the thirty-six combinations. Nevertheless, all seven males and all seven females were engaged in mounting at least once with unfamiliar individuals. Although these results seem to contradict the finding that females eluded males upon approach, there were no qualitative indications that mounting and thrusting were forced behaviors, for no female resisted the male or tried to escape. In this connection, it is noteworthy that, in only two trials of the entire experimental series, did confrontation partners fight. In contrast to the results for common marmosets, unfamiliar heterosexual pairs of titi monkeys showed affiliative and, to a limited extent, sexual behaviors. Although males were generally the initiators and females seemed to be more reluctant to interact, these behaviors undoubtedly require for their execution a certain willingness in both partners. Thus, these results indicate that, in absence of their mates, titi monkeys of both sexes were willing to interact amicably and even sexually with social alternatives. They indicate additionally that there seem to be no long-lasting effects of the absent mates on the individual's behavior.

When pairmates were present, titi monkeys showed clear evidence of conflicts of choice. In the presence of pairmates, males and females spent significantly less time in proximity (p <.001) or huddling (p <.05) with the confrontation partner. Frequency of sexual behaviors also decreased, as did the number of pairs showing sexual behaviors. For each of these measures the decrease was greater when the female's pairmate was present behind Plexiglas. In other words, the on-looking

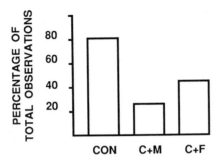

Figure 11.6 Proximity between male and female *Callicebus* confrontation pairs (mean percentage of total observations). Conditions: confrontation pair alone (CON); confrontation pair in presence of male pairmate (C+M); confrontation pair in presence of female pairmate (C+F).

male had a stronger effect on his female pairmate than did an on-looking female on her male pairmate. (Consult figure 11.6 for the general pattern of decrease.)

It is not clear whether the behavior displayed by the male pairmate distracted the rival from interacting with the female, or if the female was responsible for the behavioral changes when her mate was present during confrontations. There is evidence for both possibilities. Since the unfamiliar male was responsible for initiating sexual and affiliative interactions, it seems likely that he was engaging in fewer of these attempts when the female's mate was present. On the other hand, titi monkey females were clearly attracted by their mates and avoided unfamiliar males. Therefore, it is likely that females too were responsible, at least partly, for the changes in behavior when their mates were present. If females resolve conflicts of choice in favor of their pairmates, as expressed by drawing closer to them, and they are followed by their confrontation partners, it is likely that this will be accompanied by the expression of competition-conflict between the males. This is considered in the next section of this chapter.

Competition Conflict

Competition conflict is likely to be engendered in the confrontation situation when the pairmate is present and can observe the interaction between the confrontation pair. Allowing potential rivals unrestricted access to each other to test competition conflict is not advisable due to the risk of injurious aggression, as has been reported for captive marmosets (Epple 1975). Therefore, appraisal of competition conflict in these studies relied on attempted or intended behaviors occurring between rivals that were restrained by the intervening screen.

Competition conflict between same-sexed individuals within a heterosexual trio was also assessed indirectly by comparisons of the behavior of the confrontation pair when pairmates were present and when they were absent. Confrontations between unfamiliar heterosexual individuals in the absence of pairmates may be treated in this analysis as the baseline condition.

In the event that the general pattern of interactions between confrontation partners was altered in the presence of pairmates, there are two possible interpretations. First, the altered pattern of interactions could be the result of a conflict of choice in the confrontation partner whose pairmate was present (as shown in the preceding section). A second possibility is that the primary effect of the pairmate's presence was not on its partner, but on the rival stranger who was either inhibited or diverted by the presence and the behavior of a stranger of the same sex (its confrontation partner's pairmate), and thus involved in a competition conflict.

For *Callithrix*, two different behaviors were particularly informative in direct assessments of competition conflict between subjects and their confrontation partners' mates and family: location at dividing wire-mesh, and vocal threats. Subjects also showed other agonistic behaviors, such as tail-raised presentation, tuft-flick stare, and arch bristle. (For definitions, see Lipp 1978; Stevenson and Poole 1976). Tail-raised presentations and tuft-flick stares occurred too infrequently for statistical analysis, but it is noteworthy that males initiated all three behaviors only to mates and families of their confrontation partners when there was a possibility to interact—that is, when they were present behind wire mesh. This never occurred when the unfamiliar conspecifics were visually present but, being behind a one-way screen, were unaware of the confrontation. The males never directed these three agonistic behaviors toward their confrontation partners.

Females, in contrast, showed arch bristle under all experimental conditions, even when alone with their confrontation partners.

The different reactions of males and females to the presence of strange families resulted in a sex difference on this measure (p <.001). During the fifteen-minute confrontation period with the family of the confrontation partner present, males and females spent 36% and 33% of total observations, respectively, clinging to the wire screen. That these approaches were not motivated by affiliative tendencies is indicated by the fact that both males and females directed a high frequency of threat vocalizations toward their like-sexed rivals (means: males = 15.2; females = 14.7). Sex differences were not significant on either measure, indicating that males and females were equally involved in competition conflicts.

An interesting preliminary observation worthy of mention is that the gender of the confrontation partner led to a sexual segregation within the family. Adult family members of the same sex as the intruder were more agitated than were opposite-sexed family members, and their reaction persisted throughout the confrontation period. It should also be noted that, when any family member was confined by itself to the cage normally used for confrontations, the remaining family members did not behave any differently than they did during their daily routine.

The behavior between confrontation partners—or, more precisely, the impairment of their interactive behaviors in the presence of mates—provides less direct evidence for competition conflict in common marmosets. In encounters of mated males with unmated females, the entire array of sexual behaviors (anogenital control, mounting, and thrusting) were observed when the pair was separated and alone, that is, divided from families by opaque screens. When the subdominant female's family was visually present, sexual behaviors were clearly reduced. This suggests that the presence of the unfamiliar family had a potent influence on the males. When two-way interactions between the female's family and the confrontation pair were possible through wire mesh, sexual behaviors ceased completely. Reduction of each of the three measures of sexual behavior across exposure conditions was significant (p's <.01). These results correspond with those previously described—namely, that males spent more than one third of total observations at the dividing screen, interacting aggressively and primarily with the mated male of the group (alpha male; male parent). In confrontations between mated males and mated females, interactions largely consisted of female-initiated agonistic behavior, ranging from low intensity vocal threats to physical attacks, and these behaviors did not differ qualitatively nor quantitatively in relation to presence or absence of their mates. Nevertheless, the quality of the aggressive behaviors which the males exchanged at the wire screen left no doubt that the onlooking mate was responding to the rival and would have attacked, given the opportunity to do so.

It is noteworthy that onlooking males did not initiate any behaviors toward their mates, but concentrated all their attention on the rivals. This pattern of intense aggressive interaction between males occurred in spite of the fact that mated females were quite successful at thwarting the attempts of unfamiliar males to engage in affiliative or sexual interactions. Taken together, the data illustrate that there is competition conflict between males, although direct physical interaction was prevented by the experimental setting. One might expect that male-male interactions would differ depending on whether the intruder was with a daughter of the alpha male or its mate. This was not the case. However, we should keep in mind that the male's mate was also present in both situations, although on opposite sides of the wire mesh.

Confrontations in which marmoset females were introduced to the family room of their confrontation partners also provided evidence for competition conflict. When unfamiliar mated females (that is, subjects and onlooking pairmates) were interacting with each other at the wire screen, agonistic behaviors were comparable to those observed between the unfamiliar males. Although interactions between females were similar to those of males, they may be attributable to different underlying motivations. It is unlikely that mated females would show resistance toward unfamiliar males while responding at the same time to other females as immediate rivals for the attention of these males.

Nevertheless, we may conclude from the similarities in intrasexual interactions that both males and females were involved with their like-sex rival in competition conflict, the most likely function of which was preservation of the social and sexual integrity of the pair relationship. The high levels of intrasexual aggression in both males and females is in full accord with investigations by other authors with this species (Epple 1970; Evans 1983; Sutcliffe and Poole 1984).

Male and female *Callicebus* also exchanged agonistic gestures through the screen with the mates of their confrontation partners. The three agonistic behaviors recorded were chest rubbing, tail lashing, and attempted attacks. The first two patterns were too infrequent for statistical treatment. Subjects only showed them, however, when the mates of their confrontation partners were present, never when alone with the confrontation partner. Thus, these behaviors were clearly directed to like-sexed rivals and not to the opposite-sexed confrontation partners or their own mates. Males initiated more attempted attacks than did females, and this difference was significant ($p < .001$). However, both sexes showed intrasexual aggression, which can be construed as a manifestation of competition conflict between the two same-sexed individuals within the triad.

More indirect evidence for competition conflict in *Callicebus* is provided by changes in the interactions between confrontation partners brought about by the presence of the mate of one of them. As previously indicated, unfamiliar heterosexual pairs of titi monkeys interacted affiliatively and sexually when they were alone together. When pairmates were present, however, affiliative and sexual interactions diminished. The presence of the females' pairmate had a stronger inhibitory influence than did the presence of the males' pairmate (refer to figure 11.6). These results are in full accord with the observation that the males' behavioral reactions to the like-sexed animal were more prominent than were the females reactions. In particular, male pairmates were more agitated by the sight of their mates with a male stranger than were female pairmates under the obverse circumstances. Duetting, a remarkable and characteristic vocal behavior performed by mated titi monkeys, also suggested a greater reaction in male titi monkeys than in females upon seeing the mate with a stranger. Duetting, normally initiated by males, probably serves among other functions to strengthen the pair bond. Duets were regularly heard from pairmates, but occurred on only one occasion between confrontation partners. Duets frequently occurred following reunion of pairmates at the conclusion of testing, but this was more frequent after females had interacted with unfamiliar males while their mates—that is, the females' mates—were onlookers, than vice versa.

We may draw two conclusions from these results. First, because all interactions between confrontation partners were initiated by males, it is reasonable to assume that the major factor responsible for inhibiting sexual and affiliative interactions was the disruptive effect of competition conflict between males. Second, males were obviously more distracted by the presence of their confrontation partners' pairmates than by presence of their own mates. Females, however, were more influenced by the presence of their own mates than by the presence of the pairmates of their confrontation partners. It is entirely possible that competition conflict in male titi monkeys was so prevalent in the triadic situation, that it masked the manifestation of conflict of choice among females.

DISCUSSION

The focus of the research presented here has been on proximate causes of pair-bond behavior in two New World monkey species which both exhibit a monogamous life-style. Therefore, one might expect some behavioral congruencies when pairmates of these species have to cope with social conflicts resulting from the presence of a social rival or an alternate partner to the pairmate.

Let us first summarize results which allow one to assess manifestations of conflicts of choice. Social attraction toward mates versus strangers, as appraised by means of spatial preferences, were in full accord between species. Females clearly preferred their own mates, whereas males seemed quite attracted to unfamiliar females. Male attraction to unfamiliar females was evident in their confrontations when the females' mates were absent. Males of both species immediately approached opposite-sexed strangers, a behavior never shown by females of either species toward unfamiliar males. Males of both species also readily initiated affiliative and sexual behaviors toward unfamiliar females. Female reactions to these male overtures differed between species. *Callithrix* females reacted aggressively toward unfamiliar males. *Callicebus* females showed little or no signs of resistance, but neither did they take the initiative in these interactions. These results indicate that males of both species and *Callicebus* females interact amicably or even sexually with social alternatives if their mates are absent, whereas *Callithrix* females do not.

When pairmates were present during encounters, the aggressive behavior of *Callithrix* females toward the male stranger was unchanged while *Callicebus* females were as indifferent as before. Therefore, we may conclude that, in both species, females always prefer their mates over strangers. (For *Callicebus*, see also Mason 1971, 1974, 1975.) If there is a conflict of choice, females seem to resolve it a priori in favor of their mates. Conversely, males of both species changed their behaviors when their pairmates were present, and showed a reduction in the initiation of affiliative and sexual interactions with unfamiliar females. This can be taken as evidence of conflict of choice in males, for reduction of these behaviors could be clearly attributed to the fact that males now were occupied by interactions with their mates, resulting in quite obvious behavioral ambiguity.

Both species showed evidence of competition conflict. Males and females appeared to defend the integrity of the pair bond by behaviors that distracted same-sexed rivals or disrupted their sexual and affiliative behaviors. There were remarkable differences between species in the extent of engagement with the rival. *Callithrix* males and females were clearly more active in this respect than were *Callicebus*. Furthermore, *Callithrix* males and females engaged same-sexed rivals in competition conflict to an equivalent extent. In *Callicebus*, interventive behaviors were clearly less pronounced, and the sexes differed on these measures. Males were much more agitated and disruptive than were females, and female interventive behavior was nearly absent. These results are in accordance with findings of Cubicciotti and Mason (1978) in which females showed little evidence of jealousy behavior. An interesting question raised by the present studies is why rivals let themselves become involved in competition conflict at all. This is especially puzzling for females, because they either resisted (*Callithrix*) or reacted ambivalently (*Callicebus*) toward unfamiliar males. Yet they interacted aggressively through dividing screens with mates of these very males. Because their reactions toward the males suggest that females did not intend to monopolize their recently met confrontation partners, it seems reasonable to conclude that females simply responded in kind to the provocative interventive behaviors of the males' pairmates.

Table 11.3 attempts to summarize social tendencies in both species. This scheme can be viewed in relation to the hypotheses described in the introduction. As a consequence of the male predilection to philandering (hypothesis 2), both genders of monogamously mated species should develop jealousy-like behaviors (hypotheses 1 and 3). The functional interpretation of these behavioral strategies is that they will prevent the males from being cuckolded and the females

Table 11.3 Social Tendencies of Callithrix and Callicebus during Encounters with Strangers

	Response toward Unfamiliar Conspecifics of:		
	Opposite Sex		Same Sex
	In absence or	In Presence of Pairmate	
Callithrix			
Males	+ + +	+ -	- - -
Females	- -	- -	- - -
Callicebus			
Males	+ + +	+ -	- -
Females	+ -	+ -	-
	Conflict of Choice		Competition Conflict

+ : affiliative/sexual

- : aversive/aggressive

+ - : ambivalent

from being deserted. Therefore, highly affiliative or sexual behavior toward a stranger of the opposite sex by one member of the pair should trigger aggressive behavior toward this stranger by the other. Only *Callithrix* females reacted exactly as predicted. Their high levels of aggressive behavior toward female rivals can be seen as a complement to the philandering inclinations of the male. *Callithrix* females also behaved according to hypothesis (4), which stated that females should resist unfamiliar males and stay with the mate with whom they had already been successful in raising young (Weatherhead and Robertson 1979).

For the other three classes of subjects, some questions remain. First, it is not obvious why *Callithrix* males responded so vigorously to potential rivals when their mates resisted the unfamiliar males quite effectively. There seems to be no need for male interventive behavior in view of the strong fidelity of female *Callithrix*. The possibility of forced copulations by strange males could explain such a double security system. Yet, this functional explanation is not entirely satisfactory, because common marmoset females show a postpartum estrous and none of our individuals tested was in this state. Thus, we may presume male intrasexual aggressivity is not only motivated socially, but by selective factors in the ecology—such as resource distribution—which were not examined. Ecological influences may also be reflected in female reactions toward strange males.

It is not clear why *Callicebus* males were less agitated and less interventive than were *Callithrix* males, even though *Callicebus* females interacted sexually with strangers. Similarly, it is not clear why the reactions of *Callicebus* females to the sexual and affiliative behaviors of their pairmates with strange females was so much less than that of the *Callithrix* females in the same circumstances.

It is possible that, despite similarities between the species in general pattern of male-female associations, they differ in their basic modes of coping with social challenges. Some qualitative observations lend support to this possibility. It was striking that, during triadic situations, both sexes of *Callithrix* showed no mate-directed activities, but fully concentrated all interactions on the same-sexed rivals. In *Callicebus*, some behaviors in the triadic situations were mate-oriented, and, in general, individuals seemed not to be as overtly aggressive as *Callithrix*. Henry (1976) proposed two distinct modes of physiological response to psychological stress, namely the "flight-fight" and "conservation-withdrawal" modes. Even though these terms have been introduced and used to describe physiological patterns, they are illustrative of the behavioral patterns shown by the two species in question. *Callithrix* showed an immediate "flight-fight" response toward rivals—obviously triggered by the simultaneous presence of the mate—and this pattern of response corresponds with the remarkable levels of intrasexual aggression attributed to both sexes of this species (Epple 1970; Evans 1983; Sutcliffe and Poole 1984). In titi monkeys, responses to rivals were less spectacular. In the wild, encounters of neighboring or unfamiliar individuals are highly ritualized, and most agonistic interactions are limited to vocal exchanges (Mason 1966, 1968; Robinson 1979, 1981). Studies on captive individuals have revealed that unfamiliar pairs completely withdraw after initial interactions, and that males sometimes physically restrained, in a literal sense, their female pairmates (Anzenberger et al. 1986). Thus, this species seems to favor a behavioral pattern reminiscent of the "conservation-withdrawal" mode of reaction. Whether these different behavioral tendencies are in fact based on different physiological predispositions remains for the present a matter of speculation, although suggestive data are available (Cubicciotti et al. 1986).

Finally, I would like to discuss the general topic of the monogamous organization in the two species. We learned that males of both species interacted sexually with unfamiliar females whenever there was an opportunity, and *Callicebus* females showed no signs of resistance. Such inclinations seem barely consistent with the concept of monogamy. Moreover, Mason (1968) reported that free-living *Callicebus* females crossed territorial borders and occasionally engaged in sexual activities with neighboring males. As far as callitrichids are concerned, some recent field studies also raise doubts about their monogamous life-style (Sussman and Kinzey 1984. See also Epple et al. 1986).

When addressing this general issue, we must be aware of some inherent problems that create misunderstandings in discussions. First of all, we must make clear at which level of group structure we are arguing (Gowaty 1985; Wickler and Seibt 1983). Most misunderstandings result from the fact that demographic, motivational, functional, and genetic levels of group structure are not strictly separated in discussions. In spite of the need to take all of them into account—something not easy to achieve—it should not be expected that these different levels will turn out to be congruent in a given species, population, or group. A further problem is the sometimes unbalanced evaluation of studies conducted in the wild versus in the laboratory, although it is commonplace that both approaches have their advantages and limitations. Sussman and Kinzey (1984, 419) stated in their review article that the assumption of a monogamous life-style in callitrichids is mainly "the result of misinterpreting laboratory studies." On the other hand, data from marked, wild callitrichid populations are still so few that the alternative assumption of "small multimale groups...that tend toward polyandrous mating" (Sussman and Kinzey 1984, 445) seems premature and might be equally misinterpreting available studies. It may be mentioned that the tamarin species *Saguinus fuscicollis*, which seems to exhibit a facultatively polyandric life-style (Goldizen 1987, 1989; Terborgh and Goldizen 1985), could already have been described as facultatively polyandric on the basis of captive studies—that is to say, it could be kept occasionally in polyandric trios that were stable (Epple 1981).

The fact that gibbons and siamangs (Hylobatidae), titi monkeys (Callicebinae), and most of marmosets and tamarins (Callitrichidae) are found in nature in family-like structures containing one reproductively active pair is persuasive evidence that strong predispositions toward a monogamous life-style are prevalent. It should be noted that these predispositions are manifest under captive conditions. Moreover, they are displayed across very diverse housing conditions. Mated pairs of these taxa will not tolerate the addition of other adults to form larger reproductive units. Since housing multiple mating individuals together is not too hard to manage in most of the remaining primate species, this fact clearly illustrates the different motivational inclinations between monogamous and polygamous species. Such facts can be made visible and understandable only by focusing on the proximate causes of behavior.

REFERENCES

Abbott, D. H. (1984) Behavioral and physiological suppression of fertility in subordinate marmoset monkeys. *American Journal of Primatology* 6:169–186.

Altmann, J. (1974) Observational study of behavior: Sampling methods. *Behaviour* 49:227–267.

Anzenberger, G. (1983) *"Bindungsmechanismen in Familiengruppen von Weissbuschelaffchen* (Callithrix jacchus)." Zurich: Juris Druck and Verlag.

Anzenberger, G. (1985) How stranger encounters of common marmoset (*Callithrix jacchus jacchus*) are influenced by family members: The quality of behavior. *Folia Primatologica* 45:204–224.

Anzenberger, G. (1988) The pair bond in the titi monkey (*Callicebus moloch*): Intrinsic versus extrinsic contributions of the pairmates. *Folia Primatologica* 50:188–203.

Anzenberger, G.; Mendoza, S. P.; Mason, W. A. (1986) Comparative studies of social behavior in *Callicebus* and *Saimiri*: Behavioral and physiological responses of established pairs to unfamiliar pairs. *American Journal of Primatology* 11:37–51.

Box, H. O. (1977) Quantitative data on carrying of young captive monkeys (*Callithrix jacchus*) by other members of their family group. *Primates* 18:475–485.

Cubicciotti, D. D., III; Mason, W. A. (1978) Comparative studies of social behavior in *Callicebus* and *Saimiri*: Heterosexual jealousy behavior. *Behavioral Ecology and Sociobiology* 3:311–322.

Cubicciotti, D. D., III, Mason, W. A. (1975) Comparative studies of social behavior in *Callicebus* and *Saimiri*: Male-female emotional attachments. *Behavioral Biology* 16:185–197.

Cubicciotti, D. D., III; Mendoza, S. P.; Mason, W. A.; Sassenrath, E. N. (1986) Differences between *Saimiri sciureus* and *Callicebus moloch* in physiological responsiveness: Implications for behavior. *Journal of Comparative Psychology* 100:385–391.

Coulson, J. C.; Thomas, C. S. (1983) Mate choice in the Kittiwake Gull. In Bateson, P. (ed.), Mate Choice. Cambridge: Cambridge University Press. 361–376.

Darwin, C. (1871) *The Descent of Man, and Selection in Relation to Sex*. London: Murray John.

Epple, G. (1967) *Vergleichende untersuchungen uber Sexual-und Sozialverhalten der Krallenaffen (Hapalidae). Folia Primatologica* 7:37–65.

Epple, G. (1968) Comparative studies on vocalization in marmoset monkeys (*Hapalidae*). *Folia Primatologica* 8:1–40.

Epple, G. (1970) Quantitative studies on scent marking in the marmoset (*Callithrix jacchus*). *Folia Primatologica* 13:48–62.

Epple, G. (1974) Pheromones in primate reproduction and social behavior. In Montagna, W., Sadler, W. A. (eds.), *"Reproductive Behavior."* New York: Plenum Publishing Corporation. 131–155.

Epple, G. (1975) The behavior of marmoset monkeys (*Callithricidae*). In Rosenblum, L. A., "Primate Behavior: Developments in Field and Laboratory Research," vol. 4. New York: Academic Press. 195–239.

Epple, G. (1978) Reproductive and social behavior of marmosets with special reference to captive breeding. *Medical Primatology* 10:50–62.

Epple, G. (1981) Effect of pair-bonding with adults on the ontogenetic manifestation of aggressive behavior in a primate, *Saguinus fuscicollis*. *Behavioral Ecology and Sociobiology* 8:117–123.

Epple, G.; Küderling, I.; French, J. A. (1986) Social and sexual strategies in callitrichid monkeys. *Primate Report* 14:73.

Erb, R. (1983) *Auswirkungen auf die Gruppendynamik nach Verlust des Vaters bei Weibbuschelaffchen* (Callithrix jacchus). Unpublished masters thesis. University of Zurich.

Evans, S. (1983) The pair-bond of the common marmoset, *Callithrix jacchus jacchus*: An experimental investigation. *Animal Behaviour* 31:651–658.

Fragaszy, D. M.; Mason, W. A. (1983) Comparisons of feeding behavior in captive squirrel and titi monkeys (*Saimiri sciureus* and *Callicebus moloch*). *Journal of Comparative Psychology* 97:310–326.

Goldizen, A. W. (1987) Facultative polyandry and the role of infant-carrying in wild saddle-back tamarins (*Saguinus fuscicollis*). *Behavioral Ecology and Sociobiology* 20:99–109.

Goldizen, A. W. (1989) Social relationship in a cooperatively polyandrous group of tamarins (*Saguinus fuscicollis*). *Behavioral Ecology and Sociobiology* 24:79–89.

Gowaty, P. A. (1985) Multiple parentage and apparent monogamy in birds. In Gowaty, P. A., Mock, D. W. (eds.), "Avian Monogamy." Washington, D.C.: The American Ornitologists' Union. 11–21.

Hearn, J. P. (1983) The common marmoset (*Callithrix jacchus*). In Hearn, J. P. (ed.), "Reproduction in New World Primates." Lancaster, England: MTP Press Limited. 181–215.

Henry, J. P. (1976) Mechanisms of psychosomatic disease in animals. *Advances in Veterinary Science and Comparative Medicine* 20:115–145.

Hubrecht, R. C. (1984a) Field observations on group size and composition of the common marmoset (*Callithrix jacchus jacchus*), at Tapacura, Brazil. *Primates* 25:13–21.

Hubrecht, R. C. (1984b) Marmoset social organization and female reproductive inhibition in the wild. *Primate Eye* 24:5.

Hubrecht, R. C. (1985) Home range size and use and territorial behavior in the common marmoset, *Callithrix jacchus jacchus*, at the Tapacura Field Station, Recife, Brazil. *International Journal of Primatology* 6:533–550.

Ingram, J. C. (1977) Interactions between parents and infants, and the development of independence in the common marmoset (*Callithrix jacchus*). *Animal Behaviour* 25:811–827.

Kinzey, W. G. (1981) The titi monkeys, genus *Callicebus*. In Coimbra-Filho, A. F, Mittermeier, R. A. (eds.), "Ecology and Behavior of Neotropical Primates." Rio de Janeiro: Academia Brasileira de Ciencias. 241–276.

Kleiman, D. G. (1977a) Monogamy in mammals. *Quarterly Review of Biology* 52:39–69.

Kleiman, D. G. (1977b) "The Biology and Conservation of the Callitrichidae." Washington, D.C.: Smithsonian Institution Press.

Kleiman, D. G. (1980) The sociobiology of captive propagation. In Soule, M. E. (ed.), "Conservation Biology." Sunderland, Mass.: Sinauer Associates. 243–261.

Kleiman, D. G. (1981) Correlations among life history characteristics of mammalian species exhibiting two extreme forms of monogamy. In Alexander, D., Tinkle, D. W. (eds.), "Natural Selection and Social Behavior." New York: Chiron Press. 332–344.

Lacher, T. E.; Bouchardet da Fonseca, G.; Alves, G.; Magalhaes-Castro, B. (1981) Exudate-eating, scent-marking, and territoriality in wild populations of marmosets. *Animal Behaviour* 29:306–307.

Lienert, G. A. (1973) *Verteilungsfreie Methoden in der Biostatistik*. Meisenheim: Hain Verlag.

Lipp, H. P. (1978) Aggression and flight behaviour of the marmoset monkey *Callithrix jacchus*: An ethogram for brain stimulation studies. *Brain, Behavior and Evolution* 15:241–259.

Maier, W.; Alonso, C.; Langguth, A. (1982) Field observations on *Callithrix jacchus jacchus*. *Zeitschrift für Saugetierk* 47:334–346.

Mason, W. A. (1966) Social organization of the South American monkey, *Callicebus moloch*: A preliminary report. *Tulane Studies in Zoology* 13:23–28.

Mason, W. A. (1968) Use of space by *Callicebus* groups. In Jay, P. C. (ed.), "Primates: Studies in Adaptation and Variability." New York: Holt, Rinehart & Winston. 200–216.

Mason, W. A. (1971) Field and laboratory studies of social organization in *Saimiri* and *Callicebus*. In Rosenblum, L. A. (ed.), "Primate Behavior: Developments in Field and Laboratory Research," vol. 2. New York: Academic Press. 107–137

Mason, W. A. (1974) Comparative studies of social behavior in *Callicebus* and *Saimiri*: Behavior of male-female pairs. *Folia Primatologica* 22:1–8.

Mason, W. A. (1975) Comparative studies of social behavior in *Callicebus* and

Saimiri: Strength and specificity of attraction between male–female cagemates. *Folia Primatologica* 23:113–123.

Maynard Smith, J. (1977) Parental investment: A prospective analysis. *Animal Behaviour* 25:1–9.

Mendoza, S. P.; Mason, W. A. (1986) Parental division of labour and differentiation of attachments in a monogamous primate (*Callicebus moloch*). *Animal Behaviour* 34:1336–1347.

Menzel, C. R. (1986) An experimental study of territory maintenance in captive titi monkeys (*Callicebus moloch*). In Else, J. G., Lee, P. C. (eds.), "Primate Ecology and Conservation." Cambridge: Cambridge University Press. 133–143.

Moynihan, M. (1966) Communication in the titi monkey, *Callicebus*. *Journal of Zoology, London* 150:77–127.

Parker, G. A. (1979) Sexual selection and sexual conflict. In Blum, M. S., Blum, N. A. (eds.), "Sexual Selection and Reproductive Competition in Insects." New York: Academic Press. 123–166.

Robinson, J. G. (1979) Vocal regulation of use of space by groups of titi monkeys *Callicebus moloch*. *Behavioral Ecology and Sociobiology* 5:1–15.

Robinson, J. G. (1981) Vocal regulation in inter- and intragroup spacing during boundary encounters in the titi monkey, *Callicebus moloch*. *Primates* 22:161–172.

Rothe, H. (1975) Some aspects of sexuality and reproduction in groups of captive marmosets (*Callithrix jacchus*). *Zeitschrift für Tierpsychologie* 37:255–273.

Rothe, H. (1978) Sub-grouping behaviour in captive *Callithrix jacchus* families. In Rothe, H. Wolters, H. J. Hearn, J. P. (eds.), "Biology and Behaviour of Marmosets." Gottingen: Eigenverlag H. Rothe. 233–257.

Rothe, H.; Wolters, H. J.; Hearn, J. P. (1978) "Biology and Behaviour of Marmosets." Gottingen: Eigenverlag H. Rothe.

Rowley, I. (1983) Re-mating in birds. In Bateson, P. (ed.), "Mate Choice." Cambridge: Cambridge University Press. 331–360.

Siegel, S. (1976) *"Nicht-Parametrische Statistische Methoden."* Frankfurt: Fachbuchhandlung fur Psychologie.

Stevenson, M.; Poole, T. B. (1976) An ethogram of the common marmoset (*Callithrix jacchus jacchus*): General behaviour repertoire. *Animal Behaviour* 24:428–451.

Sussman, R. W.; Kinzey, W. G. (1984) The ecological role of the *Callitrichidae*: A review. *American Journal of Physical Anthropology* 64:419–449.

Sutcliffe, A. G.; Poole, T. B. (1984) An experimental analysis of social interaction in the common marmoset (*Callithrix jacchus jacchus*). *International Journal of Primatology* 5:591–607.

Terborgh, J.; Goldizen, A. W. (1985) On the mating system of the cooperatively breeding saddle-backed tamarin (*Saguinus fuscicollis*). *Behavioral Ecology and Sociobiology* 16:293–299.

Trivers, R. L. (1972) Parental investment and sexual selection. In Campbell, B. (ed.), "Sexual Selection and the Descent of Man." Chicago: Aldine. 136–179.

Trivers, R. L. (1985) "Social Evolution." Menlo Park, Calif.: The Benjamin/ Cummins Publishing Company.

Weatherhead, P. J.; Robertson, R. J. (1979) Offspring quality and the polygyny threshold: The "sexy son" hypothesis. *American Naturalist* 113:201–208.

Wickler, W.; Seibt, U. (1983) Monogamy: An ambiguous concept. In Bateson, P. (ed.), "Mate Choice." Cambridge: Cambridge University Press. 33–50.

Wright, P. C. (1984) Biparental care in *Aotus trivirgatus* and *Callicebus moloch*. In Small, M. (ed.), "Female Primates: Studies by Women Primatologists." New York: A. R. Liss, Inc. 59–75.

CHAPTER TWELVE

Social Conflict and Reproductive Suppression in Marmoset and Tamarin Monkeys

DAVID H. ABBOTT

Few biological urges are more insistent than those relating to procreation. We are not surprised by yet another illustration of the huge risks that an animal will take for the opportunity to mate. The anomaly is not the spawning salmon or praying mantis, giving its all for the sake of reproduction, but the individual who is reproductively mature, yet disinclined or unprepared to participate in the mating game. When the latter occurs, the how and why of it call for explanations.

Clear examples of mature individuals who experience reproductive suppression are provided by some species of New World primates, particularly

David H. Abbott—Wisconsin Regional Primate Research Center, University of Wisconsin, Madison

The author would like to thank his colleagues, including L. M. George, J. K. Hodges, O. L. Wilson, M. C. Ruiz de Elvira, S. F. Ferrari, B. R. Ferreira, J. W. Sheffield, K. T. O'Byrne, and S. F. Lunn (MRC Reproductive Biology Unit, Edinburgh) for their collaboration in some of the work reported here; I. S. Sutherland and G. R. Chambers (National Institute of Medical Research, Mill Hill, London) for designing and building the miniaturized battery-powered GnRH pumps; M. J. Llovet and the animal technical staff for their care and maintenance of the animals; T. J. Dennett and M. J. Walton for preparation of the figures and T. Grose and M. B. Foreman for typing the manuscript. The manuscript benefited from the criticism of A. P. F. Flint and P. M. Summers. This work was supported by grants from the Association for the Study of Animal Behaviour, the University of London, the Nuffield Foundation and an MRC/ARFC Programme Grant.

the tamarins and marmosets which provide the focus of this chapter. In these species, reproduction is generally restricted to a few of the mature members of the group. Social conflict within the group acts as a contraceptive, leading to the suppression of reproduction in social subordinates. The nonbreeding members of the group may include the offspring of the breeding animals, but also unrelated animals. Infertility in subordinates is maintained by both behavioral and physiological means. The level of reproductive suppression which subordinate females can experience is deeper than that of subordinate males because their ovulations can be completely inhibited. The anovulatory condition in subordinate females may be the result of two physiological causes. The more pronounced inhibitor is an increased sensitivity to estradiol negative feedback on GnRH and LH secretion. The other involves endogenous opioid peptide inhibition of GnRH secretion.

The repeatable and reliable consequences of social conflict for fertility of female marmosets and tamarins makes these monkeys particularly useful in furthering our understanding of naturally occurring infertility in ourselves and other mammals. The findings also raise questions about the adaptive significance of socially induced infertility and its possible positive or negative effects on an individual's lifetime reproductive success.

Social conflict in marmoset and tamarin monkeys, as in ourselves, produces winners and losers. This chapter is concerned primarily with the losers produced by social conflict in marmoset and tamarin monkeys and the reproductive penalties these animals incur as social subordinates.

In the wild, callitrichid monkeys live in groups of about three to fifteen individuals (Dawson 1978; Garber et al. 1984; Hubrecht 1984a; Stevenson and Rylands 1988; Sussman and Kinzey 1984; Terborgh and Goldizen 1985). Social conflict within these groups acts as an extremely powerful contraceptive. The losers, or social subordinates, do not breed, while the winners, or social dominants, freely reproduce (Abbott 1987). Such a clear-cut outcome from social conflict is rare in primates (Harcourt 1987). Literally, the winner takes all. Marmoset and tamarin monkeys, therefore, provide an excellent example of how conflict between animals can specifically and completely control reproduction.

In free-living groups of callitrichid monkeys, only one female and one or two males reproduce (Dawson 1978; Garber et al. 1984; Hubrecht 1984a; Stevenson and Rylands 1988; Terborgh and Goldizen 1985). Exceptions to this general rule are rare and transient, such as the examples of two breeding females in (1) captive groups of common marmosets, *Callithrix jacchus* (Abbott 1984) and (2) two wild groups of saddle-back tamarins, *Saguinus fuscicollis* (Terborgh and Goldizen 1985). In the laboratory, the breeding female has been identified as the socially dominant female of her group (Abbott and Hearn 1978; Epple 1967, 1975; French et al. 1984; Rothe 1975). The breeding males have been identified as (a) the socially dominant male and (b) the second-ranking male (Abbott 1986; Epple 1972). The nonbreeding animals in both captive and free-living callitrichid groups include, not only the offspring of the breeding animals, but also unrelated animals (Abbott and Hearn 1978) which, in the wild have immigrated from surrounding groups (Dawson, 1978; Scanlon et al. 1988; Sussman and Kinzey 1984; Terborgh and Goldizen 1985).

This chapter will deal mostly with captive animals and will consider the deficits imposed on the reproductive behavior and reproductive physiology of socially subordinate marmoset and tamarin monkeys. It will particularly consider the physiological mechanisms which translate subordinate status into infertility. The behavioral manifestations and consequences of social conflict in marmosets are dealt with by Anzenberger in chapter 11 of this volume. The suppression of reproduction in free-living groups of callitrichid monkeys has already been well covered by Stevenson and Rylands (1988), Sussman and Kinzey (1984), and Terborgh and Goldizen (1985).

SUBORDINATE MALES

Suppression of Sexual Behavior

Family groups. In captivity, marmosets and tamarins are commonly kept in family groups comprising a set of parents and their offspring (Hearn 1983; Tardif 1984). In marmosets, mature male offspring rank as subordinate to their fathers for as long as they remain in their original family groups (Abbott 1984; Anzenberger 1985; Rothe 1975).

Unlike their fathers, male offspring exhibit little sexual behavior, although sons have occasionally been observed mounting their mothers. All sons exhibiting sexual behavior were older than eight months and therefore past the onset of puberty. At approximately eight months of age in male marmosets, plasma testosterone concentrations start to increase above their low juvenile values of less than 3 ng/ml and the testes commence growing from their small juvenile total volume of approximately 100 mm^3 toward their final adult volume of 730 ± 25 mm^3 (Abbott and Hearn 1978). Pelvic thrusts were seen during all mounts (Abbott 1984; Stevenson and Rylands 1988). Infrequent copulations have also been observed between brother and sister in family groups (Rothe 1975; Stevenson and Rylands 1988). In Rothe's (1975) example of brother-sister mating, no intromission or ejaculation was achieved. However, it is not clear whether or not the same was true in the other instances of mounts by male offspring. The sexual abilities of male tamarin offspring remaining in their families are seemingly displayed as rarely as those of male marmoset offspring, but no specific information is yet available.

This lack of sexual activity in male offspring is not because these young males are behaviorally incapable. Abbott (1978a, 1984) has shown that male marmoset offspring, of between ten and eighteen months old, will readily show mounting behavior. However, to achieve this sexual activity, the males had to be temporarily (Abbott 1984; Anzenberger 1985) or permanently (Abbott 1978a) removed from their families and tested with unfamiliar females. The mounts observed commonly ended in the normal ejaculation of motile spermatozoa (Abbott 1984; Abbott and Hearn 1978). These matings were fertile and resulted in conceptions and full-term pregnancies (Abbott and Hearn 1978). Male offspring returned to their families after such sex tests resumed their almost celibate existence (Abbott 1984). Anzenberger (1985) took these tests one stage further and paired mature male marmoset offspring with females when the males' families were present as onlookers, but did not have access to the test cage. Males readily copulated with females when the families were not present, but copulations ceased while the males' families watched. Male tamarin offspring, of between ten and eighteen months of age, are also capable of ejaculatory mounts with females, but again only when the males were removed from their families and paired with unfamiliar females (Epple and Katz 1980). As with the young male marmosets, these matings were also fertile and resulted in conceptions and full-term pregnancies.

So, why do male marmoset and tamarin offspring show so little sexual behavior in their family groups, when they are quite capable of demonstrating it? As yet, there is no easy answer to this question. Their fathers are certainly not inhibited from demonstrating sexual behavior in the family groups. They mate almost exclusively with the breeding female (the mother of the offspring) and demonstrate unhurried and intimate ejaculatory mounts, involving licking, tonguing, and face-nuzzling behaviors (Abbott 1984; Epple 1967; Rothe 1975). However, on the rare occasions when their sons mated, the mounts were hurried and less intimate (Abbott 1984; Rothe 1975). There is no obvious inhibition imposed on the sexual behavior of these young males. There is no overt social conflict between fathers and sons when sons mount their mothers or sisters (Abbott 1984; Rothe 1975). Furthermore, when social conflict involving male offspring arose (none involved mating), either all or the majority of occasions involved brothers and not fathers (Evans, unpublished results; McGrew and McLuckie 1986; Rothe 1975).

When male offspring were removed from their families, their sexual behavior could still be inhibited by just the sight of their original family beside their new cage (Anzenberger 1985). To explain this sexual suppression in male offspring, Rothe (1975), Abbott (1984), and Anzenberger (1985) suggested that a familiarity or incest taboo may exist in marmoset and tamarin families. Male offspring may thus have a bias for copulating with females from outside the family and so help prevent inbreeding. Experiments by McGrew and McLuckie (1986) support this suggestion. These authors attempted a laboratory simulation of dispersal of animals from groups of cotton-top tamarins, *Saguinus oedipus*, and found that sons, unlike daughters, showed a significant tendency to stay with their parents and not to disperse. McGrew and McLuckie (1986) have proposed that sons may therefore inherit familial territories when their fathers die and then take in an unrelated female mate to displace the mother as the breeding female. These suggestions, however, require further testing and confirmation from field studies before they can be taken as a viable explanation of male offspring celibacy in family groups of callitrichid monkeys. Social conflict may still be found to play a subtle role in reinforcing this sexual suppression.

Peer groups and adult groups of unrelated animals.   As well as family groups, marmoset and tamarin monkeys are kept in captive groups of unrelated postpubertal or adult animals, known as *peer* or *adult groups* or sometimes just as *social groups* (Abbott 1987; Abbott and Hearn 1978; Epple 1972; Rothe 1975). In such groups, overt social conflict is responsible for the suppression of sexual behavior in subordinate males. Dominant (rank 1) males clearly impose their status on all other males in their groups—that is, the subordinates (rank 2 males and below)—through agonistic interactions with the subordinates (Abbott 1984; Abbott and Hearn 1978; Epple 1975, 1978; Rothe 1975).

Of the seven subordinate male marmosets in Rothe's (1975) three groups, only five mounted females and none of them intromitted or ejaculated in the three- to nineteen-month period of observation. Abbott and Hearn (1978) and Abbott (1984) commonly observed subordinate male marmosets mounting and ejaculating with females shortly after group formation, but after three to twelve months, subordinate male copulations were rare. However, if after two months of living in heterosexual groups, the sexes of each group were then kept apart except for observation times, the rank 2 males (subordinates), in three of the eight groups studied, exhibited ejaculatory mounts with the dominant female, but only at about half the frequency of the dominant male (Abbott 1986). In Epple's four groups of tamarin monkeys (1972, 1975), only three of the four subordinate males mounted the single female in each group, contributing just 4 to 45% of the mounts received by the females. However, it is not known if any of these were ejaculatory mounts.

Even this low frequency of sexual activity by subordinate males brought them into social conflict with the dominant males which either attacked, threatened, or displaced the copulating subordinate males and prematurely terminated their copulations (Abbott 1984;

Rothe 1975). They also prevented subordinate males from even gaining access to females (Epple 1972). If we therefore look at the overall pattern of male sexual behavior, it is not surprising that subordinate males show very low proportions of intromission and ejaculation in comparison with dominant males (figure 12.1).

Dominant males did not always attack or interrupt copulating subordinate males (Abbott 1984; Abbott and Hearn 1978; Epple 1972, 1975; Rothe 1975). There are several possible reasons for this. Dominant males may not always need to attack copulating subordinate males in order to inhibit the ability of subordinate males to ejaculate,

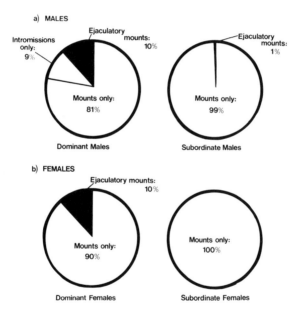

Figure 12.1 Proportion of mounts (a) exhibited by dominant and subordinate males and (b) received by dominant and subordinate female common marmosets living in captive groups. In (a), the male mounts observed were subdivided into those which ended in only the mount, or with intromission or with ejaculation. All males lived in groups of unrelated adults. In (b), the mounts received by females from males were sub-divided into those which terminated in only the mount or in ejaculation. The females lived in either family groups or in groups of unrelated adults. The data were taken from Rothe (1975) and Abbott (1986 and unpublished results).

as found in male talapoin monkeys, *Miopithecus talapoin* (Keverne 1983). Dominant males may also reduce their suppressive influence on subordinate males when no female is close to ovulation.

However, there is always the possibility that subordinate males stand a small, but finite chance of fertilizing the breeding female of the group (Abbott 1986). In free-living groups of marmosets and tamarins, where the distance between animals is much greater than in captivity, the suppressive influences of dominant males may be reduced and may thus allow subordinate males greater sexual freedom (Abbott 1984; Stevenson and Rylands 1988). Whether or not this is translated into reproductive success for subordinate males will remain for genetic fingerprinting studies to determine.

Suppression of Reproductive Physiology

The major reproductive hormones involved in maintaining spermato-genesis in male primates are illustrated in figure 12.2. Until recently, there was no clear evidence to show that this hormonal system was impaired in subordinate male marmosets or tamarins (Abbott 1984). It was assumed that subordinate males were restrained from reproduc-ing only by behavioral means. However, a more detailed look at the reproductive physiology of subordinate male common marmosets liv-ing in groups of unrelated postpubertal and adult animals has uncov-ered evidence of impaired hormonal function which could affect the quality of spermatozoa produced.

Subordinate males have significantly lower plasma testosterone concentrations than do dominant males (figure 12.3). The one excep-tion to this was the only subordinate male in the four groups studied to show ejaculatory mounts with a female (Abbott 1986). Was reduced gonadotrophin stimulation of the testes responsible for the reduced levels of circulating testosterone in subordinate males? The short answer is "yes." Only one gonadotrophic hormone—luteinizing hor-mone (LH)—can so far be measured in marmoset plasma. Subordinate males have significantly lower plasma concentrations of bioactive LH than do dominant males [6.6 + 1.4 mIU/ml (n = 3) versus 15.6 + 2.6 mIU/ml (n = 3), respectively; $p < 0.05$ Student's t-test].

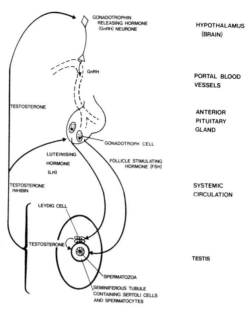

Figure 12.2   Diagrammatic representation of the major hormones involved in regulating the production of spermatozoa in male primates. GnRH is secreted from the terminals of neurons in the hypothalamus into portal blood vessels draining into the anterior pituitary gland. In the anterior pituitary gland, GnRH stimulates the synthesis and release of LH and FSH into the blood stream from gonadotrophs. LH stimulates testosterone secretion from the Leydig cells in the testis. FSH, together with testosterone, stimulates the Sertoli cells and spermatogenesis in the seminiferous tubules. Testosterone from the Leydig cells and inhibin from the Sertoli cells exert a major negative feedback inhibitory control on further secretion of LH and FSH, respectively. Testosterone also exerts an inhibitory negative feedback control on GnRH secretion. Based on De Kretser et al. (1980) and Everitt and Keverne (1986).

This result then prompted the next question. Was there reduced secretion of LH from the anterior pituitary gland in subordinate male marmosets? Possibly. The anterior pituitary gland of subordinate males is certainly less responsive to the administration of exogenous gonadotrophin releasing hormone (GnRH) in comparison to dominant males in terms of stimulated plasma LH values (figure 12.4). However, without information concerning the metabolic clearance rate of LH in dominant and subordinate males, the evidence for reduced pituitary

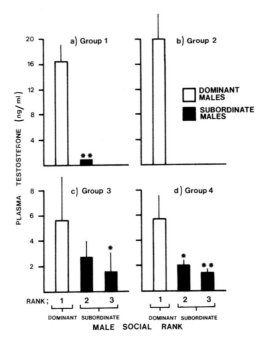

Figure 12.3 Mean (± s.e.m.) plasma testosterone concentrations in dominant and subordinate male common marmoset monkeys in peer groups of unrelated animals: (a) group 1, (b) group 2, (c) group 3 and (d) group 4. (* p<0.05, ** p<0.01 versus dominant male: Student's t-test). The males were sampled over a period of 2-4 months. Reprinted with permission from Abbott (1986).

secretion of LH in subordinate males remains circumstantial. These results suggest that reduced plasma concentrations of LH are probably responsible for the reduced plasma concentrations of testosterone in subordinate male marmosets.

In vitro experiments using testicular tissue from both dominant and subordinate male marmosets showed that tissue from subordinate males produced significantly less testosterone and androstenedione from a [^{3}H] pregnenolone precursor than testicular tissue from dominant males (Sheffield et al. 1989). This suggested that intratesticular concentrations of androgens may be lower in subordinate than in dominant males.

These endocrine deficiencies have important possible implica-

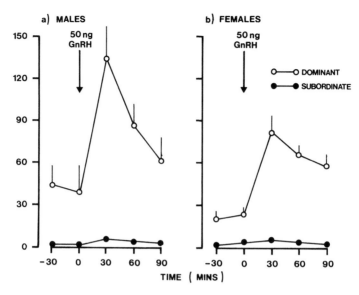

Figure 12.4 Mean (± s.e.m.) plasma LH concentrations in (a) four dominant and four subordinate male and (b) six dominant and eight subordinate female common marmoset monkeys given an i.m. injection of 50 ng GnRH at 0 min. All dominant values are significantly (p<0.05) greater than those of subordinates (Duncan's Multiple Range Test following a two-way analysis of variance for repeated measures). All animals were from peer groups of unrelated animals. Reprinted with permission from Abbott (1986).

tions for the production of normal spermatozoa. Reduced plasma LH values probably also reflect reduced plasma concentrations of follicle stimulating hormone (FSH), and therefore reduced gonadotrophin stimulation of the Sertoli cells in the sperm-producing seminiferous tubules of the testes. Since both FSH and testosterone are important hormones for the production of normal spermatozoa (De Kretser et al. 1980), deficiencies in both may lead to deficiencies in the quality and the quantity of spermatozoa in subordinate male marmosets. We are currently evaluating this possibility. Reduced plasma concentrations of testosterone may also lead to inadequacies in the sexual performance of subordinate males, in addition to the behavioral inhibition from dominant males. However, artificially elevating the circulating levels of testosterone in subordinate male talapoin monkeys did not increase their sexual behavior with females (Keverne et al. 1984). It only served to increase the aggression which subordinate males received from dominant males. Therefore, the reduced testosterone levels of subordinate male marmosets may not be an important factor in the suppression of their sexual behavior.

SUBORDINATE FEMALES

Suppression of sexual behavior

Family groups. Mature female marmoset offspring are subordinate to their mothers, in an analagous fashion to the relationship between sons and fathers (Abbott 1984; Rothe 1975). In contrast to their mothers, but like their brothers, daughters exhibit little sexual behavior. In two of seventeen families observed by Abbott (1984), the adult male mounted one of his daughters once. On both occasions, the daughters rejected the mounts. The daughters were eleven to fourteen months old, but neither ovulated nor became pregnant at that time (Abbott 1984). Rothe (1975) reported three mounts of one female offspring in a marmoset family group by her twin brother. The female was 2.75 years old, but vigorously rejected the mounts. Five further instances of brother-sister matings in marmoset families were reported by Stevenson (cited in Stevenson and Rylands 1988).

In Rothe's (1975) example of brother-sister matings, no intromission or ejaculation occurred. Whether or not the same inadequate behavior was shown in the other instances of copulations involving female offspring is not known. In family groups of cotton-top tamarins and golden lion tamarins *(Leontopithecus rosalia)*, no sexual behavior involving daughters has been observed (French et al. 1984, 1989; French and Stribley 1987; Tardif 1984; Ziegler et al. 1986).

Interestingly, when female marmoset offspring of eleven months of age or older were temporarily (Abbott 1984; Anzenberger 1985; Hubrecht 1984b) or permanently (Abbott 1978a) removed from their families and tested with unfamiliar males, they all accepted male mounts. These matings included ejaculatory mounts because pregnancies ensued (Abbott and Hearn 1978; Hubrecht 1984b). No continuing sexual behavior was noted from females returned to their families

(Abbott 1984; Hubrecht 1984b). If the female's family viewed the test, sexual behavior with the test male was greatly reduced (Anzenberger 1985). Female tamarin offspring of between ten and eighteen months of age are also capable of accepting ejaculatory mounts when they were removed from their families and housed with unfamiliar males (Epple and Katz 1980; Tardif 1984; Ziegler et al. 1986). These were also fertile matings because all females conceived.

So, female offspring in marmoset and tamarin families, in a similar fashion to their male counterparts, show little sexual behavior, despite their obvious ability to do so outside the family. They show a lack of sexual receptivity to male mounts (Abbott 1984; Rothe 1975) and, unlike their mothers (Epple 1967; Rothe 1975), may well not solicit any sexual attention from males (Epple 1967).

To explain this sexual suppression of female offspring in family groups, several authors have suggested that some form of behavioral inhibition was being imposed. This was deduced because a daughter became pregnant as a result of temporary extrafamilial matings (Hubrecht 1984b) and because some daughters in families ovulated (Abbott 1984; French and Stribley 1987; French et al. 1989; Tardif 1984), suggesting that daughters in family groups can be fertile and fail to breed only because of their lack of sexual involvement with males (Rothe 1975). As with male offspring, this behavioral inhibition may well take the form of a familiarity or incest taboo. Thus, female offspring may also have a bias toward copulating with animals of the opposite sex from outside the family.

One way in which female offspring may achieve sexual freedom is suggested by McGrew and McLuckie (1986). In a simulated dispersal of animals from groups of cotton-top tamarins in the laboratory, older female offspring were the primary dispersers from families, both in terms of leaving the home cage and of exploring new areas. Thus, female offspring may emigrate from their family groups in order to attempt to become breeding females in other groups (Abbott 1984). However, confirmation of a tendency for females to migrate between groups of free-ranging cotton-top tamarins is not yet convincing, as only 58% of all sexed transient animals were female (Neyman 1978).

Is social conflict involved in the suppression of sexual behavior in female offspring and their dispersal from the family? McGrew and McLuckie (1986) support this suggestion, but with little evidence. However, Rothe (1978) did show that, in marmosets, daughters formed a separate social subgroup from the rest of the family, which may be due to social conflict within the group. Nonetheless, when the adult female or mother in the group died or was removed (Rothe 1978) none of the remaining daughters showed sexual behavior or became breeding females. Consequently, further information is required before social conflict can be blamed completely as the cause of sexual inhibition in female offspring in marmoset and tamarin families.

Peer groups and adult groups of unrelated animals. In these types of groups, overt social conflict is the cause of suppressed sexual behavior in subordinate females. As with the males, dominant (rank 1) females openly subordinate all other females in their groups (ranks 2 and below: Abbott 1984; Abbott and Hearn 1978; Epple 1967, 1975, 1978; Rothe 1975).

This conflict among females was so severe—much more so than in family groups (Abbott 1984)—that, for 76% of newly formed groups, a subordinate female had to be removed to avoid serious injury to the subordinate (Abbott 1979). This intense severity of the conflict was greater than that seen among males (Abbott 1984) and was similarly severe in adult groups of saddleback tamarins (Epple 1975) and cotton-top tamarins (Hampton et al. 1966).

Given the severity of aggression in these groups, it was not surprising that subordinate females took little part in any sexual activity. The seven subordinate females in Rothe's (1975) three groups of marmosets received an average of only 23% of all male mounts and no ejaculations. Abbott (1986 and unpublished results) found a similarly low amount of sexual activity among subordinate female marmosets. On average, they received only 3% of all male mounts and no ejaculations. Dominant females either attacked, threatened, or displaced the copulating subordinate females (Abbott 1984; Rothe 1975). The likelihood of conflict with the dominant female probably led to the six instances, reported by Abbott (1984), of subordinate females avoiding any sexual advances by the dominant male of their group. Therefore, if we examine the overall quality of sexual interactions in which females were involved, subordinate females received only mounts from males and no ejaculations, while 10% of copulations involving dominant females ended in ejaculation (figure 12.1). The sexual activities of subordinate females are not nearly so constrained in the first three days following group formation, however, when copulations with males can end in ejaculation (Abbott 1979, 1984).

Following initial group formation, subordinate females did not receive ejaculatory mounts from males, and they showed little rejection of male mounts (figure 12.5. See also Abbott 1986). This is quite unlike the immediate rejections of mounts observed with female offspring in family groups. This would suggest that subordinate females in unrelated groups of marmosets are still receptive to male sexual behavior, whereas female offspring in family groups are not. Nonetheless, just as female offspring, subordinate females in unrelated groups showed no sexual solicitation of males and therefore no proceptivity (figure 12.6).

Dominant females did not always attack or interrupt copulating subordinate females (Abbott 1979, 1984; Rothe 1975). This may well be

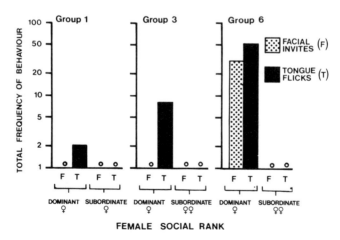

Figure 12.5 Total frequency of sexual solicitation behaviors exhibited by dominant and subordinate female common marmosets in three peer groups of unrelated animals: groups 1, 3, 6. Values for the two subordinate females in groups 3 and 6 are combined; 0 = no behavior observed during the 40h of observation per group.

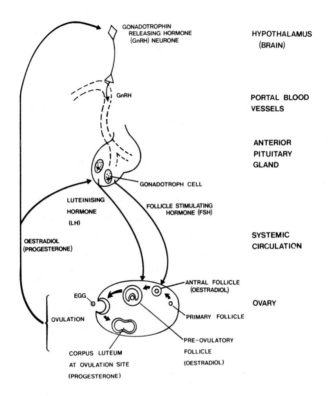

Figure 12.6 Diagrammatic representation of the important hormones involved in regulating ovulation in female primates. GnRH is secreted from the terminals of neurons in the hypothalamus into portal blood vessels draining into the anterior pituitary gland. In the anterior pituitary gland, GnRH stimulates the synthesis and release of LH and FSH from gonadotrophs into the blood stream. LH and predominantly FSH stimulate follicular development in the ovaries. Estradiol, and to a lesser extent progesterone, secreted from the developing follicles, exert a negative feedback inhibitory control of further secretion of GnRH, LH and FSH. This negative feedback control loop is overridden by the large amounts of estradiol secreted from the pre-ovulatory follicles which initiate a massive discharge of LH, and to a lesser extent FSH (and possibly GnRH) which triggers ovulation. This stimulatory effect of estradiol is known as positive feedback. The follicular tissue which remains at the ovulation site is transformed into a corpus luteum which secretes large quantities of progesterone, important for implantation and maintaining early pregnancy. If no pregnancy occurs, the corpus luteum degenerates at the end of each cycle.

due to the fact that, in marmosets and tamarins, social conflict with dominant females usually stops subordinate females from ovulating and thus renders subordinates infertile (Abbott 1984; Abbott and Hearn 1978; Epple and Katz 1980, 1984; French et al. 1984; Ziegler et al. 1986, 1987). Thus, there may be little need to impede the sexual behavior of such infertile subordinate females. The suppression of ovulation in subordinate females probably underpins most of the restricted breeding practices of marmoset and tamarin monkeys.

Suppression of Reproductive Physiology

In family groups of marmosets and tamarins, only the mother or dominant female becomes obviously pregnant and gives birth to offspring (Abbott 1984; Epple 1967, 1975; French et al. 1984, 1989; French and Stribley 1987; Hampton et al. 1966; Rothe 1975; Tardif 1984; Ziegler et al. 1987). In perhaps the majority of cases, this restriction of breeding occurs because female offspring in family groups do not ovulate (Abbott 1984; Epple and Katz 1984; Evans and Hodges 1984; French et al. 1984; Katz and Epple 1979; Ziegler et al. 1987). The main indicators of ovulation in marmosets and tamarins are elevated plasma progesterone concentrations, produced from the corpus luteum in the ovary (figure 12.6), and elevated concentrations of urinary metabolites of corpus luteum secretions, such as pregnanediol glucuronide (Evans and Hodges 1984) or conjugated estrogens (Epple and Katz 1984; French et al. 1984; Hodges and Eastman 1984; Ziegler et al. 1987). In the anovulatory examples of female offspring given here, the plasma or urinary concentrations of these hormones were very low and showed only small, acyclic variability (figure 12.7a. See also Epple and Katz 1984; French et al. 1984; Ziegler et al. 1986).

This inhibition of ovulation was not due to the immaturity of the female offspring. Most were between one and three years of age and well past the age of puberty (Abbott and Hearn 1978; Epple and Katz 1980; Katz and Epple 1979). When removed from their families and paired with males, female offspring of this age start ovulating within two weeks or more and could conceive on the first ovulation (Epple and Katz 1980, 1984; Evans and Hodges 1984; French et al. 1984; Katz and Epple 1979; Ziegler et al. 1987).

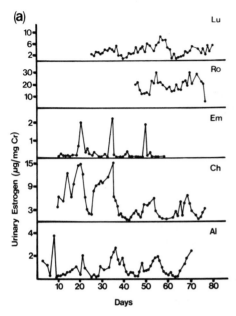

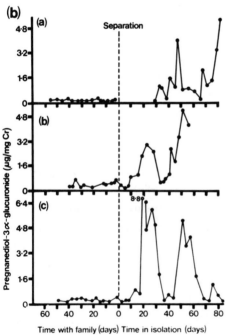

Figure 12.7 (a) Urinary estrogen concentrations in five golden lion tamarin females living in family groups. Lu, Ro and Al are breeding females and Em and Ch are daughters. The uninary hormone profiles indicate ovarian cyclicity in daughter Ch and possibly in daughter Em. Reprinted with permission from French and Stribley (1987). (b) Urinary pregnanediol glucuronide concentrations in three common marmoset daughters (a-c) remaining with their families until Day 0, when they were removed and housed singly. The urinary hormone profiles indicate a lack of ovarian cyclicity while all three daughters remained with their families. The daughters were between 400-640 days of age. Reprinted with permission from Evans and Hodges (1983).

The main source of this inhibitory influence on female offspring appears to be the breeding female herself. In an example from a family group of cotton-top tamarins, where the breeding female died, leaving only her male mate and one daughter, the daughter underwent ovarian cycles until a replacement female mate for the male was found (French et al. 1984). This replacement adult female became the new breeding female and immediately caused the suppression of ovarian cycles in the daughter. Approximately three months later, when the daughter was removed from the breeding pair, her ovarian cyclicity recommenced immediately (figure 12.8).

The chief physiological culprit for the ovarian inhibition in female offspring appears to be the suppression of pituitary LH secretion (figure 12.6). As there is no follicle stimulating hormone (FSH) assay for marmosets or tamarins, the suppressed secretion of this gonadotrophic hormone from the pituitary is assumed to accompany that of LH. Basal plasma concentrations of bioactive LH were continually low and acyclic in female offspring (8.2 ± 2.9 mIU/ml: n = 9) in comparison to breeding females during the follicular phase (13.3 ± 5.3 mIU/ml: n= 6), periovu-

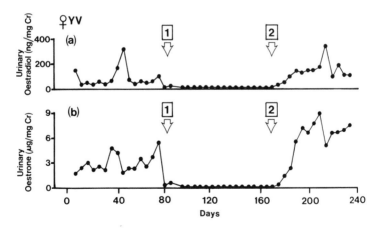

Figure 12.8 Urinary concentrations of (a) estradiol and (b) estrone in a daughter cotton-top tamarin monkey (YV). From days 1-82, female YV was housed alone with her father. At arrow 1 (day 82) an unrelated cycling adult female was introduced to the father-daughter pair. At arrow 2 (day 167), female YV was removed and housed with an unrelated adult male.

latory stage (101.6 ± 28.8 mIU/ml: n = 5) and luteal phase (45.5 ± 7.1 mIU/ml: n = 10) of the ovarian cycle (Abbott et al. 1988). Therefore, it is easy to understand why there is suppression of ovarian function in daughters (see figure 12.6). Without gonadotrophic stimulation, follicular development in the ovary will be inadequate and ovulation will not occur. The suppression of LH secretion may be the result of suppressed secretion of GnRH from the hypothalamus. Some evidence in support of this is provided by the poor LH rise induced by 200 ng GnRH, given intramuscularly, to female offspring in direct contrast to the large rise seen in breeding females in the follicular phase (26.2 ± 6.0 mIU/ml versus 40.0 ± 13.3 mIU/ml, after thirty minutes, respectively: $p < 0.001$) (Abbott unpublished results).

However, not all female offspring in family groups were anovulatory. In common marmosets (Abbott 1984), cotton-top tamarins (Tardif 1984) and golden lion tamarins (French and Stribley 1987, in press) examples of ovulatory daughters have been found in normal family groups (see figure 12.7b). In common marmosets and cotton-top tamarins, ovulatory daughters were found in approximately 50% of the families examined. In common marmosets and golden lion tamarins, the daughters' cycles were of a similar duration to that of breeding females (Abbott 1984; French and Stribley 1987; French et al. 1989). As none of these daughters copulated or became pregnant, it appears that behavioral inhibitory mechanisms play the paramount role in maintaining infertility in otherwise fertile female offspring.

Hormonal studies of subordinate females in peer or adult groups of unrelated animals have been carried out only on common marmosets monkeys (Abbott 1984; Abbott and Hearn 1978; Abbott et al. 1981). Subordinate females in these marmoset groups were mostly anovulatory, as illustrated by figure 12.10a. In the ovaries of such acyclic subordinates, follicular development was suppressed, producing a relatively uniform population of small, immature antral follicles (Abbott et al. 1981; Harlow et al. 1986). However, four of the twenty-eight subordinates studied by Abbott et al. (1988) underwent one or two inadequate ovulatory cycles. Plasma progesterone concentrations in these four females remained above 10 ng/ml or the ovulation criterion for only one to ten days (figure 12.9b), in contrast to the nineteen to twenty days for the normal luteal phase of the cycle (Harlow et al. 1983). Since marmoset and tamarin monkeys provide the most extreme primate examples of social contraception (Abbott 1987; Harcourt 1987), studies have concentrated on the acyclic subordinate female marmoset in order to elucidate the physiological mechanisms which biologically translate behavioral subordination, a product of social conflict, into infertility.

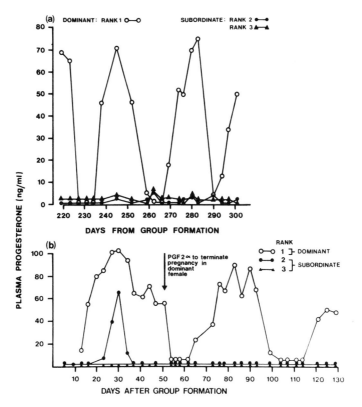

Figure 12.9 Two examples of plasma progesterone concentrations in dominant and subordinate female common marmosets in peer groups of unrelated animals: (a) depicts the typical hormonal profile of a cyclic dominant female with two acyclic subordinates and (b) depicts the rare occasion of an inadequate ovarian cycle shown by a subordinate female. An early pregnancy in the dominant female was terminated by an i.m. injection of a synthetic analogue of prostaglandin F2α (Summers et al., 1985). (a) reprinted with permission from Abbott (1988) and (b) from Abbott (1986).

So what is the initial physiological change responsible for the anovulatory condition of subordinate females? The ovulatory failures suggested that a reduced gonadotrophic stimulation of the ovaries was the initial physiological cause for the anovulatory condition, similar to that found in acyclic daughters in family groups. This was indeed the case. Subordinate females had low, acyclic plasma concentrations of bioactive LH in comparison to the higher and cyclical pattern of LH found in ovulatory dominant females (figure 12.10). Furthermore, when subordinate females were tested with exogenous GnRH, they showed a reduced ability to secrete LH from the pituitary gland in comparison to dominants (figure 12.4b) (Abbott et al. 1981, 1988).

The pituitary and ovarian failures were readily reversed by either behavioral or hormonal means. The behavioral reversal was accomplished through the experimental manipulation of the social sta-

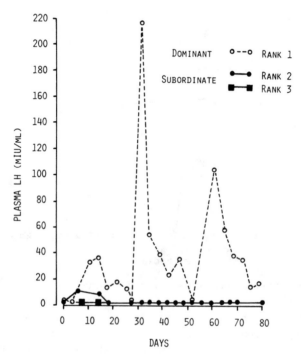

Figure 12.10 Plasma LH concentrations in the dominant (O − − O) and two subordinate (●—●, ■ − − ■) female common marmosets in a typical peer group of unrelated animals. Reprinted with permission from Abbott (1987).

tus of subordinate females. Five long-term subordinate females were removed from their groups and housed singly (Abbott et al. 1988). Plasma LH and estrogen concentrations then started to rise within one to seven days of the removal of the females from their groups, and ovulations commenced after nine days or more. Figure 12.11 illustrates a typical example from one of the five females, and emphasizes the rapidity of the hormonal activation. Abbott (1984) demonstrated similar reversals of infertility when two subordinate females were removed from groups and were paired separately with males. When isolated females were returned to subordinate social status in new groups, plasma LH, estrogen, and progesterone concentrations fell rapidly, and ovulations ceased. In three of the five females, the rapid fall in plasma progesterone concentrations produced obviously curtailed luteal phases. Figure 12.11 provides a typical example from one of those three. This result suggested that marmoset *corpora lutea* are dependent on pituitary LH secretion, similar to the findings in female rhesus monkeys (Hutchison and Zeleznik 1984). The gonadotrophic dependency has since been confirmed by Hodges and colleagues (1988), using a GnRH antagonist to bring about the premature demise of *corpora lutea* in cycling female marmosets.

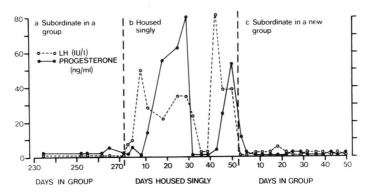

Figure 12.11 Plasma progesterone and LH concentrations in a female common marmoset (224W) while (a) subordinate in a peer group (established nine months previously), (b) housed singly (for 53 days), and (c) during and after becoming subordinate in a new peer group. Reprinted with permission from Abbott (1989).

The hormonal reversal of socially induced infertility was accomplished by leaving subordinate females in their groups and giving them long-term GnRH treatment. The long-term GnRH treatment was administered via mini osmotic or battery-powered pumps housed in a lightweight backpack fitted to subordinate females (Abbott 1987; Ruiz de Elvira and Abbott 1986). Such elaborate systems of administration were necessary because GnRH must be given in pulsatile manner, approximately once every one to two hours, to induce and maintain normal pituitary gonadotrophic secretion in primates (Belchetz et al. 1987). Continuous administration is not effective. In subordinate female marmosets, GnRH therapy induced elevated plasma LH concentrations in all thirteen treated subordinate females as exemplified by figure 12.12. The GnRH therapy also induced plasma estrogen secretion, increased ovarian follicular growth and development, ovulation and a short-term pregnancy. On the cessation of the GnRH treatment, all induced hormonal secretion and ovarian function stopped (Abbott 1987, 1989).

These experimental reversals of subordinate female infertility convincingly demonstrated the unusually high degree of dependence which female marmosets have on dominant or nonsubordinate social status for a normal fertile life. They also demonstrated the direct and rapid physiological consequences of female social conflict in this monkey and directly implicated the suppression of endogenous GnRH

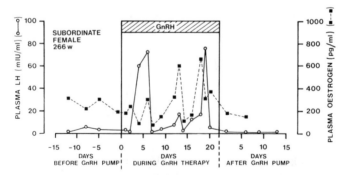

Figure 12.12   Plasma LH and estrogen concentrations in subordinate female common marmoset 266W before, during and after GnRH treatment via a mini-osmotic pump (1 µg GnRH every 2.4h, s.c.).

secretion from the hypothalamus as the cause of the ovarian and pituitary failures. Given this conclusion, the next logical step was to investigate known inhibitors of GnRH and LH secretion to find out which were involved in this particular case of reproductive suppression. Elevated circulating concentrations of cortisol and prolactin and low body weight occur under conditions of stress arising out of various forms of conflict, and all of these have been implicated in the suppression of GnRH and LH secretion in primates and women (Dubey and Plant 1985; Kaplan et al. 1986; Keverne 1979; Schilling and Perret 1987; Yen 1986). However, in the case of acyclic, subordinate female marmosets, plasma cortisol and prolactin concentrations were not elevated above those of cyclic dominant females, and the body weights of dominant and subordinate female marmosets were the same (Abbott 1984, 1986; Abbott et al. 1981).

The clue as to what was inhibiting GnRH and LH secretion in subordinate female marmosets came from ovariectomized females. In normal cycling females, ovarian steroid hormones, such as estradiol, exert a negative feedback control on GnRH and LH secretion (figure 12.7. See also Everitt and Keverne 1986). Removal of this ovarian control, following ovariectomy, results in a sustained and vastly elevated secretion of GnRH and LH. In ovariectomized subordinate female marmosets, this rise in LH secretion—and, therefore, most probably a rise in GnRH secretion (Knobil, 1980)—was significantly attenuated. Plasma LH concentrations in dominant females rose to a postovariectomy maximum of 71.4 ± 8.7 mIU/ml (n = 3) within a week, whereas plasma LH concentrations in subordinate females took two to three weeks to reach their postovariectomy maximum of 37 ± 10.0 mIU/ml (n = 5) which was well below that of the dominants. This differential in plasma LH concentrations was still evident one year after ovariectomy [dominants (n = 4): 63.6 ± 4.0 mIU/ml versus subordinates (n = 3): 14.2 ± 1.7 mIU/ml] (Abbott 1988). The important point about these findings is that LH secretion—and probably GnRH secretion—in subordinate females does increase when the ovaries are removed. It therefore implies that some secretion from the ovaries was responsible for the LH (and GnRH) suppression. The prime candidate for the inhibitor secreted by the ovaries was estradiol, exerting its effects on GnRH and

LH secretion through an increased negative feedback sensitivity, operating at the level of the hypothalamus or pituitary, or both. While this
proposal may seem surprising because the ovarian function of subordinate female marmosets is suppressed, the ovaries of subordinate
female marmosets do, in fact, secrete measurable, but small amounts
of estradiol. Intact female subordinates have plasma estrogen concentrations of 86.5 ± 18.0 pg/ml (n = 4). After ovariectomy, this drops to
48.3 ± 9.5 pg/ml (n = 3) (Abbott unpublished results). However, both
these levels are well below the 150 to 300 pg/ml levels of estradiol
found in cycling female marmosets during the follicular phase of their
ovarian cycles (Hodges et al. 1987). Nevertheless, the low levels of circulating estradiol, secreted from the ovaries of intact subordinate
female marmosets, certainly appear sufficient to sustain an increased
negative feedback suppression of LH secretion.

Consequently, could a low-dose implant of estradiol differentially
suppress LH (and GnRH) secretion in subordinate but not dominant
ovariectomized females? The short answer was "yes." The estradiol
implant that was used produced circulating levels of approximately 350
pg/ml (Abbott 1988), well below the 1 ng/ml levels previously used to
suppress LH secretion in ovariectomized female marmosets (Hodges
1978). Plasma LH concentrations in ovariectomized subordinate females
fell to their previously low intact levels within one week (3.1 ± 1.1
mIU/ml: n = 4), while those in dominant females remained unchanged
at 55.4 ± 13.0 mIU/ml (n = 4). This situation remained unaltered during
the next 2.5 months while the estradiol implants remained in place.

Following the removal of the estradiol implants, plasma LH concentrations in subordinates returned to their intermediate level of 15 to
35 mIU/ml. These results suggested that increased sensitivity to the
negative feedback effects of estradiol may well be the cause of suppressed LH (and GnRH) secretion in subordinate female marmosets.

However, this is not the whole story. Even when the inhibitory
effect of ovarian estradiol is removed from subordinate females, there
is still evidence of continuing LH suppression, although not to the
degree seen in intact subordinates. This is because, even in long-term
ovariectomized subordinate females, plasma LH concentrations
remain well below those of ovariectomized dominant females (Abbott

1988). While this remaining suppression of LH (and GnRH) may reflect the negative feedback inhibition of nonovarian estradiol (such as from the adrenal glands), it could also reflect a separate neural component of GnRH and LH suppression which operates independently of an increased sensitivity to estradiol negative feedback. Such a dual inhibitory mechanism seems to operate, at least in the seasonally anestrus ewe, for which a steroid-independent neural mechanism inhibits GnRH and LH, in conjunction with a sensitized estradiol negative feedback mechanism (Meyer and Goodman 1986).

The possible existence of such a steroid-independent suppression mechanism in subordinate female marmosets has been suggested by two different experiments. In the first, five ovariectomized subordinate female marmosets were isolated from their social groups, housed singly for seven weeks and then were returned to subordinate social status. Plasma LH concentrations in four of the five separated females rose within thirty days to the concentrations found in dominant ovariectomized females. Figure 12.13 illustrates a typical example. Nevertheless, within thirty days of returning these ovariectomized females to subordinate social status, plasma LH concentrations had returned to their previously low levels. Plasma estradiol concentrations remained low throughout. This finding at least suggested the possibility that social subordination was bringing a steroid-independent suppression mechanism into play in inhibiting LH secretion in subordinate female marmosets, independently of the negative feedback action of ovarian estrogen.

The second experiment involved the use of the opiate receptor blocker, naloxone. Endogenous opioid peptides have been implicated directly and indirectly with the suppression of hypothalamic GnRH secretion (Bicknell 1985; Grossman et al. 1981). The use of naloxone to stimulate LH secretion in women with hypothalamic amenorrhea associated the endogenous opioid peptides with GnRH-related infertility (Quigley et al. 1980; Yen 1986). Opioids have also recently been linked with the social suppression of reproduction as increased concentrations of ß-endorphins were found in the cerebrospinal fluid of subordinate male talapoin monkeys (Martensz et al. 1986). Therefore, naloxone was given to both ovariectomized and intact dominant and subordinate

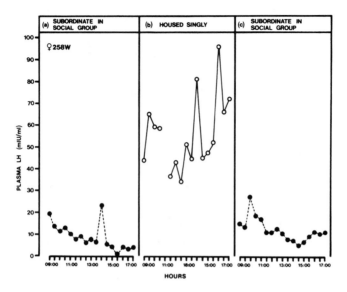

Figure 12.13 Plasma LH concentrations in an ovariectomised female common marmoset (258W): (a) when subordinate in a peer group (●—●), (b) while housed singly for 30 days (○—○) and (c) after returning to subordinate status in her original peer group for 30 days (● – – ●). Blood samples were taken every 30 min between 0900 and 1700 h on the days concerned. Reprinted with permission from Abbott (1988).

female marmosets to find out whether or not endogenous opioid peptides played a role in the suppression of GnRH in subordinates.

In ovariectomized females, naloxone produced a significant increase in plasma LH concentrations in subordinates but not in dominants (figure 12.14). In intact females, the opposite was achieved. Naloxone induced LH rises in dominant females, in the luteal phase of their ovarian cycle, and no LH rise in subordinates.

These results can be interpreted as follows: With the removal of ovarian estradiol negative feedback in ovariectomized subordinates, the remaining GnRH and LH inhibition appears to be due to endogenous opioid activity. In ovariectomized dominants, this does not apply. In this latter respect, ovariectomized dominant marmosets are similar to ovariectomized rhesus monkeys in that endogenous opioids normally play no significant role in suppressing GnRH output in the

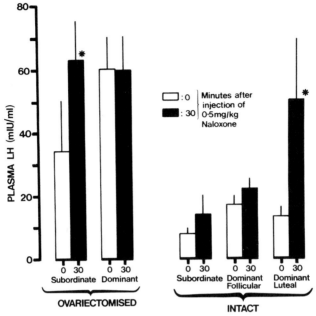

Figure 12.14 Mean (± s.e.m.) plasma LH concentrations in six ovariec-tomized and five intact dominant female common marmosets, and four ovariectomised and five intact subordinate female common marmosets at 0 and 30 min. after an i.v. injection (ovariectomized females) or an i.m. injection (intact females) of 0.5 mg/kg of naloxone. *p<0.05 versus 0 min (Duncan's Multiple Range Test following a two-way analysis of variance for repeated measures).

ovariectomized state (Kesner et al. 1986). In intact subordinate female marmosets, the intense inhibition provided by the increased sensitiv-ity to estradiol negative feedback either overrides or masks the opioid involvement in GnRH and LH suppression because naloxone fails to elicit an LH rise (figure 12.14). Naloxone also only induced an LH rise in intact dominants in the luteal phase. This latter result is similar to that found in female rhesus monkeys and women, in which endoge-nous opioid peptides seem to be most obviously functionally involved in GnRH control during the luteal phase, when blood progesterone levels are high (Ropert et al. 1981; Van Vugt et al. 1984). Thus, endoge-nous opioid involvement in the social suppression of reproduction in subordinate female marmosets can only be functionally demonstrated in ovariectomized subordinates.

Taking the results from the social manipulation of ovariectomized subordinates together with those from the naloxone studies, it is possible that there may be two neuroendocrine components of reproductive suppression in subordinate female marmosets. One is manifest as an increased sensitivity to estradiol negative feedback. The other is manifest as an ovarian steroid-independent, opioid-related inhibition of GnRH. It will be interesting to see whether or not these are two independent suppression pathways or whether they merely reflect individual components of a single suppression mechanism. Whatever the outcome, the subordinate female marmoset now provides an excellent opportunity to determine the neuroendocrine mechanisms which control GnRH secretion in a primate.

One exciting new possibility for the physical activator of these neuroendocrine changes in subordinate females is a pheromone or chemical signal from the dominant female. This suggestion is built on a case study of one subordinate female saddleback tamarin, *Saguinus fuscicollis* (Epple and Katz 1984). This female was taken out of her family group and paired with a male. Up until this time, the daughter's urinary estradiol concentrations were low, indicating no ovulatory cycles (figure 12.15). She was twenty-eight months old and well past puberty. During the first six weeks of pairing, plates from the daughter's family cage, mainly scent marked by the mother, were transferred daily into the isolated pair's cage.

Throughout this scent transfer time, urinary estradiol concentrations in the paired female were erratic and did not resemble the cyclic patterns normally shown by breeding females. However, when the scent transfer was stopped, the female immediately commenced normal cyclic patterns of urinary estradiol. Normally, daughters of this age exhibit cyclical patterns of urinary estradiol immediately on separation from their family. No intermediate maturation phase is observed. This is the first evidence indicating that the reproductive inhibition of subordinate female marmosets and tamarins may be brought about by a chemical signal from the dominant female. In this instance, the chemical signal involved was not powerful enough to

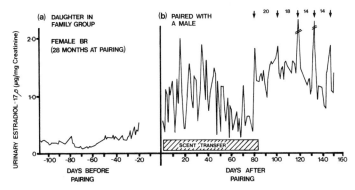

Figure 12.15 Urinary estradiol concentrations in a saddleback tamarin daughter (a) living with her family, (b) after being paired with a male away from her family but receiving scent-marked plates from the family daily and after the scent transfer stopped. The arrows indicate peaks of uninary estradiol at normal, cyclic intervals. Reprinted with permission from Epple and Katz (1984).

completely maintain ovarian inactivity. However, this may be due to the small amount of scent transferred in this experiment. A chemical signal from dominant male dwarf mouse lemurs *(Microcebus murinus)* has certainly been identified as a reproductive suppressant for subordinate male dwarf mouse lemurs (Schilling and Perret 1987). Consequently, further work must be carried out to confirm whether or not a similar system operates among female marmosets and tamarins. Rylands (1985) has suggested that, in the wild, the frequent scent marking behavior of breeding female marmosets over the gouged feeding holes on their group's feeding trees could serve to transmit a contraceptive-like pheromone to all other females in their group.

THE PAIR BOND AND SHARED OFFSPRING CARE: THE KEYS TO COMPLETE REPRODUCTIVE SUPPRESSION IN SUBORDINATES?

The clear behavioral and hormonal sources of infertility in subordinate male and female marmosets and tamarins are unusually restrictive for primate species (Abbott 1987; Harcourt 1987). Why is this so? This question is simple, but finding an answer is difficult.

For a start, only one male and one female (the dominant pair) in any captive group of marmosets or tamarins displays a pair bond

(Abbott and Hearn 1978; Anzenberger 1985; Epple 1975; Evans and
Poole 1984; Rothe 1975). The pair bond is reflected by the animals'
coordinated, synchronized, and intimate social and sexual behavior.
The dominant pair actively prevent any pair-bond formation between
any other male and female combination in their group. Furthermore,
once they are pair-bonded to a particular male, female marmosets are
highly aggressive toward the sexual advances of other males (Anzenberger 1985). Consequently, the exclusivity of the pair bond to the
dominant male and female of a marmoset or tamarin group could well
explain the sexual inadequacies of subordinates. Subordinates are simply blocked from developing any relationship with the opposite sex.
This, in itself, may form a key part of the stress of social subordination
which leads to suppressed sexual behavior and physiological inhibition. Such a block on close relationship bonding with other animals is
now being considered as an important component underlying sexual
and reproductive inadequacies in subordinate talapoin monkeys (Keverne cited by Harcourt 1987).

Why develop a pair bond and a virtually monogamous reproductive system? The answer to this question may lie in the shared care of
the twin infants among the breeding male and other animals in the
group. Marmoset and tamarin females normally give birth to twins
(Epple 1975; Hearn 1983). Females also appear to require helpers to aid
with the rearing of their young, if the young are to have the maximum
chance of surviving (Sussman and Kinzey 1984). The best evidence for
this comes from a field study of moustached tamarin monkeys, *Saguinus mystax* (Garber et al. 1984; Sussman and Garber 1987). All the adult
males and the nonbreeding adult females helped carry and care for the
offspring of the breeding female. Immature animals helped to a limited
extent as well. The twin offspring of the breeding female were both
more likely to survive in the groups with larger numbers of adults than
in groups with fewer adults (Abbott 1987). Apparently, the breeding
female benefited from the reproductive inactivity of the other adult
females in her group, which would otherwise compete to obtain the
help of the few adult group members in rearing their young.

Similar evidence of a reduced maternal load comes from captive
studies of the *Callithrix* and *Saguinus* species. When other group mem-

bers played a significant part in carrying infants, both parents showed reduced but successful infant care activities (Kleiman 1985). Under these circumstances, female helpers cannot reproduce. Otherwise, they would no longer be available to aid the original breeding females and, moreover, would require helpers themselves to successfully rear their offspring. In such a free-for-all situation, successful rearing of offspring might seldom take place.

Given this scenario, subordinate females might actually benefit from delaying their actively reproducing phase of life until they themselves inherit the breeding position or attain it in another group (Carr and MacDonald 1986; MacDonald and Carr 1989; Wasser and Barash 1983). Such severe constraints on reproduction are not placed on subordinate males, which might explain their less obvious suppression of reproduction. Certainly, total infertility among subordinates only seems to occur in mammalian species when breeding females require help to rear their young. The ecological conditions which probably play a powerful role in determining the one-female breeding systems of marmoset and tamarin monkeys, and the extreme suppression of reproduction observed in subordinate females, are beyond the scope of this chapter. Such ecological considerations are dealt with by Sussman and Garber (1987), Sussman and Kinzey (1984), and Terborgh (1983).

DISCUSSION

Infertility is probably the major consequence of becoming subordinate during social conflict in marmoset and tamarin monkeys. These monkeys are not unique in this respect among the primates. They just express this contraceptive trait to an unusually extreme degree. Suppressed reproduction among subordinate female primates is manifest in many ways in both captive and free-living species. Subordinate females show deficient physiological responses (talapoin monkey: Keverne et al. 1982), delayed first conception (rhesus monkey, *Macaca mulatta*: Meikle et al. 1984; Wilson et al. 1983; Japanese macaque, *Macaca fuscata*: Sugiyama and Ohsawa, 1982) reduced numbers of offspring born (gelada baboon, *Theropithecus gelada*: Dunbar 1989; vervet monkey, *Cercopithecus aethiops*: Whitten 1983; rhesus monkey: Wilson et al. 1978; several species: Harcourt 1987) and increased infant mortality (rhesus monkey: Wilson et al. 1983). Suppressed reproduction among subordinate male primates is depicted by reduced sexual activity (talapoin monkey: Keverne et al. 1978, 1984; squirrel monkey, *Saimiri sciureus:* Coe and Levine) and lowered blood testosterone levels (as in the dwarf mouse lemur: Schilling and Perret 1987).

Social contraception is also not just a condition experienced by primates. It occurs in other mammals. The extreme form of social contraception practised by the marmosets and tamarins is found, as predicted, in species where the breeding female requires help in rearing her offspring, such as in the silver-backed jackal, *Canis mesomelas* (Moehlman 1983); dwarf mongoose, *Helogale parvula* (Rood 1980); prairie vole, *Microtus ochrogaster* (Carter et al. 1986); and naked mole-rat, *Heterocephalus glaber* (Brett 1986; Faulkes et al. 1990; Jarvis 1981). It is interesting to speculate that all these species may have developed similar blocks to reproduction as those found in marmosets and tamarins, in order to exert such extreme suppression of reproduction on subordinates.

The specific and repeatable reproductive inhibition found in female marmoset and tamarin monkeys may have particular relevance to ourselves. Some forms of infertility in women are caused by stress (Yen 1986). Certainly one such condition is similar to that seen in female marmosets: GnRH secretion is deficient and the women have no ovulatory cycles in psychogenic hypothalamic amenorrhea (Yen 1986). Physiological investigations of the natural reproductive suppression imposed on female marmosets and tamarins by social conflict may usefully advance our understanding of the physiological disturbances underlying stress-related infertility in ourselves and other mammals.

REFERENCES

Abbott, D. H. (1978a) Hormones and behavior during puberty in the marmoset. In Chivers, D. J., Herbert, J. (eds.), "Recent Advances in Primatology," vol. 1. London: Academic Press. 497–499.

Abbott, D. H. (1978b) The physical, hormonal and behavioral development of the common marmoset, *Callithrix jacchus jacchus.* In Rothe, H.,

Wolters, H. J., Hearn, J. P. (eds.), "Biology and Behaviour of Marmosets." Göttingen: Eigenverlag Hartmut Rothe. 99–106.

Abbott, D. H. (1979) The sexual development of the common marmoset monkey, *Callithrix jacchus jacchus*. Doctoral thesis. University of Edinburgh.

Abbott, D. H. (1984) Behavioral and physiological suppression of fertility in subordinate marmoset monkeys. *American Journal of Primatology* 6:169–186.

Abbott, D. H. (1986) Social suppression of reproduction in subordinate marmoset monkeys *(Callithrix jacchus jacchus)*. In De Mello, M. T. (ed.), "*A Primatologia No Brasil*," vol. 2. Brasilia: *Sociedade Brasileira de Primatologia*. 1–16.

Abbott, D. H. (1987) Behaviorally mediated suppression of reproduction in female primates. *Journal of Zoology, London* 213:455–470.

Abbott, D. H. (1988) Natural suppression of fertility. *Symposium of the Zoological Society of London* 60:7–28.

Abbott, D. H. (1989) Social suppression of reproduction in primates. In Standen, V., Foley, R. A. (eds.), "Comparative Socioecology: The Behavioral Ecology of Humans and Other Mammals." Oxford: Blackwell Scientific Publications. 285–304.

Abbott, D. H.; Hearn, J. P. (1978) Physical, hormonal and behavioral aspects of sexual development in the marmoset monkey, *Callithrix jacchus*. *Journal of Reproduction and Fertility* 53:155–166.

Abbott, D. H.; Hodges, J. K.; George, L. M. (1988) Social status controls LH secretion and ovulation in female marmoset monkeys *(Callithrix jacchus)*. *Journal of Endocrinology* 117:329–339.

Abbott, D. H.; McNeilly, A. S.; Lunn, S. F.; Hulme, M. J.; Burden, F. J. (1981) Inhibition of ovarian function in subordinate female marmoset monkeys *(Callithrix jacchus jacchus)*. *Journal of Reproduction and Fertility* 63:335–345.

Anzenberger, G. (1985) How stranger encounters of common marmosets *(Callithrix jacchus jacchus)* are influenced by family members: The quality of behavior. *Folia Primatologica* 45:204–224.

Belchetz, P. E.; Plant, T. M.; Nakai, Y.; Keoch, E. J.; Knobil, E. (1978) Hypophysial responses to continuous and intermittent delivery of hypothalamic gonadotrophin-releasing hormone. *Science* 202:631–633.

Bicknell, R. J. (1985) Endogenous opioid peptides and hypothalamic neuroendocrine neurones. *Journal of Endocrinology* 107:437–446.

Brett, R. A. (1986) The ecology and behaviour of the naked mole-rat *(Heterocephalus glaber Ruppell) (Rodentia: Bathyergidae)*. Doctoral thesis. University of London.

Carr, G. M.; MacDonald, D. W. (1986) The sociality of solitary foragers: A model based on resource dispersion. *Animal Behaviour* 34:1540–1549.

Carter, C. S.; Getz, L. L.; Cohen-Parsons, M. (1986) Relationships between social organization and behavioral endocrinology in a monogamous mammal. *Advances in the Study of Behavior* 16:109–145.

Coe, C. L.; Levine, S. (1981) Psychoneuroendocrine relationships underlying reproductive behavior in the squirrel monkey. *International Journal of Mental Health* 10:22–42.

Dawson, G. A. (1978) Composition and stability of social groups of the tamarin, *Saguinus oedipus geoffroyi*, in Panama: Ecological and behavioral implications. In Kleiman, D. G. (ed.), "The Biology and Conservation of the *Callitrichidae*." Washington, D.C.: Smithsonian Institution Press. 23–37.

De Kretser, D. M.; Kerr, J. B.; Rich, K. A., Risbridger, G.; Dobos, M. (1980) Hormonal factors involved in normal spermatogenesis and following the disruption of spermatogenesis. In Steinberger, A., Steinberger, E. (eds.), "Testicular Development, Structure and Function." New York: Raven Press. 107–115.

Dubey, A. K.; Plant, T. M. (1985) A suppression of gonadotrophin secretion by cortisol in castrated male rhesus monkeys *(Macaca mulatta)* mediated by the interruption of hypothalamic gonadotrophin-releasing hormone release. *Biology of Reproduction* 33:423–431.

Dunbar, R. I. M. (1989) Reproductive strategies of female gelada baboons. In Rasa, A., Vogel, C., Voland, E. (eds.), "The Sociobiology of Sexual and Reproductive Strategies." London: Chapman Hall. 74–92.

Epple, G. (1967) *Vergleichende untersuchungen über sexual und sozialverhalten der krallenaffen (Hapalidae). Folia Primatologica* 7:37–65.

Epple, G. (1972) Social behavior of laboratory groups of *Saguinus fuscicollis*. In Bridgewater, D. D. (ed.), "Saving the Lion Marmoset." West Virginia: The Wild Animal Propagation Trust. 50–58.

Epple, G. (1975) The behavior of marmoset monkeys *(Callithricidae)*. In Rosenblum, L. A. (ed.), "Primate Behavior: Developments in field and laboratory research" vol. 4. New York: Academic Press. 195–239.

Epple, G. (1978) Notes on the establishment and maintenance of the pair bond in *Saguinus fuscicollis*. In Kleiman, D. G. (ed.), "The Biology and Conservation of the *Callitrichidae*." Washington, D.C.: Smithsonian Institution Press. 231–237.

Epple, G.; Katz, Y. (1980) Social influences on first reproductive success and related behaviors in the saddle-back tamarin *(Saguinus fuscicollis, Callitrichidae). International Journal of Primatology* 1:171–183.

Epple, G.; Katz, Y. (1984) Social influences on estrogen excretion and ovarian cyclicity in saddle-back tamarins *(Saguinus fuscicollis). American Journal of Primatology* 6:215–227.

Evans, S.; Hodges, J. K. (1984) Reproductive status of adult daughters in family groups of common marmosets *(Callithrix jacchus jacchus)*. *Folia Primatologica* 42:127–133.

Evans, S.; Poole, T. B. (1984) Long-term changes and maintenance of the pair-bond in common marmosets, *Callithrix jacchus jacchus*. *Folia Primatologica* 42:33–41.

Everitt, B. J.; Keverne, E. B. (1986) Reproduction. In Lightman, S. L., Everitt, B. J. (eds.), "Neuroendocrinology." Oxford: Blackwell Scientific Publications. 472–537.

Faulkes, C. G.; Abbott, D. H.; Jarvis, J. U. M (1990) Social suppression of ovarian cyclicity in captive and wild colonies of naked mole-rats, *Heterocephalus glaber*. *Journal of Reproduction and Fertility* 88:559–568.

French, J. A.; Abbott, D. H.; Snowdon, C. S. (1984) The effect of social environment on estrogen excretion, scent marking and sociosexual behavior in tamarins *(Saguinus oedipus)*. *American Journal of Primatology* 6:155–167.

French, J. A.; Inglett, B. J.; Dethlefs, T. D. (1989) Reproductive status of non-breeding group members in captive social groups of lion tamarins *(Leontopithecus rosalia)*. *American Journal of Primatology* 18:73–86.

French, J. A.; Stribley, J. A. (1987) Synchronization of ovarian cycles within and between social groups in golden lion tamarins *(Leontopithecus rosalia)*. *American Journal of Primatology* 12:469–478.

French, J. A.; Stribley, J. A. (in press) The reproductive behavior of the golden lion tamarin. In Kleinman, D. G. (ed.) "The Golden Lion Tamarin: A Study in Conservation." Washington, D.C.: The Smithsonian Institution Press.

Garber, P. A.; Moya, L.; Malaga, C. (1984) A preliminary field study of the moustached tamarin monkey *(Saguinus mystax)* in North Eastern Peru: Questions concerned with the evolution of a communal breeding system. *Folia Primatologica* 42:17–32.

Grossman, A.; Moult, P. J. A.; Gaillard, R. C.; Delitalia, G.; Toff, W. D.; Rees, L. H.; Besser, G. M. (1981) The opioid control of LH and FSH release: Effects of a met-enkephalin analogue and naloxone. *Clinical Endocrinology* 14:41–47.

Hampton, J. K.; Hampton, S. H.; Landwehr, B. T. (1966) Observations on a successful breeding colony of the marmoset, *Oedipomidas oedipus*. *Folia Primatologica* 4:265–287.

Harcourt, A. H. (1987) Dominance and fertility among female primates. *Journal of Zoology, London* 213:471–487.

Harding, R. S. O. (1980) Agonism, ranking and the social behavior of adult male baboons. *American Journal of Physical Anthropology* 53:203–216.

Harlow, C. R.; Gens, S.; Hodges, J. K.; Hearn, J. P. (1983) The relationship between plasma progesterone and the timing of ovulation and early embryonic development in the marmoset monkey *(Callithrix jacchus)*. *Journal of Zoology* 201:273–282.

Harlow, C. R.; Hillier, S. G.; Hodges, J. K. (1986) Androgen modulation of follicle stimulating hormone-induced granulosa cell steroidogenesis in the primate ovary. *Endocrinology* 119:1403–1405.

Hearn, J. P. (1983) The common marmoset *(Callithrix jacchus)*. In Hearn, J. P. (ed.), "Reproduction in New World Primates." Lancaster, England: MTP Press, Ltd. 181–215.

Hodges, J. K. (1978) Effects of gonadectomy and estradiol treatment on plasma luteinizing hormone concentrations in the marmoset monkey, *Callithrix jacchus*. *Journal of Endocrinology* 76:271–281.

Hodges, J. K.; Eastman, S. A. K. (1984) Monitoring ovarian function in marmosets and tamarins by the measurement of urinary estrogen metabolites. *American Journal of Primatology* 6:187–197.

Hodges, J. K.; Cottingham, P. G.; Summers, P. M.; Yingnan, L. (1987) Controlled ovulation in the marmoset monkey *(Callithrix jacchus)* with human chorionic gonadotrophin following prostaglandin-induced luteal regression. *Fertility and Sterility* 48:299–305.

Hodges, J. K.; Green, D. I.; Cottingham, P. G.; Sauer, M. J.; Edwards, C.; Lightman, S. L. (1988) Induction of luteal regression in the marmoset monkey *(Callithrix jacchus)* by a gonadotrophin-releasing hormone antagonist and the effects on subsequent follicular maturation. *Journal of Reproduction and Fertility* 82:743–752.

Hubrecht, R. C. (1984a) Field observations on group size and composition of the common marmoset *(Callithrix jacchus jacchus)* at Tapacura, Brazil. *Primates* 25:13–21.

Hubrecht, R. C. (1984b) Marmoset social organization and female reproductive inhibition in the wild. *Primate Eye* 24:5.

Hutchison, J. S.; Zeleznik, A. J. (1984) The rhesus monkey corpus luteum is dependent on pituitary gonadotrophin secretion throughout the luteal phase of the menstrual cycle. *Endocrinology* 115:1780–1786.

Jarvis, J. U. M (1981) Eusociality in a mammal: Cooperative breeding in naked mole-rat colonies. *Science* 212:571–573.

Kaplan, J. R.; Adams, M. R.; Koritnik, D. R.; Rose, J. G.; Manuck, S. B. (1986) Adrenal responsiveness and social status in intact and ovariectomized *Macaca fascicularis*. *American Journal of Primatology* 11:181–193.

Katz, Y.; Epple, G. (1979) The coming of age in female *Saguinus* (marmoset monkeys). In "Proceedings of the 61st Annual Meeting of the Endocrine Society." Abstract No. 473.

Kesner, J. S.; Kaufman, J. M.; Wilson, R. C.; Kuroda, G.; Knobil, E. (1986) The effect of morphine on the electrophysiological activity of the hypothalamic luteinizing hormone-releasing hormone pulse generator in the rhesus monkey. *Neuroendocrinology* 43:686–688.

Keverne, E. B. (1979) Sexual and aggressive behaviour in social groups of talapoin monkeys. In "Sex, Hormones, and Behavior." Ciba Foundation Symposium (New Series), vol. 62. Amsterdam: Excerpta Medica. 271–297.

Keverne, E. B. (1983) Endocrine determinants and constraints on sexual behavior in monkeys. In Bateson, P. P. G. (ed.), "Mate Choice." Cambridge: Cambridge University Press. 407–420.

Keverne, E. B.; Eberhart, J. A.; Meller, R. E. (1982) Dominance and subordination: Concepts or physiological states? In Chiarelli, O. (ed.), "Advanced Views in Primate Biology." Berlin: Springer-Verlag. 81–94.

Keverne, E. B.; Meller, R. E.; Martinez-Arias, A. M. (1978) Dominance, aggression and sexual behaviour in social groups of talapoin monkeys. In Chivers, D. J., Herbert, J. (eds.), "Recent Advances in Primatology," vol. 1. London: Academic Press. 533–547.

Keverne, E. B.; Eberhart, J. A.; Yodyinguad, U.; Abbott, D. H. (1984) Social influences on sex differences in behavior and endocrine state of talapoin monkeys. *Progress in Brain Research* 61:331–347.

Kleiman, D. G. (1985) Paternal care in New World Primates. *American Zoologist* 25:857–859.

Knobil, E. (1980) The neuroendocrine control of the menstrual cycle. *Recent Progress in Hormone Research* 36:53–88.

MacDonald, D. W.; Carr, G. M. (1989) Food security and the rewards of tolerance. In Standen, V., Foley, R. A. (eds.), "The Behavioral Ecology of Humans and Other Mammals." Oxford: Blackwell Scientifics. 75–99.

Martensz, N. D.; Vellucci, S. U.; Herbert, J.; Keverne, E. B. (1986) ß-endorphin in the cerebrospinal fluid of male talapoin monkeys related to dominance status in social groups. *Neuroscience* 18:651–658.

McGrew, W. C.; McLuckie, E. C. (1986) Philopatry and dispersion in the cotton-top tamarin, *Saguinus (O) oedipus*: An attempted laboratory simulation. *International Journal of Primatology* 7:401–422.

Meikle, D. B.; Tilford, B. L.; Vessey, S. H. (1984) Dominance rank, secondary sex ratio and reproduction of offspring in polygamous primates. *American Naturalist* 124:173–188.

Meyer, S. L.; Goodman, R. L. (1986) Separate neural systems mediate the steroid-dependent and steroid-independent suppression of tonic luteinizing hormone secretion in the anoestrus ewe. *Biology of Reproduction* 35:562–571.

Moehlman, P. D. (1983) Socioecology of silver-backed and golden jackals,

Canis mesmomelas and *C. auvenus*. In Eisenberg, J. F., Kleiman, D. G. (eds.), "Recent Advances in the Study of Mammalian Behavior." Special Publication 7 of the American Society of Mammalogists. 423–453.

Neyman, P. R. (1978) Aspects of the ecology and social organization of free-ranging cotton-top tamarins (*Saguinus oedipus*) and the conservation status of the species. In Kleiman, D. G. (ed.), "The Biology and Conservation of the Callitrichidae." Washington, D.C.: The Smithsonian Institution Press. 39–71.

Quigley, M. E.; Sheehan, K. L.; Casper, R. F.; Yen, S. S. C. (1980) Evidence for an increased dopaminergic and opioid activity in patients with hypothalamic hypogonadotrophic amenorrhea. *Journal of Clinical Endocrinology and Metabolism* 50:949–954.

Rood, J. P. (1980) Mating relations and breeding suppression in the dwarf mongoose. *Animal Behaviour* 28:143–150.

Ropert, J. F.; Quigley, M. E.; Yen, S. S. C. (1981) Endogenous opiates modulate pulsatile luteinizing hormone release in humans. *Journal of Clinical Endocrinology and Metabolism* 52:583–585.

Rothe, H. (1975) Some aspects of sexuality and reproduction in groups of captive marmosets *(Callithrix jacchus)*. *Zeitschrift für Tierpsychologie* 37:255–273.

Rothe, H. (1978) Sub-grouping behaviour in captive *Callithrix jacchus* families. In Rothe, H., Wolters, H. J., Hearn, J. P. (eds.), "Biology and Behaviour of Marmosets." Göttingen: Eigenverlag Hartmut Rothe. 233–257.

Ruiz de Elvira, M. C.; Abbott, D. H. (1986) A backpack system for long-term osmotic minipump infusions into unrestrained marmoset monkeys. *Laboratory Animal* 20:329–334.

Rylands, A. B. (1985) Tree-gouging and scent-marking by marmosets. *Animal Behaviour* 33:1365–1367.

Scanlon, C. E.; Chalmers, N. R.; Monteiro, M. A. (1988) Changes in size, composition and reproductive conditions of wild marmoset groups (*Callithrix jacchus jacchus*) in north-east Brazil. *Primates* 29:295–305.

Schilling, A.; Perrett, M. (1987) Chemical signals and reproductive capacity in a male prosimian primate *(Microcebus murinus)*. *Chemical Senses* 12:143–158.

Sheffield, J. W.; Abbott, D. H.; O'Shaugnessy, P. J. (1989) The effects of social rank on [^{3}H] pregnenolone metabolism and androgen production by the common marmoset testis. *Serono Symposium Review* 20, 1:200.

Stevenson, M. F.; Rylands, A. B. (1988) The marmoset monkeys, genus *Callithrix*. In Mittermeier, R. A., Rylands, A. B., Coimbra-Filho, A., Fonsela, G. A. B: "Ecology and Behavior of Neotropical Primates," vol. 2. Washington, D.C.: World Wildlife Fund. 131–222.

Sugiyma, Y.; Ohsawa, H. (1982) Population dynamics of Japanese monkeys with special reference to the effect of artificial feeding. *Folia Primatologica* 39:238–263.

Summers, P. M.; Wennick, C. J.; Hodges, J. K. (1985) Cloprostenol-induced luteolysis in the marmoset monkey *(Callithrix jacchus)*. *Journal of Reproduction and Fertility* 73:133–138.

Sussman, R. W.; Garber, P. A. (1987) A new interpretation of the social organization and mating system of the *Callitrichidae*. *International Journal of Primatology* 8:73–92.

Sussman, R. W.; Kinzey, W. G. (1984) The ecological role of the *Callitrichidae*: A review. *American Journal of Physical Anthropology* 64:419–449.

Tardif, S. D. (1984) Social influences on sexual maturation of female *Saguinus oedipus oedipus*. *American Journal of Primatology* 6:199–209.

Terborgh, J. (1983) "Five New World Primates. A Study in Comparative Ecology." Princeton, N.J.: Princeton University Press.

Terborgh, J.; Goldizen, A. W. (1985) On the mating system of the cooperatively breeding saddle-backed tamarin *(Saguinus fuscicollis)*. *Behavioral Ecology and Sociobiology* 16:293–299.

Van Vugt, D. A.; Lam, N. Y.; Ferin, M. (1984) Reduced frequency of pulsatile luteinizing hormone secretion in the luteal phase of the rhesus monkey: Involvement of endogenous opiates. *Endocrinology* 115:1095–1101.

Wasser, S. K.; Barash, D. P. (1983) Reproductive suppression among female mammals: Implications for biomedicine and sexual selection theory. *Quarterly Review of Biology* 58:513–538.

Whitten, P. L. (1983) Diet and dominance among female vervet monkeys *(Ceropithecus aethiops)*. *American Journal of Primatology* 5:139–159.

Wilson, M. E.; Gordon, T. P.; Bernstein, I. S. (1978) Timing of births and reproductive success in rhesus monkey social groups. *Journal of Medical Primatology* 7:202–212.

Wilson, M. E.; Walker, M. L.; Gordon, T. P. (1983) Consequences of first pregnancy in rhesus monkeys. *American Journal of Physical Anthropology* 61:193–210.

Yen, S. S. C. (1986) Chronic anovulation due to CNS—hypothalamic—pituitary dysfunction. In Yen, S. S. C., Jaffe, R. B. (eds.), "Reproductive Endocrinology. Physiology, Pathophysiology and Clinical Management," 2d ed. Philadelphia: W. B. Saunders Company. 500–545.

Ziegler, T. E.; Savage, A.; Snowdon, C. T. (1986) Behavioral and hormonal correlates of sexual maturation in female cotton-top tamarins *(Saguinus o. oedipus)*. *Primate Report* 14:79.

Ziegler, T. E.; Savage, A.; Scheffler, G.; Snowdon, C. T. (1987) The endocrinology of puberty and reproductive functioning in female cotton-top tamarins *(Saguinus oedipus)* under varying social conditions. *Biology of Reproduction* 37:618–627.

CHAPTER THIRTEEN

Biological Antecedents of Human Aggression

LIONEL TIGER

Both the affiliative and aggressive elements in human behavior have an evolutionary history. Analytical confusion between biology and physiology, and between cognition and culture, have produced broad misunderstanding of the evolutionary preparedness of the human species to form complex affiliations and antipathies. Among other consequences, this confusion has resulted in a bias toward study of sophisticated prosocial behavior rather than antisocial behavior. It has also tended to focus discussion of the biology of aggressive behavior on internal secretions and other clearly physiological phenomena, rather than on symbolic and social interactions, although these rely as much on evolved capacities as does physiology.

One of these evolved capacities relates to kinship. Among humans, kinship has characteristically been defined in terms of social conflict. Kin are those one is with, as opposed to nonkin, who are those one is against. This basic distinction is understandable because the brain which creates kinship categories evolved to act on perceived distinctions, not to think about them. When societies grow too large for kinship as the organizing principle, religious or political ideologies of various forms take their roles. However complex such ideologies may be, they are, nevertheless, biologically predictable products of a species committed to generate both affiliative and separatist/agonist social networks.

Lionel Tiger—Department of Anthropology, Rutgers University, New Brunswick, N.J.

It is by now commonplace for scholars of human biology to note that the discussion about whether or not human aggression has a biological component is invariably marked by the display of generous aggressive assertiveness. This behavior is emitted with equal prodigality by both camps: those who inform us that human aggression is the result of bad, unequal, or immoral social systems which, in the Rousseauist mode betray Man's natural goodness; and those who opine that, lurking within the human heart, is a well-tuned engine of mayhem—the destructive innard of the killer ape. Perhaps the most well-clarified renditions of these themes were sounded in the by-now classical arguments of Robert Ardrey's *African Genesis* of 1961 on the one hand and Ashley Montagu's rejoinder collection *Man and Aggression* of 1968.

There is little need to rehearse these themes in a publication reflecting the sophistication of research and conceptualization which has, in good measure, benefited from the clarifying extremism of these views and which has ,in a synthetic manner, sought to neither exclude nor embrace arguments on programmatic lines. However, it may be worthwhile to glance briefly at the impact of these initial points of departure on the scientific journeys which have been taken. This will be in itself highly synthetic, an overview rather than a bibliographic account.

A BIAS TOWARD THE POSITIVE

First, it may help to establish a contrast in the broad approach to the biology of what may be described as prosocial behavior in comparison with antisocial behavior. For example, the initiative of Bowlby (1958) in describing mother-infant bonding, and subsequent application of the notion to very early experience in hospitals and elsewhere by Klaus and Kennel (1976) has been, by and large, accepted as a reasonable practical extrapolation of biology to policy, even in such precincts as the United States Congress as it considers possible legislation to promote family effectiveness in the face of economic claims made on parents. As a result, many hospitals boast birthing rooms, more newborns than ever before are presented to their mothers immediately at birth, neonates are more likely than ever before to share rooms with their mothers rather than a barracks with nineteen other infants.

There are strong images of the favorable nature of the bonding relationships between newborn and mother—a bonding dependent on a biological predisposition and yielding basic benefits in social and even medical tonus for both mother and child. While there was some controversy about this, it was surprisingly limited, particularly in view of the relatively expensive practical outcome of the hypothesis which is that hospitals had to make investments in new facilities, new training and programs, and individuals had to relax their dependence on drugs as a way of confronting the trauma of parturition.

While in retrospect it is clear this should have been neither surprising nor should it have required scientific innovation from so gregarious a mammal as *Homo sapiens*—scientists somewhat laboriously having to prove the obvious—nevertheless, it is interesting that the whole matter stimulated little of the acrimony and assertiveness associated with the nature-nurture question when applied to human aggression. Here, there seems to be implicit the possibility that non-scientific considerations are involved—that, in a sense, the discussion has become politicized, and that the scientific positions which commentators take on the subject reflect broader values and beliefs than specifically those integral to analyzing data about behavior in the context of acceptable hypotheses.

Why? One must assume that what people attend to is at least some indication of their basic profiles of interest. Hence, the chronic glare of emphasis on the biology of human aggression may belie not only the vexatious nature of the scientific controversy, but the interesting, even captivating, impact of aggressive behavior on human consciousness and self-consciousness.

My principal purpose in this essay will be to suggest that a good deal of the difficulty surrounding this analysis has rested on a confusion between biology and physiology, and culture and cognition. The orienting notion is that the human capacity for cognitive conduct of aggressive encounter—and for creating the group conditions for potential and real conflict—is as biological a phenomenon as the evidently inescapable association, under particular circumstances, between gonadal secretions and bellicose behavior. That is to say, the relatively conventional, traditional view of this was that the animating organs of human aggression were metaphorically—and even practically—below the belt. However, as with so many other systems of human behavior, aggression may find its origin or management above the neck, as one of the higher order functions, not as a low-level process with an endocrinologically determined result.

This may appear to reassure those who conclude that human behavior is under control of potentially ratiocinative intellectual processes. However, on the contrary, this adds another level of phylogenetically modulated influences on human behavior. In effect, this argument turns the most cultural of human accomplishments into yet another alloy of various forces including those which some thought to be primordial. While enthusiasts of the nurture school might regard such primordial factors as irrelevant to contemporary circumstances, it is nevertheless worthwhile to explore in greater detail the possibility that, just as in organic evolution, little or nothing is lost but rather abetted, modulated, or reinterpreted as to function. Similar considerations apply to cognitive patterns so that even cognition about the most modern of events or objects may also involve, as it were, lower strata of the archeology of thought. Indeed, should this be so, then the title of this chapter—Biological Antecedents of Human Conflict—should more strictly be Biological Elements of Human Conflict, on the broad principle that, if forms of conflict behavior were sufficiently successful in the past to help capture genetic continuity, then such forms must, in some manner, endure as part of the available repertoire of the species or, in some way, be retained as genotypical information.

CONSPECIFIC NONRANDOMNESS

We can begin considering this by acknowledging that living forms above the simplest do not respond to conspecifics randomly. Rather, they exert preferences and maintain aversions according to particular categories of choice. The most obvious preference is for close kin. Assertions about kin recognition have even included such creatures as bats and tadpoles who are evidently able to discern kin from nonkin.

So there is positive nonrandomness in that animals choose to be close in space and time with particular conspecifics. Presumably, few would doubt that such sturdy and understandably functional choices reside in a general sense at the very heart of the biosocial nature of a species.

It is also technically reasonable to further claim that, the more complicated the social repertoire and life of the species in question, the more likely are preferences to be exerted. Otherwise, what is the complexity for, if not to enhance particular ends of the animals, and to permit them to do this in a congenial social environment? Thus, social preference—conspecific nonrandomness—facilitates the process of dealing with the external environment and its dangers and challenges within a social world of particular beings. These individuals share some general commitment to the assembled group and are willing and able to cooperate in varying degrees in achieving ends. This is why one striking development in modern primatology has been the demonstration of individual differences among primates and yet the ordered nature of their social processes, particularly those associated with kin and especially the matrilineage. So, for example, Strum's account of her extended work with baboons emphasizes both the individuality of the animals she came to know and the subtle but discernible manner in which social process was mediated by social bonds (Strum 1987). In a similar description of her work with chimpanzees, Goodall is able, also almost novelistically, to tell the story of the social lives of animals who, to an earlier generation of primatologists, or less persistent ones today, would seem to display less precision and continuity in their social relationships than, in fact, they appear to have (Goodall 1986).

Presumably there is little disagreement with this rendition. There may be more disagreement with the contrasting assertion that, just as animals pick and choose in ordered ways those with whom they wish to associate, then it may be inescapable that they also pick and choose those with whom they do not wish to associate, and that this, too, occurs within a framework of discernible order. Just as their preferences will vary in intensity and importance to their lives, so also will their aversions. Those aversions—which are severe—may go beyond avoidance, and they may produce active conflict. It is such aversions which I want to examine here because they may reveal what real empirical and conceptual bridges exist between the conflicts among nonhuman primates and our own conflicts.

In so doing, I will also consider briefly the issue of social competence, in an effort to suggest that skill in aversion and conflict may be an underestimated feature of the repertoire of the primates. It may have been particularly underestimated among humans, perhaps because of the apparent bias toward prosocial behavior alluded to at the beginning of this comment. The preeminent tests of human skill—such as the IQ and Miller analogies—are principally concerned with technical skills not social ones. They are definitely—and perhaps inescapably—biased in favor of the literate modes of communications characteristic of industrial societies. The experience of nonhuman primates may be irrelevant to the conduct of industrial societies. Nevertheless, if recalcitrant elements persist in our species—in this archeology of cognition and communication—then it may well be that patterns of behavior which are at great remove from the industrial society may be precisely those which, because of their very exoticism, have some coercive impact on how we behave and how we evaluate this behavior.

A greater understanding of the nature and significance of conflict—both in other species and in our own—may also be helpful to members of human society seeking to create the so-called "social furniture" appropriate for comfortable community in crowded settings filled with many strangers. Fox and I once described social scientists who were unwilling to consider the possible biology of aggression as "the Christian Scientists of sociology," because we saw similarities in the empirical realism of the two approaches, particularly the preference for locating the causation of the problem in question in psychosocial or spiritual realms, not physical (Tiger and Fox 1989). Of course, there has been ample evidence of the fallacy of misplaced concreteness by persons basing their broad explanations of a general behavior in specific organic operations. Nevertheless, as a scientific strategy, there is a reasonable possibility that any major and basic behavioral system—such as conflict—will implicate mechanisms researchable by crossspecific investigation.

THE FALSE PRIMACY OF THE SQUISHY BITS BELOW THE BELT

First, it seems necessary to consider the proposal that one reason for the antipathy to or confusion about biological explanations is that they have tended to focus on what is essentially physiological processes. That is, the dominating explanatory emphasis has been to relate specific behavioral events to specific physiological, organic ones. For example, there is an extensive literature on the relations between endocrinological secretions and aggressive behavior, in other species and in *Homo sapiens*. These range from the classic establishing work, such as Beach (1965) to the more recent considerations by Rheinisch and Sanders (1982) about the possible impact of administering barbiturates to pregnant females carrying male offspring. Evidently, some feminization of the profile of behavioral responses occurs in rodents, rabbits, and, speculatively, in humans. The interplay between particular bodily squishy bits and specific events need flow not only from the body to the behavior, but it can go the other way around. I sought to indicate this premise to social scientists by pointing out to them the hitherto underestimated contribution they could make to the natural sciences whose awareness of social structure may have been unduly simple-minded (Tiger 1975).

Both natural and social scientists have suffered from the academic canyon between the natural and social sciences—in effect, from the gulf created by Durkheim and his sociological followers who saw, in any form of reductionism, a threat to their effort to establish autonomous social sciences in the competitive academic arena. This also happened to coincide with the full emergence of the industrial system which has, in general, found the modern environmentalist social sciences congenial to its profound challenge to all forms of nature, including a human nature which should not be permitted to interfere with contemporary processes—and which, in addition, may be excluded, almost by definition, from analyses of social life. (See The Psycho-Industrial Complex in Tiger 1987.)

Notwithstanding this, now we are aware that the relationship between behavior and physiology can be complex and genetically consequential, as has been argued by Wasser and Isenberg who have sought to relate such an important genetic and physiological event as reproductive failure to the largely social stimulus of psychosocial stress and even broader psychosocial dynamics (Wasser and Isenberg 1986). Or, as with Madsen (1985), who revealed that the position which an individual occupies in a sociopolitical structure appears to have an effect on the secretion of circulating serotonin, which in turn has an impact on individual equanimity and prudent response to stress. (This is perhaps the first clear finding in humans of the neurophysiological impact of changes in social position and structure.) Broadly, this style of work reflects real progress in clarifying the relationships between behavior and organic processes. Certainly, the findings about the somatic correlates of aggression have been among the most interesting and influential in the biosocial field. The work also possesses the real advantage and attraction of relying on stimuli, mechanisms, and outcomes which can be explored among a variety of species, from rodents to humans, with some confidence that there is a legitimate comparison to be made and even equations to be drawn. Certainly, there is a reason for confidence and energy in continuing the study of the relationship between aggression and various squishy bits.

But my concern here is not with those somatic processes, which have, in any event, been fully detailed elsewhere—for example by Davies (1987). My principal effort will be to suggest the value of what may be a new stage of this general analysis, namely, an effort to relate aggressive behavior to social bonding patterns—the conspecific nonrandomness to which I earlier referred. This depends upon a number of basic steps. (1) The brain in its higher cortical functions is involved. (2) The brain evolved principally to act not to think. (3) The social map called kinship locates people in a landscape of ingroups and outgroups. (4) There is no reason to believe natural selection yielded individuals expert only at prosocial behavior—the principle of nonrandomness demands there be some individuals nonrandomly disliked or detested. Therefore, there was some premium for those individuals able to learn antipathy and act accordingly. (5) Finally, contemporary cultural norms for scientific testing of intelligence and general competence offer scant opportunity and challenge to individuals adept at aggressive encounter. That is, the community broadly seeks to reward patterns of interpersonal skill which do not reflect the broad range of evolved human social capacities. (This may be desirable as far as social policy is concerned but should not be confused with objective analysis.) It is parenthetically interesting that substantial coin can be earned by purveyors of assertiveness training who broadly claim that people are inadequately aggressive in pursuing their own interests, and must have aggressive energies liberated from the restrictions of oversocialized congeniality.

The relation of this to aggressive behavior of other primates lies in the similarities and distinctions which can be drawn in the various contexts which have been identified. For example, presumably few primatologists would expect a competent animal to entirely eschew aggressive endeavor. So if tests of competence or intelligence were created for primates, they would presumably include a component measuring an animal's capacity to identify an appropriate opponent in a particular situation and interact in whatever effortful manner—with whatever assertiveness—was necessary to solve the problem from the point of view of the animal being tested. It is unclear whether testing of humans is sufficiently secular as far as aggression is concerned for that sort of presupposition to be tolerable to professional testmakers. For them, prosocial behavior and even attitudes may be a sine qua non of occupational acceptability and of good citizenship generally. Accordingly, an important primate-like antecedent of human aggressive behavior may be overlooked.

I am saying nothing more here than, among the psychological testing trades, a bias exists against identifying aggressive skillfulness and treating it as seriously, for example, as mathematical problem-solving ability. However, human beings evolved through social processes in which aggressive assertion is likely to have been as important if not more so than mathematical calculation. Yet, the communal self-consciousness reflected in the testing professions hardly suggests this. It seems likely that the controversies surrounding the biology of aggression both reflect and fuel this strongly held and consistent bias against behavior defined as antisocial. So the bias toward the positive has had not only practical implications but also scientific ones.

Let me examine a concrete case in which both scientific and popular biases have contributed to inadequate understanding of conflict situations. It is clear that abuse and homicide of children is widely and generally censured. It is seen as unusually reprehensible, and certainly without redeeming social value. Who would disagree with this in popular terms?

However, speaking scientifically, the issue becomes more interesting once it is discovered, as Daly and Wilson (1988) have reported, that what may be at issue is not child abuse but more specifically stepchild abuse, particularly by stepfathers. The general tendency is sharp—adults are far more likely to abuse children who are not their biological own. Citing Canadian data, Daly and Wilson (1988) indicate that children under the age of five are forty times more likely to suffer abuse if they live with a stepparent than with natural parents. Their other data also demonstrate the relationship between aggression against children and kin connectedness.

It so happens that similar assertions have been made about the aggressive interactions of other animals. For example, Ghiltieri (1988) has described such behavior among the chimpanzees of Kibale Forest in Uganda. Hrdy (1979) found similar trends among the langurs. It has also been shown that wild stallions usurping a harem will so harass pregnant mares that they will abort. This accomplishes infanticide in effect, as well as causing the mares to begin cycling and, hence, become available for impregnation by the new male (Berger 1983).

Here, then, is a clear biological element of aggression which is if not prosocial at least pro-genetic from the point of view of the infant-abuser. Perhaps there is no clear biological *mechanism* at work here, but there is a fairly evident genetic *effect*. Certainly, the nonhuman animals involved in these drastic interactions are unaware of their genetic meaning.

However, to a clear extent, they act as if they were, and so do humans. But even the scientists studying the matter did not comprehend the genetic infrastructure of the abusive event until surprisingly recently. Indeed, the explanation of the matter had to await the refinement of genetic analysis. Now the behavior of both humans and other animals can be understood in a new light and more clearly—and in a dramatic way at that, because few events are so difficult to comprehend as the abuse or murder of a defenseless immature being. Yet, there can be no process as central to any species as its kinship arrangements and the elaborate series of preferences and antipathies which are at the heart of these arrangements. So, if most perpetrators of this form of violence do not appreciate fully or even partly the sources of their animus—and if scientists have gained this understanding only recently (and even so, controversy remains)—then we are entitled to ask: What other types of conflict may be subject to, or at least affected by, a similar order of biological rule? At the very least, scientists should now have to assume formally that there will be explicit non-randomness in social behavior as far as *both* affection and antipathy are concerned. Once more, we turn to Daly and Wilson for an indication of how coercive is kin relatedness in determining who will be victimized by murderers. "Cohabitants...not blood relatives of the killer are more than eleven times as likely to be murdered as cohabitant kin. Spouses are clearly the principal victims, but if spouses are removed from the analysis, the result is little changed. Unrelated cohabitants still endure more than eleven times the risk experienced by blood kin, and the effect is still highly significant." (Daly and Wilson 1988, 23).

WHEN MATTERS COME TO A HEAD

I want to conclude this brief comment by considering forms of conflict which appear to be specifically human—those conflicts which are ideological, religious, national, ethnic, class-based, or otherwise dependent on groups of participants decisively different in scale, and usually form, from the relatively small kin-based groups whose conflicts crossspecies data may help us understand (Tiger 1990).

Succinctly, the steps of argument here are:

1. Large-scale human conflict depends on symbolic determination of who is friend and who is foe. Hence, the overall guidance of conflict processes is essentially cerebral and, presumably, quintessentially hominid.

2. But the brain which does this work evolved to act, not to think, and the principal arena for selectively advantageous action had

to remain the same kinship matrix in which small-scale conflict also takes place.

3. The hominid mechanism which permits the translation from small- to large-scale enables us to produce ideologies or religions, for example, which provide symbolic linkages between otherwise unaffiliated people. Even so, kinship terminology is often used, as in Brother, Sister, Pope (Father). This offers strong hint of the effective emotional and motivational factors which surface in these situations.

4. Thereafter, once the symbolic kin group is established, the people involved may—and usually do—expend much energy in purely symbolic conflict. There may be doctrinal disputes of varying degrees of intensity, efforts to achieve national, class, ethnic, or other hegemony; and symbolic insults or assaults may well be treated as seriously as real ones. Occasionally, there may be real conflict when the symbolic forms no longer suffice. Then, the costs in economic and human terms of such conflict may well dwarf the specific economic value of whatever loss or insult stimulated the onset of real conflict.

In this circumstance the squishy bit which is significantly involved is the uniquely human brain, not the hormonal or other physiological systems which, in varying degrees, hominids share with other primates. Clearly, neither the Pentagon nor the Kremlin depend for their smooth function on the elevated testosterone levels of soldiers. Instead, a highly cerebralized system of alliances, defenses, conjectures, projections, and external scientific and internal bureaucratic imperatives become essential features of an almost exquisitely non-physiological version of elemental or primordial conflict.

CONCLUSIONS

I have implied that, whether or not there are biological antecedents to human conflict, there are certainly biological elements which may take physiological and/or cerebral forms. I did not address the important issue of mechanism. How does kin-recognition occur? What are the neurophysiological correlates of strong belief? Are there discernible physiological consequences of destroying or prevailing over an opponent? Consider the triumph displays of winning athletes and teams. Are there linking networks which unite basic genetics and complex kin groups? (See Rushton 1986.)

Mechanisms notwithstanding, it appears responsible to describe how contemporary human conflicts may reflect patterns of affiliation and disaffiliation which proved of selective relevance during the evolution of our species. These may still be occasionally reiterated even though they may no longer have any evident function in securing gene-survival. They may well, in fact, imperil such survival.

REFERENCES

Ardrey, R. (1961) "African Genesis." New York: Atheneum.

Beach, F. (ed.) (1965) "Sex and Behavior." New York: John Wiley.

Berger, J. (1983) Induced abortion and soical factors in wild horses. *Nature* 303:59–61.

Daly, M.; Wilson, M. (1988) "Homicide." New York: Aldine de Gruyter.

Davies, J. C. (1987) Aggression: Some definition and some physiology. *Political Life Science* 6:1.

Ghiltieri, M. P. (1988) "East of the Mountains of the Moon: Chimpanzee Society in the African Rain Forest." New York: The Free Press.

Goodall, J. (1986) "The Chimpanzees of Gombe: Patterns of Behavior." Cambridge, Mass.: Harvard University Press.

Hrdy, S. (1979) Infanticide among animals: A review, classification, and examination of the implications for the reproductive strategies of females. Ethology and Sociobiology 1:13-40.

Klaus, M.; Kennell, J. H. (1976) "Maternal-Infant Bonding." St. Louis, Mo.: Mosby.

Madsen, D. (1985) A biochemical property relating to power seeking in humans. *The American Political Science Review* 79:448-457.

Montagu, A. (1968) "Man and Aggression." New York: Oxford University Press.

Rheinisch, J. M.; Sanders, S. (1982) Early barbituarate exposure: The brain, sexually dimorphic behavior and learning. *Neuroscience and Biobehavioral Reviews* 6:311-319.

Rushton, J. P. (1986) Gene-culture coevolution and genetic similarity theory: Implications for ideology, ethnic nepotism, and geopolitics. *Political Life Science* 4:144-148.

Strum, S. (1987) "Almost Human: A Journey Into the World of Baboons." New York: Random House.

Tiger, L. (1975) Somatic factors and social behavior. In Fox, R. (ed.), "Biosocial Anthropology." New York: John Wiley & Sons.

Tiger, L. (1987) "The Manufacture of Evil: Ethics, Evolution, and the Industrial System." New York: Harper & Row.

Tiger, L. (1990) The cerebral bridge from friend to foe. In Falger, V. (ed.), "The Sociobiology of Violence." London: Croon Helm.

Tiger, L.; Fox, R. (1989) "The Imperial Animal." 2d ed. New York: Henry Holt and Co.

Wasser, S.; Isenberg, D. (1986) Reproductive failure among women: Pathology or adaptation? *Journal of Psychosomatic Obstetrics and Gynecology* 5: 153-175.

CHAPTER FOURTEEN

Conflict as a Constructive Force in Social Life

DAVID M. LYONS

The final chapter of this book addresses a view toward social conflict that is not often considered in animal behavior research. Although we are accustomed to thinking about conflict as a divisive, inherently destructive aspect of social life, many influential theories of human development and social change have treated conflict as a crucial turning point for better or for worse—as a moment of adapting or not adapting, progressing or regressing.

Building on this perspective, this chapter considers circumstances in which conflicts are likely to proceed in a constructive direction with mutually beneficial outcomes for all concerned parties. Following a schematic outline of the distinguishing characteristics of constructive social conflicts, examples of constructive conflicts drawn from three traditionally distinct levels of analysis are discussed. Those levels are conflicts within the individual, conflicts between individuals, and conflicts within and between groups. At each level, the analysis of constructive conflicts in humans, monkeys, and apes raises a number of issues relevant to a broader understanding of primate social conflict.

David M. Lyons—California Regional Primate Research Center, University of California, Davis

I am grateful to Tom Geen, Bill Mason, and Sally Mendoza for their constructive comments and suggestions. Financial support was provided by Grant RR00169 from the National Institutes of Health, Division of Research Resources, and a National Research Service Award HD07293 from the National Institute of Child Health and Human Development.

Given a world of limited resources, opportunities for personal choice, and some degree of interpersonal dependence, social conflicts are inevitable. In a world of limited resources, there will always be situations in which individuals initially want the same thing and must necessarily settle for different things. As long as there are opportunities for choice, there will also be individuals who initially want different things but, for reasons of mutual interdependence, must ultimately settle for the same thing. Thus, conflicts are a pervasive aspect of social life. They can never be eliminated nor avoided for very long in societies of human (Coombs 1987) and nonhuman primates (see Mason in chapter 2 of this volume).

It is equally obvious that only very few of the conflicts that occur in organized societies, social institutions, or interpersonal relationships actually culminate in chaos, catastrophe, or destruction. Indeed, many influential theories of human development and social change view conflict as a crucial turning point for better *or* for worse (Coser 1956; Deutsch 1971; Erikson 1968; Freud 1930; Simmel 1955). According to John Dewey, for example, it is not a matter of whether conflicts in human societies will continue to occur, "but whether they shall be characterized chiefly by uneasiness, discontent, and blind antagonistic struggles, or whether intelligent direction may modulate the harshness of conflict, and turn the elements of disintegration into a contructive synthesis" (1922, 129). This point of view is also evident outside of Western cultures. In China, for instance, the written character for conflict consists of two superimposed symbols, one for opportunity and one for danger (Hocker and Wilmot 1985).

The notion that conflicts represent inevitable turning points— vital moments of adapting or not adapting, progressing or regressing—raises a question that is central to this essay. Under what circumstances are conflicts likely to proceed in a constructive direction with mutually positive outcomes for all parties, and in what circumstances are conflicts likely to proceed in a destructive direction with negative outcomes for either one concerned party or for all?

In exploring this question, I will provide a selective overview of ideas and findings that have emerged in research on the constructive functions of conflict in human development, interpersonal relationships, and organized groups. Then, I will consider this work with respect to research on nonhuman primates. Because the destructive aspects of social conflict are well-known in animal behavior research, in this essay I will focus chiefly on those circumstances in which conflict represents a potentially constructive force in the social lives of monkeys and apes.

CONSTRUCTIVE AND DESTRUCTIVE CONFLICTS

Social conflicts tend to be episodic. They generally begin with some initial incompatibility among the actions, goals, or interests of two or more participants, and they end when the incompatibilities have been resolved, at least temporarily, through some form of accommodation, compliance, or lack of opposition on the part of either one or both participants. Deutsch (1971) identified the following characteristics as distinguishing between constructive and destructive processes of resolving conflictual episodes in the social lives of human beings.

Task Orientation. In constructive conflicts, incompatible goals and conflicting interests are viewed as a mutual problem to be resolved by joint problem-solving. In destructive conflicts, the solution for resolving incompatibilities is generally viewed as being of the type in which winners impose their needs, desires, or interests upon losers.

Scope. In constructive conflicts, there tends to be a limiting of the scope of the incompatiblities and conflicting interests, while in destructive conflicts, there tends to be an expansion of the scope of the incompatibilities. The social conflict escalates to become a matter of general discord, rather than a point of contention confined to a particular issue at a given time and place. Expansion of the conflict increases its significance and intensifies emotional involvement in it, making constructive resolutions more difficult to achieve.

Perceptions of the Other Party. In constructive conflicts, there is a tendency to recognize common interests and similarities, and to minimize the salience of differences. Destructive conflicts tend to increase sensitivities to threats and differences, minimize the perception of commonalities, and stimulate a sense of complete oppositeness.

Attitudes toward the Other Party.   Constructive conflicts involve cautiously cooperative, trusting, and friendly attitudes. Destructive conflicts involve exploitative, suspicious, and hostile attitudes, and are generally associated with a tendency to react negatively to another's actions, interests, or values.

Influence Attempts.   In constructive conflicts, attempts to influence the other party take the form of persuasive processes of negotiation and involve an open and honest exchange of relevant information. Each party is open to information which it receives from the other, and each party also provides honest information to the other. In destructive conflicts, influence attempts take the form of coercive processes and deceit. Each party is interested in information received from the other but, in turn, withholds relevant information, or provides irrelevant, discouraging, or misleading information to the other.

Outcomes.   Constructive conflicts lead to positive outcomes of mutual satisfaction and mutual net gain for all participants. Destructive conflicts end either with negative outcomes of mutual dissatisfaction and loss, or with outcomes in which one party gains while the other party loses.

The outline just sketched highlights the principal differences between constructive and destructive conflicts, with each considered in its pure or ideal form. Of course, in real life settings, any given conflictual episode may simultaneously involve both constructive and destructive elements. Nevertheless, this conceptual framework provides a useful starting point for considering the constructive functions of conflict in humans, monkeys, and apes. The discussion is organized around three traditionally distinct levels of analysis—conflicts within the individual, conflicts between individuals, and conflicts within and between groups.

INTRAINDIVIDUAL CONFLICTS IN SOCIAL DEVELOPMENT

Intraindividual conflicts occur whenever two or more incompatible tendencies of roughly equal strength are aroused simultaneously within an individual. Although this form of conflict is generally equated with a state of ambivalence, uncertainty, or psychological disequilibrium (see Mason in chapter 2 of this volume), intraindividual conflicts are not necessarily aversive or destructive. On the contrary, intraindividual conflicts serve essential constructive functions in many influential theories of human development.

For Piaget (1970), the essential conflict in cognitive development is a lack of fit between previously established modes of understanding and newly discovered properties of objects and events. In the process of resolving these incompatibilities or contradictions between previously understood appearances and a newly discovered reality, old cognitive schemas are modified or abandoned, and new ones are developed. The growing child thus advances toward more complex levels of understanding. Many of the standardized problems employed by Piaget to reveal qualitative shifts between stages of cognitive development relied upon creating in children a sense of contradiction between appearances and reality. Building on the notions of contradiction and conflict, Reigel (1975) proposed the addition of a fifth and final stage to Piaget's stage theory of cognitive development—that is, the stage of dialectical operations. In this stage, a person is said to accept contradiction as the basis of thought. Accordingly, such a person realizes that contradiction and conflict are both essential and constructive, and does not regard them as necessarily catastrophic or destructive.

Intraindividual conflict also plays a central role in theories of personality development and individual identity. Freud, for example, envisioned three conflicting aspects of the human psyche (Hall 1954). Ego was viewed as negotiating the perpetual conflict between what the individual wants to do as determined by primitive impulses or drives (the province of the id), and what the individual ought to do in light of acquired values and standards of social conduct (the province of the superego). Whereas Freud formulated his theories around conflicts pertaining primarily to issues of sexuality, Erikson's elaborations on psychoanalytic theories of human development consider conflicts of a more general nature. Examples of intraindividual conflict which Erikson refers to as crises of psychosocial development include issues such as "trust" versus "mistrust," "self-certainty" versus "self-consciousness," and "intimacy" versus "isolation" (1968, 94).

Both Freud and Erikson share the view that human development progresses through a series of qualitatively different stages with each one involving a basic issue of conflict. The constructive resolution of these conflicts is considered to be essential to becoming a mature, well-adjusted adult.

While their theories of human development differ in many ways, Piaget, Riegel, Freud, and Erikson all postulate a developmental process in which the constructive resolution of intrapsychic conflicts across the life span leads to progressively more complex levels of psychological functioning. However, only those conflicts whose scope and complexity is somewhat greater—but not too much greater—than the present psychological complexity of the individual are likely to be constructively resolved, and thus lead to more functionally advanced developmental outcomes. If forced to deal with conflicts that are too complex—or if conflicts are not constructively resolved—psychological development may become fixated or even retarded. Thus, these theorists envision a process in which human development is paced by the progressive resolution of intraindividual conflicts of gradually increasing scope and complexity. A similar theme can be discerned in theories of social development in nonhuman primates (Mason 1968, 1971, 1986; Sackett 1965).

Mason (1971) identified two major functional systems that are pertinent to the social development of primates: the filial (mother-directed) and the exploitative (other-directed) developmental programs. Both systems are present throughout life, but their relative prominence changes as development proceeds. These developmental programs may thus be viewed as the basis of two complementary—and sometimes conflicting—trends in social development.

The filial or mother-directed trend is the most clearly focused and the easiest to characterize. This trend is inferred from strongly motivated contact-seeking behaviors—such as clinging, rooting, and sucking—that serve the obvious function of promoting the survival of the newborn monkey or ape. These behaviors also share the characteristic of being most likely to occur under conditions of high psychophysiological arousal, and their performance leads to a reduction in arousal (Mason and Berkson 1962; Hill et al. 1973; Mendoza et al.

1978). Early in life, arousal reduction appears to be an important factor in the formation of filial attachments or emotional bonds. Although infantile contact patterns, such as clinging, occur progressively less often over the course of the life span, they never disappear entirely, and they presumably retain their primitive ability to reduce arousal.

The second major developmental trend becomes increasingly prominent as the growing infant shifts its attention from the mother to the surrounding environment. The exploitative, other-directed trend is inferred from a gradual increase in the frequency of behaviors such as visual exploration, object play, social play, and various investigatory activities. In contrast to the filial behaviors, which occur under conditions of high arousal and are arousal reducing, exploitative behaviors occur under conditions of low or moderate arousal, and their performance leads to further increments of arousal. Exploitative, other-directed behaviors are stimulation-seeking activities. Their function early in life is to support the developing primate's move away from its mother and into the larger world.

The actions associated with filial and exploitative tendencies in development are, therefore, incompatible. Intraindividual conflicts can occur whenever these tendencies are aroused simultaneously in the developing primate, but they are especially likely to occur in early childhood when the relative strengths of the filial and exploitative tendencies are most nearly equal (Mason 1971). It is during this period of life that encounters with novel, moderately arousing objects (or events) generally elicit conflicting tendencies to either move away from the mother and engage the object, or withdraw from the object and reestablish contact with the mother. Earlier in life, when the novelty or salience of an object is beyond the range to which the developing infant is accustomed, initial encounters produce withdrawal, clinging, and other filial, mother-directed behaviors. However, repeated exposure generally leads to a reduction in arousal and, in moderately arousing situations, when an object's novelty or salience is not too much greater than that to which the infant is accustomed, exploitative or other-directed behaviors are more likely to be expressed. Over the course of development, the gradual resolution of recurring conflicts between these opposing filial and exploitative tendencies normally has the cumulative effect of broadening the range and complexity of objects which the infant tolerates and approaches freely. The developing primate moves away from its mother to investigate its surroundings. In matters such as the acquisition of information about sources of solid food, the recognition of environmental sources of danger, and the development of species-typical patterns of social interaction—such as play, sex, social dominance, and the like—intraindividual conflicts of gradually increasing scope and complexity thus provide potentially constructive opportunities for exploration and learning. These processes are crucial for personal growth and the development of higher levels of psychosocial complexity in both human and nonhuman primates.

Whether or not constructive opportunities for growth and psychosocial development are realized in moments of intraindividual conflict depends, of course, on many factors. For example, the early rearing history of the developing primate is of obvious importance. It is now widely recognized that monkeys and apes raised without social companions are deficient in social responses and appear to be generally uninterested in exploring novel or variable objects and events (Harlow and Harlow 1962). This failure to show more developmentally advanced levels of social and exploitative behaviors is almost certainly traceable, in part, to limitations in the variety and amount of experience with moderately arousing situations in early development (Mason 1968, 1971; Sackett 1965). When suddenly confronted later in life with situations that arouse conflicting tendencies to either investigate or withdraw, the isolate-reared primate often withdraws and performs autistic-like self-clasping, huddling, and digit-sucking behaviors characteristic of the arousal reducing filial tendency that is normally prominent much earlier in development. Descriptively, it appears as if the scope and complexity of conflicting filial and exploitative tendencies suddenly and simultaneously elicited in moderately arousing situations far exceeds the isolate-reared primate's ability for constructive resolution. The individual regresses into developmentally primitive, self-directed, and arousal-reducing activities.

While this conclusion is clearly speculative, it is noteworthy that chimpanzee males reared in social isolation and then gradually introduced into more complex situations have, as adults, engaged in constructive, biologically effective patterns of sexual behavior with experienced females (R. K. Davenport, Jr., cited in Mason 1968). Constructive outcomes are also evident in isolate-reared rhesus monkeys that have been socially rehabilitated as a consequence of living with younger conspecifics (Suomi et al. 1972). At one year of age, these isolate-reared monkeys were virtually indistinguishable from their socially reared cagemates in the amount of exploratory behavior, the complexity of play behavior, and the appearance of elementary sexual behavior. That these beneficial effects in rhesus monkeys have been achieved through exposure to younger, but not older, monkey "therapists" suggests that the process of overcoming intraindividual conflicts to engage in constructive opportunities for exploration, social learning, and the development of species-typical patterns of social interaction can be influenced early in life by the attitudes and actions of other parties.

Social influences on the constructive resolution of intraindividual conflicts may also occur in adolescence when conflicts such as those described by Smuts (1985) for young free-ranging female olive baboons are likely to occur in a wide variety of nonhuman primates. Immature female baboons generally avoid most adult males, and they never engage in sexual transactions with them. This state of affairs begins to change at four to five years of age. The female reaches menarche, and begins to show sexual interest in adult males. Gradually, the adolescent female finds herself attracted to individuals that she previously avoided and, according to Smuts, her subsequent transactions with males "reflect this tension between sexual desire and fear." Whether or not these internally conflicted tendencies in adolescent females are resolved in a constructive fashion may, in turn, be influenced by the attitudes and actions of the males. Subtle social initiatives that allowed females to set the pace in their transactions with males appeared to be more likely to lead to the development of long-term mutually positive social relationships than did other tactics that were adopted by males (Smuts 1985, 204–205).

Mothers are especially likely to have an important impact on the constructive resolution of intraindividual conflicts that occur over the course of social development. Throughout infancy and early child-hood, mothers are the principal target of their offsprings' arousal-reducing behaviors. Consequently, the presence of a reassuring mother may serve to reduce levels of arousal, thereby promoting in the developing infant exploitative tendencies to leave her side to investigate and learn about novel objects or social events (Ainsworth and Bell 1970; Lyons et al. 1988). Furthermore, since most nonhuman primate mothers carry their infants with them in the course of their daily travels, the mother's own activities often place her developing infant in novel situations that are most likely to elicit conflicting ten-dencies to investigate or withdraw. A totally permissive maternal style in situations of this sort might prematurely allow the infant to engage in risky or dangerous situations with potentially destructive out-comes. On the other hand, a maternal style of total restrictive vigilance could prevent the infant from engaging in constructive opportunities for valuable exploration and social experience. The problem of achiev-ing a balance between these extremes involves more than the infant's best interests. Maternal attentiveness imposes demands on the mother that may also conflict with her other needs and interests, such as procuring food (see Andrews et al. in chapter 9 of this volume) or establishing and maintaining social relationships with other members of her social group (Simpson 1988).

Intraindividual conflicts marked by ambivalence, uncertainty, or a vascillation between mutually exclusive courses of action are clearly a pervasive aspect of social development over the life span. Surpris-ingly, few systematic studies of social ontogeny have been conducted explicitly from the standpoint of the developing primate's naturally recurring problem of having to cope with social situations in which opposing behavioral tendencies are aroused simultaneously. For psy-chologists traditionally concerned with studying behavior in con-trolled experimental settings, the investigation of intraindividual con-flict in free-ranging social situations has had little appeal. Ethologists once showed a keen interest in intraindividual conflict as exemplified in behavioral phenomena such as ambivalence, redirection, and dis-placement activities (Baerends 1976; Hall 1964; Tinbergen 1952). How-ever, in the last two decades, this field has increasingly shifted its attention toward the study of sociobiological strategies of competition and cooperation. Among the many good reasons for studying intraindividual conflict, however, is the likelihood that internal conflicts within the individual lead to interpersonal conflicts between individuals that make it difficult to work out durable, mutually satisfying, and cooperative relationships (Deutsch, 1971; see also Mason in chapter 2 of this volume).

INTERINDIVIDUAL CONFLICTS IN SOCIAL RELATIONSHIPS

Broadly defined, interindividual conflicts occur whenever the expectations, goals, or interests of two or more parties are incompatible. Incompatibilities, in turn, are inferred when the actions, requests, or demands of one party are at least initially resisted or opposed by another. Accordingly, interindividual conflict is a feature of transactions between individuals, rather than an aspect of any single individual's behavior. This dynamic, transactional perspective toward social conflict is evident in contemporary studies of naturally occurring interindividual conflicts in human social relationships (Goodwin 1982; Maynard 1985; Shantz 1987; Vuchinich 1984), and is in keeping with an important theme in this book on social conflict in nonhuman primates. (See Mason in chapter 2 and Mason et al. in chapter 8 of this volume.)

Although conflicts are an unavoidable source of tension in interpersonal relationships, there are several reasons for concluding that interindividual conflicts are not always divisive or destructive.

First, conflicts over the means for achieving mutually desired outcomes are a fairly common aspect of cooperative activities in human and nonhuman primates. In monogamous species in which parental care is shared by both the mother and the father, parents with congruent interests in raising their offspring to maturity might occasionally find themselves at odds, for example, when they both attempt to retrieve their infant from danger at precisely the same moment. Conflicts of this sort are not necessarily divisive nor destructive so long as both parties are motivated to select the most effective means of achieving their mutual objective rather than each of them persisting stubbornly with the course of action which they initially pursued (Deutsch 1971).

Second, social conflicts in nonhuman primates can lead to the establishment of rules, norms, and systems of mutual expectations that serve to reduce uncertainty and promote social cohesion in interpersonal relationships. (See Mendoza in chapter 4 and Menzel in chapter 10 of this volume.) In the human case, conflict has likewise been viewed as a source of law (Coser 1956; Twining and Miers 1976) and social order (Williams 1970).

Finally, through social conflict, individuals acquire an informed sense of changes in another's goals, interests, and values (Coser 1956; Maynard 1985; Vuchinich 1984). Thus, moments of social conflict may provide potential opportunities for negotiating constructive adjustments in long-term interpersonal relationships that better accommodate each party's changing needs and interests.

That conflict is not necessarily divisive is also indicated by empirical evidence on behavioral and affective states following moments of interindividual conflict. Dawe (1934) found that about 75% of two hundred quarrels observed among children were followed by little or no upset, and that play was usually resumed. The responses to most conflicts in this study were coded as *cheerful*. Similarly, Houseman (1972) reported that children's social behavior after a conflict was not significantly different from behavior before a conflict. Parallel play and cooperative play occurred after 82% of the observed conflicts, and affective states after conflicts were coded significantly more often as being positive (16%) or neutral (83%) than negative (1%). In a review of the literature on postconflict behavior in seven species of monkeys and apes, de Waal described situations in which the probability of reassuring forms of bodily contact between individuals was greater following their dyadic conflicts than in control observations. (See chapter 5 of this volume.)

Whereas the studies described here indicate that social relationships generally persist despite moments of conflict, de Waal goes one step further in suggesting that "mild antagonism does not disturb [social] bonds, but actually makes them stronger" (de Waal 1989, 15; see also Kummer 1968; Rosenblum and Harlow 1963). This is not to say that individuals engage in conflict for the sake of strengthening their interpersonal relationships, but that, under certain conditions, constructive consequences may emerge from the inevitable conflicts that occur in social life. To avoid confusion then, some distinctions are required.

In the simplest sense, the constructive functions of conflict are essentially equivalent to mutually beneficial consequences or effects. A functional analysis of interindividual conflict might then answer the question of what is social conflict good for? As already discussed, interindividual conflicts may have directly beneficial effects in strengthening social bonds, establishing rules of social conduct, instigating beneficial changes in ongoing social relationships, and so on.

Another possibility is that conflicts per se are not directly beneficial, but that constructive processes of resolving conflict serve beneficial functions in preserving valued social relationships. In children's social conflicts, for example, Hartup and colleagues found that conflictual episodes between mutual friends—as compared with those between neutral associates—did not occur less often, differ in length, or differ in the situations that instigated them. Conflicts among friends, however, tended to be less intense, were less often accompanied by physical and/or verbal aggression (26% versus 42%), and more frequently resulted in equal or nearly equal outcomes (67% versus 31%). Continued association was also more likely following conflicts between friends (Hartup et al. 1988). These findings corroborate the idea that conflicts in ongoing relationships are more likely to be resolved constructively and with mutually beneficial pay-offs for all concerned parties when the parties have less at stake in the conflict than they have in maintaining their ongoing relationship (Deutsch 1971).

From an evolutionary perspective, beneficial pay-offs are generally viewed in terms of a currency of inclusive fitness (see Silk in chapter 3 of this volume). An evolutionary analysis of conflict resolution in interpersonal relationships might then attempt to answer the question of through what specific fitness pay-offs does natural selection act to maintain (or at least not eliminate) a disposition or tendency for individuals to constructively resolve conflicts in a cooperative, egalitarian fashion?

Hand (1986) discusses an example from her research on animal social conflict. In their day-to-day activities, mated pairs of gulls must resolve conflicts over food encountered while foraging together. When conflicts are induced during the egg-laying period by throwing choice food items directly between mated pairs, Hand found that individuals did not threaten or displace their mates. They resolved conflicts over food using egalitarian first-come-first-served or sharing conventions. Food sharing or deference to their mates by males seemed rather surprising since the larger, more aggressive males could have easily displaced the smaller females, and they apparently did so in other circumstances. Hand accounted for these egalitarian modes of conflict resolution during the egg-laying period in terms of their fitness payoffs. Food appears to be a critical resource for egg production. In the evolution of food-sharing behavior, breeding males that aggressively competed for food and did not defer to their mates were presumably less successful reproductively than those who did defer.

Egalitarian modes of conflict resolution may thus reflect an evolutionary heritage in which it was (and remains) more costly for individuals to pursue competitive strategies in conflicts with their partners than it is to adopt egalitarian conventions for resolving conflicts. One reason for this is that the payoff structure of conflict is radically different in social relationships based on an element of mutual interdependence. In competitive conflicts, the pay-offs are such that each party generally benefits from the other's losses and suffers if the other gains. In relationships of mutual interdependence, however, individuals often benefit when their partner profits and suffer when their partner loses. Of course, in an evolutionary analysis, there is no implication that individuals actually perceive the potential pay-offs that might be achieved in moments of conflict, but only that the mutually beneficial consequences that are realized through constructive egalitarian modes of conflict resolution contribute jointly to each animal's reproductive success.

Analyses of conflict in human relationships are generally more concerned with individual perceptions, motives, and intentions. For humans, conflicts are usually viewed as social (rather than evolutionary) problems, and the strategies for resolving conflicts are chosen by reason (as opposed to natural selection) to maximize the satisfaction of personal needs, desires, and interests. A psychosocial analysis of conflict resolution in interpersonal relationships might then attempt to answer the question of are individuals able to directly perceive opportunities for mutually beneficial outcomes in moments of conflict, and then act on this knowledge to achieve constructive, equitable solutions in resolving their conflicts?

There is experimental evidence to suggest that, if people have a preexisting problem-solving orientation toward conflict, they are more likely to resolve conflicts in a constructive, mutually beneficial manner than if they have a competitive winner-versus-loser orientation toward conflict. In one experiment, different pairs of children played a game designed to generate an element of interindividual conflict (Deutsch 1971). The same game was played under one of two different instructions. The "chicken" instructions identified the game as one that separates people into two groups—those who give under pressure and those who do not. The "problem-solving" instructions identified the game as one that distinguishes between people who can arrive at a solution to a problem that will bring maximum benefits to both of the players, and those who cannot work out this solution. In both conditions, the children were told that it was important for each of them to earn as much money as they could. The children also played the game for real money in amounts that were significant to them. The outcomes observed in these experimentally induced interindividual conflicts are striking. The "chicken" instructions resulted in a substantial mutual loss for nine of ten pairs of children. The "problem-solving" instructions resulted in a substantial mutual gain for all but three pairs.

In a slightly different study of adults, Deutsch (1958, 1960) found that, when people adopted a mutually cooperative orientation toward conflict, this outlook gave rise to friendly, trusting attitudes and honest communication in moments of conflict. A mutually competitive orientation, in contrast, led to suspicious, hostile attitudes and either no communication or misleading communication.

Researchers studying social conflict in nonhuman primates often assume that monkeys and apes view their own conflicts from a strictly self-interested standpoint of winning or losing competitive contests,

and the aggressive potential of primates has been thoroughly documented. When relationships between individuals are based solely on
self-interest, power, and control, a competitive orientation toward
conflict in which winners aggressively impose their needs and desires
upon losers may be appropriate. Contemporary research clearly indicates, however, that social relationships in nonhuman primates are
often multiplex, involving aspects of both self-interest and mutual
interdependence (see Mason in chapter 2 of this volume). For many
monkeys and apes, long-term, mutually beneficial relationships represent valuable commodities that are worth preserving (Cheney et al.
1986; Kummer 1978), and these are the very conditions that favor constructive problem-solving mechanisms for resolving conflicts in an
equitable manner.

Evolutionary and psychosocial approaches toward the study of
social conflict indicate two general classes of problem-solving mechanisms that may lead to constructive, mutually beneficial modes of conflict resolution. The process of natural selection may produce equitable
solutions for resolving social conflict insofar as animals with heritable
dispositions or tendencies to resolve conflicts with their partners in an
egalitarian fashion may, under certain conditions, perpetuate their
own kind more successfully than those who do not (Hand 1986). Psychosocial studies of problem-solving, on the other hand, are more
often concerned with cognitive skills, such as the abilities to process
information accurately, to take various perspectives, to consider alternative courses of action, to anticipate consequences, and to evaluate
outcomes. The strategies used to achieve interpersonal goals in
moments of conflict have been studied as a sociocognitive form of
problem-solving in children (Krasnor and Rubin 1983; Shantz 1987),
and a sociocognitive perspective toward conflict resolution has some
interesting implications for research on nonhuman primates as well.
Chance was among the first to suggest that the demands of resolving
recurring conflicts in the social lives of monkeys and apes may have
provided a constructive impetus for the evolution of complex cognitive abilities in primates (Chance 1966; Chance and Mead 1952. See
also Cheney et al. 1986; Humphrey 1976).

CONFLICTS WITHIN AND BETWEEN GROUPS

Group conflicts have long been a central concern for students of human social behavior. Throughout history, conflicts among religious sects, ethnic groups, socioeconomic classes, and nations or states have been so pervasive that the possibility of a society free of conflict is generally described as *utopian*—meaning literally, from the Greek, *not of this place*). Although we are accustomed to thinking of these conflicts as being inherently disruptive or dysfunctional, many constructive functions of group conflict have been discussed by sociologists (Coser 1956; Simmel 1955). Naturally, most of these are peculiar to human societies, but the following possibilities seem relevant to nonhuman primates.

Conflicts in human societies maintain boundary lines between groups and enhance social cohesion within groups by drawing individuals together to confront a common enemy (Coser 1956; Sherif 1956; Simmel 1955). In nonhuman primates, territorial conflicts that lead to a spacing out of groups might be thought of as benefiting all participants by preserving the availability of ecological resources, reducing predation risks, or preventing the spread of epidemic diseases. It has also been suggested that conflicts between groups of nonhuman primates function to strengthen social cohesion and maintain social order within groups (Bernstein et al. 1974; Hall 1964). Aggressive conflicts between groups of hamadryas baboons, for example, sometimes appear to start when intragroup aggression is redirected toward members of a neighboring group, thus alleviating internal dissension and preserving social cohesion within a group (Kummer 1968). In a similiar situation, de Waal (1989) observed a group of long-tailed macaques run to a pool of water to jointly threaten their own images. According to de Waal, the need for a common enemy in this captive group of monkeys was apparently so great that an imaginary one was fabricated.

Conflicts also bring together, in a cooperative venture, groups that might otherwise have little to do with one another because of divergent interests or even mutual antagonisms. By giving rise to temporary coalitions created solely for the purpose of fighting for a common cause, social conflicts bind together diverse elements of human society (Coser 1956; Simmel 1955). At an international level, for example, the war against Nazi Germany brought together an alliance of nation states with the most diverse, if not antagonistic, interests and values, including democratic capitalist states, states that were capitalist but not democratic, and Stalinist Russia, which was neither capitalist nor democratic. At a national level, otherwise unrelated special interest groups often band together on a temporary basis to influence the outcomes of conflicts that arise over public policy. The formation of cooperative, multimale coalitions in conflicts over access to estrous females might be taken as an even more simple example in nonhuman primates (Bercovitch 1988; Noë 1990).

Conflicts may provide an outlet for the release of tension between antagonists within human groups, thus serving as a constructive stabilizing function that contributes to group unity (Simmel 1955). However, not all within-group conflicts are likely to have constructive consequences. Only those in which concern goals, values, or interests that do not disrupt the basic consensus upon which the group's organization is founded are likely to end constructively (Coser 1956). In his research on nonhuman primates, de Waal (1989) interprets the frequent but relatively minor hostilities that occur within groups of stump-tailed macaques along these lines.

Conflicts within groups may also give rise to rules and norms of social conduct that benefit all participants by reducing the occurrence of aggressive and potentially damaging power struggles over limited resources. In human societies, this is exemplified in the growth of legal, judicial, and authoritarian institutions concerned with creating, evaluating, and enforcing laws (Coser 1956; Twining and Miers 1976). Although nothing so complex as this exists in societies of monkeys and apes, simple social dominance hierarchies have been viewed as serving a similar function. In transactions over contested resources, it generally appears as if subordinate group members follow the rule that they will defer to dominant members of the group. The overall effect is a reduction in disruptive, competitive conflicts in the group as a whole. A reduction in internal conflict is presumably beneficial to all group members, and this outcome is often said to be a functional consequence of social dominance hierarchies (Bernstein and Gordon, 1974; Hall 1964). Others have suggested that a reduction in aggressive conflict has consequences that are beneficial in different ways for dominants (such as access to resources without risk of injury) and subordinates (perhaps reduction in risk of injury and the possibility of securing desired resources elsewhere), and that social dominance hierarchies are a consequence of the reduction in fighting rather than vice versa (Hinde 1975). Clearly, there are opportunities for progress in understanding the constructive functions of conflict in groups of nonhuman primates.

CONCLUSIONS

In considering the constructive functions of social conflict, I have simplified the problem by focusing on the question of what is social conflict good for? Examples of the beneficial consequences of conflict are easy to come by, and those discussed in this essay are probably familiar to most primatologists. Recognizing that social conflicts may have mutually beneficial consequences raises two additional questions that have received much less attention in primatological research. Are primates able to directly perceive opportunities for mutually beneficial outcomes in moments of conflict, and then act on the basis of this knowledge to achieve constructive solutions in resolving their conflicts? Also, through what specific fitness pay-offs has natural selection acted to promote the evolution of psychosocial dispositions and sociocognitive abilities for resolving conflicts in a constructive, mutually beneficial fashion? These issues are clearly relevant to an understanding of social conflict in both human and nonhuman primates, but they are only imperfectly understood. It is perfectly clear, however, that our understanding of the constructive aspects of primate social conflict will continue to remain obscure so long as conflict is operationally identified in terms of aggression, and conceptually equated with competitive contests in which winners impose their needs, desires, and interests upon losers. Aggression is quite obviously destructive, and competitive conflicts tend to be inherently divisive. To consider conflict solely as a destructive, inherently divisive phenomenon is to overlook its enormous potential as a constructive force in social life.

REFERENCES

Ainsworth, M. D. S.; Bell, S. M. (1970) Attachment, exploration, and separation: Illustrated by the behavior of one-year-olds in a strange situation. *Child Development* 41:49–67.

Baerends, G. P. (1976) The functional organization of behaviour. *Animal Behaviour* 24:726–738.

Bercovitch, F. B. (1988) Coalitions, cooperation and reproductive tactics among adult male baboons. *Animal Behaviour* 36:1198–1209.

Bernstein, I. S.; Gordon, T. P. (1974) The function of aggression in primate societies. *American Scientist* 62:304–311.

Bernstein, I. S.; Gordon, T. P.; Rose, R. M. (1974) Factors influencing the expression of aggression during introductions to rhesus monkey groups. In Holloway, R. L. (ed.), "Primate Aggression, Territoriality, and Xenophobia." New York: Academic Press. 211–272.

Chance, M. R. A. (1966) Resolution of social conflict in animals and man. In de Reuck, A. V. S., Knight, J. (eds.), "Ciba Foundation Symposium on Conflict in Society." London: J. and A. Churchill, Ltd. 16–35.

Chance, M. R. A.; Mead, A. P. (1952) Social behaviour and primate evolution. *Society for Experimental Biology Symposium* 7:395–439.

Cheney, D.; Seyfarth, R.; Smuts, B. (1986) Social relationships and social cognition in nonhuman primates. *Science* 234:1361–1366.

Coombs, C. H. (1987) The structure of conflict. *The American Psychologist* 42:355–363.

Coser, L. (1956) "The Functions of Social Conflict." Glencoe, Ill.: The Free Press.

Dawe, H. C. (1934) An analysis of two hundred quarrels of preschool children. *Child Development* 5:139–157.

Deutsch, M. (1958) Trust and suspicion. *Journal of Conflict Resolution* 2:265–279.

Deutsch, M. (1960) The effect of motivational orientation upon trust and suspicion. *Human Relations* 13:123–139.

Deutsch, M. (1971) Conflict and its resolution. In Smith, C. G. (ed.), "Conflict Resolution: Contributions of the Behavioral Sciences." Notre Dame, Ind.: University of Notre Dame Press. 36–57.

de Waal, F. B. M. (1989) "Peacemaking among Primates." Cambridge, Mass.: Harvard University Press.

Dewey, J. (1922) "Human Nature and Conduct." New York: Henry Holt.

Erikson, E. H. (1968) "Identity, Youth and Crisis." New York: W. W. Norton and Company.

Freud, S. (1930) "Civilization and Its Discontents." London: Hogarth Press.

Goodwin, M. H. (1982) Processes of dispute management among urban black children. *American Ethnologist* 9:76–96.

Hall, C. S. (1954) "A Primer of Freudian Psychology." New York: World.

Hall, K. R. L. (1964) Aggression in monkey and ape societies. In Carthy, J. D., Ebling, F. J. (eds.), "The Natural History of Aggression." New York: Academic Press. 51–64.

Hand, J. L. (1986) Resolution of social conflicts: Dominance, egalitarianism, spheres of dominance, and game theory. *Quarterly Review of Biology* 61:201–220.

Harlow, H. F.; Harlow, M. K. (1962) Social deprivation in monkeys. *Scientific American* 207:136–146.

Hartup, W. W.; Laursen, B.; Stewart, M. I.; Eastenson, A. (1988) Conflict and the friendship relations of young children. *Child Development* 59:1590–1600.

Hill, S. D.; McCormack, S. A.; Mason, W. A. (1973) Effects of artificial mothers and visual experience on adrenal responsiveness of infant monkeys. *Developmental Psychobiology* 6:421–429.

Hinde, R. A. (1975) The concept of function. In Baerends, G., Beer, C., Manning, A. (eds.), "Function and Evolution of Behaviour." Oxford: Clarendon Press. 3–15.

Hocker, J. L.; Wilmot, W. W. (1985) "Interpersonal Conflict." 2d ed. Dubuque, Iowa: William C. Brown.

Houseman, J. (1972) "An ecological study of interpersonal conflicts among preschool children." Unpublished doctoral dissertation. Detroit, Mich.: Wayne State University. University microfilms no. 73–12533.

Humphrey, N. K. (1976) The social function of intellect. In Bateson, P. P. G., Hinde, R. A. (eds.), "Growing Points in Ethology." Cambridge: Cambridge University Press. 303–317.

Krasnor, L. R.; Rubin, K. H. (1983) Preschool social problem solving: Attempts and outcomes in naturalistic interaction. *Child Development* 54: 1545–1558.

Kummer, H. (1968) "Social Organization of Hamadryas Baboons." Chicago: University of Chicago Press.

Kummer, H. (1978) On the value of social relationships to nonhuman primates: A heuristic scheme. *Social Science Information* 17:687–705.

Lyons, D. M.; Price, E. O.; Moberg, G. P. (1988) Social modulation of pituitary-adrenal responsiveness and individual differences in behavior of young domestic goats. *Physiology and Behavior* 43:451–458.

Mason, W. A. (1968) Early social deprivation in nonhuman primates: Implications for human behavior. In Glass, D. C. (ed.), "Biology and Behavior: Environmental Influences." New York: Rockefeller University Press. 70–101.

Mason, W. A. (1971) Motivational factors in psychosocial development. In Arnold, W. J., Page, M. M. (eds.), "Nebraska Symposium on Motivation." Lincoln: University of Nebraska Press. 35–67.

Mason, W. A. (1986) Early socialization. In Benirschke, K. (ed.), "Primates, the Road to Self-Sustaining Populations." New York: Springer-Verlag. 321–329.

Mason, W. A.; Berkson, G. (1962) Conditions influencing vocal responsiveness of infant chimpanzees. *Science* 137:127–128.

Maynard, D. W. (1985) On the functions of social conflict among children. *American Sociological Review* 50:207–223.

Mendoza, S. P.; Smotherman, W. P.; Miner, M. T.; Kaplan, J.; Levine, S. (1978) Pituitary-adrenal response to separation in mother and infant squirrel monkeys. *Developmental Psychobiology* 11:169–175.

Noë, R. (1990) A veto game played by baboons: A challenge to the use of the Prisoner's Dilemma as a paradigm for reciprocity and cooperation. *Animal Behaviour* 39:78–90.

Piaget, J. (1970) Piaget's theory. In Mussen, P. H. (ed.), "Carmichael's Manual of Child Psychology," vol 1, 3d ed. New York: Wiley. 703–732.

Riegel, K. F. (1975) Adult life crises: A dialetic interpretation of development. In Datan, N., Ginsberg, L. H. (eds.), "Life-Span Developmental Psychology: Normative Life Crises." New York: Academic Press. 99–128.

Rosenblum, L. A.; Harlow, H. F. (1963) Approach-avoidance conflict in the mother-surrogate situation. *Psychological Reports* 12:83–85.

Sackett, G. P. (1965) Effects of rearing conditions upon the behavior of rhesus monkeys *(Macaca mulatta)*. *Child Development* 36:855–868.

Shantz, C. U. (1987) Conflicts between children. *Child Development* 58:283–305.

Sherif, M. (1956) Experiments in group conflict. *Scientific American* 195:54–58.

Simmel, G (1955) "Conflict." Wolff, K. H. (trans.), Glencoe, Ill.: The Free Press.

Simpson, M. J. A. (1988) How rhesus monkey mothers and infants keep in touch when infants are at risk from social companions. *International Journal of Primatology* 9:257–274.

Smuts, B. B. (1985) "Sex and Friendship in Baboons." New York: Aldine.

Suomi, S. J.; Harlow, H. F.; McKinney, W. T. (1972) Monkey psychiatrists. *American Journal of Psychiatry* 128:41–46.

Tinbergen, N. (1952) Derived activities: Their causation, biological significance, origin and emancipation during evolution. *Quarterly Review of Biology* 27:1–32.

Twining, W.; Miers, D. (1976) "How to Do Things with Rules." London: Weidenfeld and Nicolson.

Vuchinich, S. (1984) Sequencing and social structure in family conflict. *Social Psychology Quarterly* 47:217–234.

Williams, R. M., Jr. (1970) Social order and social conflict. *Proceedings of the American Philosophical Society* 114:217–225.

Author Index

Subject Index

ACTH (adrenocorticotropic hormone), 180
Adrenals, 179, 186, 194
Affiliation, 107, 120, 138, 155, 316–318, 320, 385
  affiliative behaviors, 55, 90, 98, 102, 112–114, 124, 128, 132, 237, 244, 250, 292, 303, 312–314, 319, 322, 373
  attraction, 8, 85, 119, 127, 130, 275, 282, 285, 309, 310, 314, 319, 395
Age changes, 3, 145, 146, 151, 153, 166, 167, 206, 207, 211–213, 215, 216, 219, 221, 230, 232, 264, 271, 298, 343, 390–395
  adulthood, 150, 342
  maturation, 53, 54, 64, 65
  puberty, 55, 175, 334, 347
  senility, 148, 163, 164
Aggression, 4–7, 14, 18, 19, 21, 23–25, 49, 50, 56, 57, 59, 60, 64, 72, 92, 97, 100, 102, 105, 111, 113–115, 117, 120, 121, 124, 126, 127, 129, 130, 132, 134, 135–139, 146, 147, 150, 154, 158, 165–169, 175, 176, 178, 182, 183, 187, 189–191, 194, 197, 198, 207, 219, 292, 298, 311, 313, 317, 318, 320–323, 337, 342, 344, 345, 362, 374–376, 378, 382, 383, 399, 400, 403–405
  aggressive behaviors, 20, 31, 58, 63, 85, 88, 90, 91, 94–96, 98, 118, 123, 155, 162, 163, 266, 273, 283, 336, 373, 380, 381
  sources of, 22, 63, 153, 379
  winner/loser, 85, 93, 94, 106, 189, 402
Agitation, 30, 231, 266, 268, 269, 276, 279, 280, 283, 316, 320
Agonism, 15, 20, 21, 85, 88, 90, 93, 95–97,

100, 104, 106, 111, 120, 176, 195, 218, 303, 311, 315–318, 323, 336, 373
Alliances, 3, 26, 49, 55, 60, 72, 139, 145, 158, 160, 163, 165–167, 182, 183, 188, 195, 384, 403, 404
Alouatta (howler monkey), 53, 55, 60, 61, 63, 65, 256
Ambiguity, 2, 23, 29, 39, 40, 93, 263, 269, 312, 320
Ambivalence, 2, 13, 23, 29, 32, 35–37, 39, 40, 261, 262, 283, 321, 391, 396
Androstenedione, 340
Arousal, 30, 114, 138, 247, 392–394, 396
Attachment. *See also* Social bond
  filial bond, 393
  mother–infant bond, 374, 375
  pair bond, 295, 296, 298, 300, 318–320, 361, 362
  security of, 244, 250, 251
Autonomic nervous system, 104, 105, 172, 177, 178, 181, 186, 187, 220

Baboon. *See Papio*; *Theropithecus*
Bats, 376
Birth, 362
  interbirth interval, 53, 54, 65, 298
Body contact, 104, 117–121, 125, 135, 136, 139, 270, 271
  affiliative, 111, 114, 127, 128, 138, 218, 237, 261, 263, 267, 312, 393, 398
  aggressive, 95, 97, 123, 261, 266, 273, 283
  male-female, 274
  mother-infant, 206, 207, 215, 231, 235, 236, 238–241, 243, 245, 247, 249, 250

415